Driving Affordable and Clean Energy Through AI and Intelligent Systems

Gunjan Gupta
Cape Peninsula University of Technology, South Africa

Sandeep Gupta
https://orcid.org/0000-0002-3734-3723
Graphic Era University, Dehradun, India

Tarun Varshney
https://orcid.org/0000-0002-4047-7351
Sharda University, Greater Noida, India

Tianyi Han
Tsinghua University, China

IGI Global
Scientific Publishing
Publishing Tomorrow's Research Today

Vice President of Editorial — Melissa Wagner
Director of Acquisitions — Mikaela Felty
Director of Book Development — Jocelynn Hessler
Production Manager — Mike Brehm
Cover Design — Jose Rosado

Published in the United States of America by
IGI Global Scientific Publishing
701 East Chocolate Avenue
Hershey, PA, 17033, USA
Tel: 717-533-8845 | Fax: 717-533-7115
Website: https://www.igi-global.com E-mail: cust@igi-global.com

Library of Congress Cataloging-in-Publication Data

Names: Gupta, Gunjan, 1986- editor | Gupta, Sandeep, 1984- editor | Varshney, Tarun, 1978- editor | Han, Tianyi, 1993- editor
Title: Driving affordable and clean energy through AI and intelligent systems / edited by: Gunjan Gupta, Sandeep Gupta, Tarun Varshney, Tianyi Han.
Description: Hershey PA : IGI Global Scientific Publishing, [2026] | Includes bibliographical references. | Summary: "This book aims to explore how intelligent systems and AI can be effectively leveraged to enhance engineering practices in alignment with global goals for clean energy sustainability"-- Provided by publisher.
Identifiers: LCCN 2025043717 (print) | LCCN 2025043718 (ebook) | ISBN 9798337370415 hardcover | ISBN 9798337370422 paperback | ISBN 9798337370439 ebook
Subjects: LCSH: Renewable energy sources | Clean energy | Artificial intelligence--Industrial applications | Sustainable development | Power resources--Costs
Classification: LCC TJ808 .D75 2026 (print) | LCC TJ808 (ebook)
LC record available at https://lccn.loc.gov/2025043717
LC ebook record available at https://lccn.loc.gov/2025043718

British Cataloguing in Publication Data
A Cataloguing in Publication record for this book is available from the British Library.

Table of Contents

Detailed Table of Contents

Antenna systems having small size, higher efficiency and stable radiation behaviour at terahertz frequencies are need for today's scenario for sustainable energy communication network. Proposed work suggests the design and optimization of DRA antenna for THz microwave frequency which is based on the parameters like voltage standing wave ratio (VSWR), axial ratio, and total efficiency. The optimized DRA possesses a value of VSWR that is less than 2.0 at the entire operating frequency that makes it an ideal match of the impedance. The antenna axial ratio is approximately 1.2 dB, which ensures the stability of circular polarization, suitable to the capacity of wireless connections of high capacity. Moreover, the design will achieve a maximum overall efficiency of about 85, which is very optimum in the context of energy conscious and low loss communication systems. Because of these features, the proposed THz DRA can be applied in smart grid successfully. Sustainable wireless, IoT devices which are energy-efficient and communication.

Motivated by growing worries about urgent environmental issues such as global warming and the rapid depletion of fossil fuel supplies, people all over the world are turning to cleaner, more sustainable alternatives. Among various energy saving technologies, wind energy conversion system (WECS) is among the most prominent and rapidly evolving technologies and playing a critical role in renewable power generation. During generation variable speed wind turbine (VSWT) present numerous issues including wind variability, pitch regulation, and rotor speed fluctuation. To meet rising demand and reduce the susceptibility of VSWT, an effective control system is required which assures the efficient power production. Recently, researchers are focusing on advances in the field of artificial intelligence (AI) which improves both sustainability and effectiveness of electricity production in WECS. Accordingly, this chapter presents an extensive review of existing literature on the application of Artificial Intelligence (AI) based control strategies for the operating zones of VSWT in WECS.

The increasing global importance of sustainable and "green" approaches has transformed renewable energy integration into a priority for multiple sectors, including the industrial sector. Industrial processes and activities that are typically energy-intensive and pose both challenges and opportunities for the use of on-site Renewable Energy (RE) sources such as solar, wind, geothermal, and biomass. In this chapter, we introduce the theme of incorporating renewable energy into industrial processes by analyzing the technical, economic, and regulatory aspects. An extensive examination of the current renewable energy literature, encompassing various technological solutions, challenges to integration, and case studies on successful applications, is presented. Furthermore, the role of EMS, smart grids, clean energy systems, and energy storage in the optimal utilization of renewable energy is also discussed in this chapter.

Buildings are responsible for nearly 35% of global energy consumption, with heating and cooling accounting for a significant share. As urbanization and population growth accelerate, energy demand in cities continues to rise, leading to increased CO_2 emissions, higher electricity costs, and greater stress on power grids. To address these challenges, real-time control systems and building automation have shown strong potential in improving energy efficiency in the built environment. In particular, optimized Heating, Ventilation, and Air Conditioning (HVAC) control systems can significantly reduce energy consumption, lower operational costs, and mitigate environmental impacts. This study investigates the application of machine learning models—including Linear Regression, Random Forest, Support Vector Machine (SVM), Long Short-Term Memory (LSTM), and Gated Recurrent Unit (GRU)—to predict HVAC performance. The models are evaluated using Mean Squared Error (MSE) and R^2 score, with the goal of identifying the most effective approach for enabling data-driven energy optimization in buildings.

to affordable, reliable, sustainable, and modern energy.Coming from a systems perspective, the way we can have at the relationships between AI and clean energy systems is something we're now looking at in this chapter.

In a dual 2-level inverter system using two separate isolated DC voltage sources, a hybrid method incorporating both Space Vector Pulse Width Modulation (SVPWM) and alternate inverter PWM technique along with swarm intelligence is adopted. This blend aims to specifically reduce the overall harmonic distortion. When comparing the Unified SVPWM with an alternative inverter PWM, the former demonstrates superior performance across a wide spectrum of speeds. Unified SVPWM demonstrates to display a lower Total Harmonic Distortion (THD) when compared with SVPWM by means of alternative inverter PWM with the help of swarm technology. Due to recommended attributes like efficiency and precise controllability, the said inverters discover is in the state of multidisciplinary problem, where in decision making tools may be required. Swarm Intelligence tool (PSO) is selected to handle the problem so that a decision making for the energy efficiency. The outcome of this research is to design 2-Level Inverter for the recent technology of USV-PWM for AC Machine along with Swarm Intelligence.

Industry 5.0 revolutionizes industrial energy management by fusing human–AI collaboration within cyber-physical ecosystems to enable adaptive, resilient, and sustainable operations. This proposal introduces a Hybrid Q-learning + Rule-based Energy Management System (EMS) that unites reinforcement learning's optimization capabilities with rule-based security to harmonize distributed energy resources in a grid-connected hybrid renewable system. Experimental validation on a realistic microgrid demonstrates superior performance—Hybrid Q+Rule EMS achieves an operational profit of $0.47, while the conventional rule-based EMS incurs a $1.44 cost. The improvement stems from a 37% reduction in grid imports, $1.67 grid export revenue, and 100.4% renewable penetration. Wind power utilization rises to 89.6% (vs. 75%), with smart battery arbitrage transforming the system from cost-saving to profit-generating. This hybrid framework establishes a new paradigm for intelligent, sustainable, and economically resilient industrial energy management.

This paper presents the optimal placement and sizing of Distributed Generation (DG) using the Random Drift Particle Swarm Optimization (RDPSO) algorithm. The study focuses on minimizing power losses, thus improving profiles of the voltage and maintaining stability of the system. The proposed methodology combines analysis of power flow through the Backward Forward Sweep (BFS) approach with RDPSO to ascertain optimal DG configurations. The proposed strategy is tested on the IEEE 33-bus system. The results obtained have shown a significant improvement in voltage stability and a reduction of power losses. Simulations with one, two, and three DGs highlight the scalability and effectiveness of the approach. The study underscores the potential of RDPSO in optimising DG placement, providing a workable way to enhance the dependability and efficiency of contemporary distribution networks.

The limited energy storage capacity of batteries in wireless sensor networks that is use for various different remote areas presents a significant constraint, limiting network lifespan. Energy harvesting (EH) systems offers a promising solution to overcome this limitation. While various energy sources exist, solar energy (clean and green energy) is widely utilized for energy harvesting due to its flexibility, maturity, and external power availability. The aim of this chapter is to extract effective solar energy from harvesting technology that uses solar energy with maximum power to give explanation for restricted cell life issue of wirelesssensor nodes. This research will provide enhanced and effective energy harvesting by solar method for wireless

Preface

The global demand for affordable, reliable, and clean energy continues to intensify as nations work toward sustainable development and climate resilience. In this evolving landscape, artificial intelligence (AI) and intelligent systems have emerged as powerful enablers, offering unprecedented capabilities to optimize energy generation, enhance system efficiency, and support informed decision-making. This edited volume, Driving Affordable and Clean Energy Through AI and Intelligent Systems, brings together contemporary research and practical advancements that illustrate the transformative potential of intelligent technologies across diverse energy domains.

The chapters in this book address a broad range of applications in which AI, optimization, and computational intelligence significantly improve energy systems and infrastructure. The opening chapter introduces intelligent design strategies for THz dielectric resonator antennas, highlighting their relevance in next-generation sustainable communication networks. Subsequent contributions explore the role of AI in wind energy through advanced control of variable-speed turbines and the integration of renewable energy in industrial operations using intelligent frameworks.

Several chapters focus on enhancing the performance of energy systems through smart optimization and control. These include intelligent HVAC systems based on machine-learning and deep-learning models, hybrid optimization for standalone solar–wind–battery systems, and swarm-intelligence-driven inverter design for AC machine control. Additional research investigates hybrid Q-learning for human–AI collaborative energy management in Industry 5.0 environments, as well as the optimal placement and sizing of distributed generation using advanced particle swarm optimization methods. Emerging trends in electric mobility and intelligent energy scheduling are also covered. Chapters on quantum-computing-based state-of-charge estimation for electric vehicle batteries and intelligent control strategies for optimal EV charging scheduling reflect the growing importance of AI in transportation energy systems. Solutions for optimizing solar energy use and wireless sensor networks using fuzzy logic further showcase the depth and diversity of intelligent approaches presented in this volume.

This book is intended for researchers, engineers, and practitioners seeking insights into the intersection of intelligent technologies and sustainable energy systems. We express our gratitude to all contributors for their impactful work and to our institutions and colleagues for their continued support. We hope this volume inspires further innovation and contributes to advancing affordable and clean energy solutions worldwide.

CHAPTER OVERVIEWS

Chapter 1: Intelligent Design and Optimization of THz Dielectric Resonator Antennas for Sustainable Energy Communication Networks. This chapter presents the design and optimization of a Dielectric Resonator Antenna (DRA) tailored for THz microwave frequencies, emphasizing metrics such as voltage standing wave ratio (VSWR), axial ratio, and overall efficiency. Antenna systems specifically designed for terahertz (THz) frequencies are critical for establishing sustainable energy communication networks. The optimized DRA exhibits a VSWR of less than 2.0 across the entire operating frequency range, ensuring optimal impedance matching. The axial ratio of approximately 1.2 dB confirms stability in circular polarization, making it suitable for high-capacity wireless communication. Furthermore, the design aims for an overall efficiency of nearly 85%, highlighting its energy efficiency and low-loss attributes. These features are instrumental in enhancing data transmission reliability, reducing power wastage, and ensuring polarization purity. Consequently, the proposed THz DRA is well-suited for applications in smart grid communications, energy-efficient Internet of Things (IoT) devices, and sustainable wireless systems.

Chapter 2: Application of Artificial Intelligence to Wind Energy for the Control of Variable Speed Wind Turbine. This chapter presents an extensive review of existing literature on the application of Artificial Intelligence (AI) based control strategies for variable speed wind turbines (VSWT) in wind energy conversion systems (WECS). Motivated by urgent environmental concerns such as global warming and fossil fuel depletion, there is a global shift towards cleaner energy alternatives, with WECS being a prominent technology in renewable power generation. The chapter addresses the challenges faced by VSWT, including wind variability, pitch regulation, and rotor speed fluctuations, and emphasizes the need for effective control systems to enhance power production efficiency. Researchers are increasingly focusing on AI advancements to improve the sustainability and effectiveness of electricity generation in WECS.

Chapter 3: Renewable Energy Integration in Industrial Operations with Intelligent Systems. This chapter introduces the integration of renewable energy into industrial processes, highlighting the technical, economic, and regulatory aspects involved. It reviews current literature on renewable energy technologies, challenges to integration, and successful case studies. The chapter also discusses the importance of energy management systems, smart grids, and energy storage for optimizing renewable energy use. It offers insights into future developments and technologies for efficient industrial decarbonization, along with economic perspectives, cost-benefit analysis, financial modeling tools, and ROI analysis. Additionally, it examines policy and government incentives that can help businesses leverage resources for renewable energy integration.

Chapter 4: Intelligent HVAC Control Systems Based on Machine Learning and Deep Learning Models. This chapter discusses the significant energy consumption of buildings, which accounts for about 35% of global energy use, primarily for heating and cooling. With rising energy demand in urban areas due to population growth and urbanization, CO2 emissions and power grid strain are increasing. To address these challenges, the chapter proposes a real-time building automation system to improve energy efficiency through optimized HVAC control. It examines various machine learning models, including linear regression, random forest, support vector machine, long short-term memory, and gated recurrent unit, to enhance HVAC performance. The effectiveness of these models is evaluated using mean squared error and R^2 as performance indicators. Results from real-time simulations demonstrate that the hybrid model combining GRU and RF outperforms existing strategies, achieving a maximum energy saving of 24.29%.

Chapter 5: Next-Generation Control for Standalone SPV–WECS– BESS Using Hybrid Optimization. This chapter focuses on hybrid renewable energy sources, specifically wind and solar, integrated with a battery energy storage system (BESS) to enhance power quality, stability, and supply continuity. It discusses the advanced control architecture that incorporates key components, including incremental conductance (InC), maximum power point tracking (MPPT), a bidirectional DC-DC converter (BDC), a battery management system, and a voltage source converter (VSC) with an active power filter (APF). The chapter also details the use of a hybrid PSO-GWO optimization technique for tuning a PID controller under dynamic conditions. The system is modeled in MATLAB/Simulink to demonstrate improved performance. This method enhances the system reliability and efficiency of vital off-grid applications, including rural electrification and islanded microgrids.

Chapter 6: Accelerating Clean and Affordable Energy Solutions Through Artificial Intelligence. This chapter explores the dynamic relationship between artificial intelligence (AI) and sustainable energy, highlighting the urgent demand for reliable energy solutions amid industrialization and urbanization. It discusses how AI can address cost, reliability, and sustainability challenges in the energy sector through advanced technologies like machine learning, deep learning, and digital twin technology. These innovations enable better forecasting of power surges, grid stabilization, and improved operational efficiency. Additionally, the chapter emphasizes the importance of considering policy, ethics, and cybersecurity to ensure that these technologies promote universal access to sustainable energy without compromising environmental integrity.

Chapter 7: Design of 2-Level Inverter System with USV-PWM for AC Machine using Swarm Intelligence. This chapter discusses a dual 2-level inverter system utilizing two isolated DC voltage sources, employing a hybrid method that combines Space Vector Pulse Width Modulation (SVPWM) and an alternate inverter PWM technique, enhanced by swarm intelligence. The primary goal of this approach is to minimize overall harmonic distortion. It highlights that Unified SVPWM outperforms alternative inverter PWM across various speeds, exhibiting lower Total Harmonic Distortion (THD) when integrated with swarm technology. The inverters are positioned within a multidisciplinary context, necessitating decision-making tools for efficiency and controllability. Particle Swarm Optimization (PSO) is chosen as the decision-making tool to enhance energy efficiency. The chapter concludes with the design of a 2-Level Inverter utilizing the latest USV-PWM technology for AC machines, supported by swarm intelligence.

Chapter 8: A Hybrid Q-Learning Decision Support Framework for Human–AI Interactive Energy Management in Industry 5.0. This chapter discusses the transformative impact of Industry 5.0 on industrial energy management through the integration of human-AI collaboration in cyber-physical ecosystems, promoting adaptive, resilient, and sustainable operations. It presents a Hybrid Q-learning + Rule-based Energy Management System (EMS) that combines the optimization strengths of reinforcement learning with the security of rule-based approaches to manage distributed energy resources in a hybrid renewable system effectively. Experimental results from a realistic microgrid show that the Hybrid Q+Rule EMS significantly outperforms traditional rule-based systems, achieving an operational profit of $0.47 compared to a $1.44 cost for conventional methods. Key improvements include a 37% reduction in grid imports, $1.

Chapter 9: Optimal Placement and Sizing of Distributed Generation Using Random Drift Particle Swarm Optimization. This chapter presents the optimal placement and sizing of Distributed Generation (DG) using the Random Drift Particle Swarm Optimization (RDPSO) algorithm, focusing on minimizing power losses and improving voltage profiles while maintaining system stability. The methodology integrates power flow analysis through the Backward Forward Sweep (BFS) approach with RDPSO to determine optimal DG configurations, tested on the IEEE 33-bus system. Results indicate significant improvements in voltage stability and reductions in power losses, demonstrating the scalability and effectiveness of the approach with one, two, and three DGs. The study highlights RDPSO's potential in optimizing DG placement, enhancing the reliability and efficiency of modern distribution networks.

Chapter 10: Optimization and Control of Solar Energy System by Fuzzy Logic for Wireless Sensor Network. This chapter addresses the limitations of battery energy storage in wireless sensor networks, particularly in remote areas, which restricts network lifespan. It explores energy harvesting (EH) systems as a viable solution, with a focus on solar energy due to its clean and flexible nature. The chapter aims to enhance solar energy extraction through photovoltaic technology, specifically targeting the issue of limited cell life in wireless sensor nodes. It proposes a method that integrates maximum power point tracking (MPPT) and fuzzy logic controls to improve energy harvesting efficiency. The research emphasizes optimizing various factors, including solar panel types and MPPT efficiency, to achieve better load voltage, current, delivered power, and overall performance.

Chapter 11: Quantum Computing based State of Charge Estimation of Electric Vehicle Battery. This chapter introduces a novel quantum computing approach to state of Charge (SoC) estimation using the Qiskit framework, addressing issues of model uncertainty, sensor noise, and nonlinearities found in traditional methods. It describes how classical battery parameters, such as voltage and capacity, are mapped onto quantum states through a parameterized Ry gate. The resulting quantum states are represented on the Q-sphere and state vector representations, highlighting the probabilistic nature of quantum estimation. The chapter suggests that hybrid quantum-classical battery management systems could outperform traditional models under certain constraints and discusses the implications for battery electrochemistry and quantum information science in the context of future smart energy networks.

Chapter 12: Optimal Scheduling of Electric Vehicle Charging Using Intelligent Control Strategies. This chapter introduces a reinforcement learning framework for managing Electric Vehicle (EV) charging within a MATLAB and Simulink microgrid. The model encompasses renewable energy generation, varying residential and commercial demands, and distributed EV charging stations. Utilizing Deep Q-Network and Proximal Policy Optimization algorithms, the framework determines charging actions based on system metrics such as base load, electricity pricing, renewable energy forecasts, and state of charge. The reward system is designed to minimize energy costs, improve the peak-to-average ratio, and maximize renewable energy usage. Simulation results indicate that Proximal Policy Optimization converges more rapidly, achieves greater rewards, offers smoother load profiles, and enhances state-of-charge management compared to adaptive DQN, linear programming, and heuristic approaches. Moreover, the framework shows potential scalability for extensive EV fleets and supports reliable, low-carbon energy operations.

We extend our sincere gratitude to all the contributors whose expertise and dedication have made this volume possible. Their research and practical insights have enriched this work, shaping it into a comprehensive guide for driving affordable and clean energy through AI and intelligent systems. We also recognise the support of our institutions and colleagues, whose encouragement and assistance have been invaluable in completing this book.

Gunjan Gupta
Cape Peninsula University of Technology, South Africa

Sandeep Gupta
Graphic Era Deemed to be University, Dehradun, Uttarakhand, India

Tarun Varshney
SSES, Sharda University, Greater Noida, UP, India

Tianyi Han
Tsinghua University, Beijing, China

Chapter 1

Intelligent Design and Optimization of THz Dielectric Resonator Antennas for Sustainable Energy Communication Networks

Satya Prakash Singh
Faculty of Sciences, University of Allahabad, Prayagraj, India

Prabhat Chandra Shrivastava
https://orcid.org/0000-0003-0766-6422
Faculty of Sciences, University of Allahabad, Prayagraj, India

Prashant Kumar
Bhagalpur College of Engineering, Bhagalpur, India

Manish Tiwari
MNNIT Allahabad, India

Shadab Hussain
MathCo, Chicago, USA

ABSTRACT

Antenna systems having small size, higher efficiency and stable radiation behaviour at terahertz frequencies are need for today's scenario for sustainable energy com-

DOI: 10.4018/979-8-3373-7041-5.ch001

munication network. Proposed work suggests the design and optimization of DRA antenna for THz microwave frequency which is based on the parameters like voltage standing wave ratio (VSWR), axial ratio, and total efficiency. The optimized DRA possesses a value of VSWR that is less than 2.0 at the entire operating frequency that makes it an ideal match of the impedance. The antenna axial ratio is approximately 1.2 dB, which ensures the stability of circular polarization, suitable to the capacity of wireless connections of high capacity. Moreover, the design will achieve a maximum overall efficiency of about 85, which is very optimum in the context of energy conscious and low loss communication systems. Because of these features, the proposed THz DRA can be applied in smart grid successfully. Sustainable wireless, IoT devices which are energy-efficient and communication.

INTRODUCTION

The phenomenal development of wireless communication technologies in the past few decades has played a significant role in transforming contemporary information systems and smart infrastructures. Due to the growing number of connected devices, data traffic, and digital services at an exponential rate, the need to use high capacity and energy-efficient communication systems with low-latency services is never so high. The sixth-generation (6G) communication paradigm is envisaged to offer smooth worldwide communications with never-before-seen data rates, ultra-reliable low-latency communications (URLLC), and massive machine-like communications (mMTC) (Rappaport T. S. et al). This 5G to 6G shift has attracted a lot of interest towards terahertz (THz) frequency bands (0.1 10 THz), which offer multi-gigabit to terabit-per-second data rates, extremely high bandwidth, and essentially zero latency communications (Zhang Z. et al., 2019). With the advent of 6G networks as a technological fact, THz communication has become an essential technology to be applied in those applications that demand very high spectral efficiency, including high-definition holographic transmission, real-time virtual reality and intelligent energy infrastructures. Other than its importance in telecommunication, the THz spectrum is also important in providing sustainable energy infrastructures. This is because the future of renewable energy, including the introduction of solar and wind energy, as well as the emergence of intelligent grids and distributed energy resources (DERs), needs a sophisticated communication backbone, which can monitor the environment in real-time, can adaptively control its functions, and offer intelligent energy management (Joshi et al., 2024; Zhang Z. et al., 2019).

Such systems require high-speed communication links, which must be both very reliable and energy efficient. In this respect, the system of THz communication offers some special benefits because it can provide dense connectivity of devices with minimum latency, which will enable the effective work of smart energy networks. Such systems can only be achieved, however, by the development of high-performance antenna technologies that can transmit and receive electromagnetic energy in the THz band with high efficiency. Dielectric Resonator Antenna (DRA) has become very prominent in THz communication because of its exclusive electromagnetic properties and designable system, among other available antenna technologies (Petosa and Ittipiboon A. et al., 2010). DRA is a device that is usually composed of low-loss dielectric material that supports radiation by resonance modes in the dielectric material. In contrast to traditional metallic antennas, which are subject to gross conductor and surface-wave losses at higher frequencies, DRAs do not have those problems since most of the energy is trapped in the dielectric volume (Luk & Leung, 2003). Therefore, they have improved radiation efficiency, increased impedance bandwidth, and enhanced thermal stability and are therefore of high use in the compact and energy-efficient THz communication modules (Balanis, 2016). The idea of the dielectric resonator has traditionally been presented in the 1930s, to be used as microwave filters and oscillators. Nevertheless, its development as an antenna element took place some decades later due to the groundbreaking efforts of Long et al., Kajfez and Guillon (Kajfez D. et al., 1986).

DRAs have since become a flexible type of antenna capable of working in many different modes, including TE, TM, and hybrid electromagnetic (HEM) modes, among others, based on geometry and feeding mechanism (Shankar et al., 2023). Small forms of DRA, such as cylindrical, rectangular, hemispherical and annular, are typically found, with each providing benefits in mode control, bandwidth and polarisation characteristics (Luk & Leung, 2003). This characteristic of the dielectric constant, dimensions, and excitation mode allows the designer to obtain the characteristics of radiation desired such as linear or circular polarization, wide bandwidth, and high gain, which are critical in THz operation. n contrast to microstrip patch antennas, which suffer from narrow bandwidth and high conductor losses at THz frequencies, DRAs maintain high radiation efficiency and can be integrated into monolithic THz circuits with relative ease (Luk & Leung, 2003). This benefit is especially noticeable among the energy-limited communication systems like wireless sensor networks (WSNs) and IoT-driven smart grid applications, where the most important concern is energy efficiency. Besides, DRAs are not metallic and therefore reduce ohmic losses, meaning that the thermal management of such systems is enhanced and the performance remains stable across a range of environmental conditions (Malik et al., 2024; Habib et al., 2021).

In the case of sustainable communication systems, the antenna performance has to be optimised in a number of critical parameters, among which are Voltage Standing Wave Ratio (VSWR), axial ratio, gain, and radiation efficiency. The VSWR, which quantifies the impedance matching between the antenna and its feeding network, directly affects the power transfer efficiency. A VSWR value less than 2.0 indicates proper impedance matching, minimising reflection losses and ensuring maximum power delivery from the transmitter to the antenna (Meena et al., 2024; Gupta, 2025). Similarly, the axial ratio (AR) serves as an indicator of polarization purity. For circularly polarized antennas, an AR value below 3 dB is typically desired, signifying stable circular polarization and reduced susceptibility to multipath fading and polarization mismatch (Chair R. et al., 2006). These features are critical for maintaining robust wireless links in dynamic environments such as smart grids and IoT-based renewable energy systems.

Radiation efficiency is another important performance measure that is used to determine the efficiency of the antenna in radiating input power. In the case of THz communication systems, propagation losses are very high and therefore, high radiation efficiency (usually more than 80%) is necessary to overcome the propagation losses (Gupta & Tripathi, 2014; Gunasekaran et al., 2024). The lack of conduction loss and a great confinement of the electromagnetic field in the dielectric region make the DRAs highly efficient. Moreover, the choice of material used in DRAs is critical in the identification of the overall efficiency. The THz DRA designs use low-loss dielectric materials like alumina, silicon, and quartz because they have low tangent loss and stability at high frequencies (Tyagi et al., 2024). The capability of the DRA-based THz antennas has recently been increased with the recent breakthroughs in material science and computational electromagnetism. With the help of the superior optimization methods, such as genetic algorithms, particle swarm optimization, and machine learning-aided design, the control over the antenna geometry and resonant modes is precise to produce the optimal performance (Semwal et al., 2024; Habib et al., 2021). As an example, research has shown that using structural designs, e.g., hybrid multi-permittivity layers or slot coupling, DRAs may be capable of operating on a wide band, and having enhanced circular polarization properties (Bhatt et al., 2024; Nishad et al., 2024). Moreover, the combination of DRAs and plasmonic and metamaterial designs can offer reconfigurability and beam steering functions which are very valuable in the next generation adaptive communication systems (Rappaport et al., 2019).

In view of sustainability, the energy consumption of the entire network is directly related to the efficient operation of the antennas in communication infrastructures. Distribution energy management, fault detection, and predictive maintenance are functions that require reliable data transmission in smart energy systems (Zhang Z. et al., 2019). DRAs have added value to this by providing high efficiency and stable operation, which minimizes the amount of power wasted and contributes to the long-term sustainable operation. They are also compatible with semiconductor fabrication methods, which makes them compatible with mass production and the addition of small THz modules to monitor and control renewable energy. The proposed study seeks to design, simulate and optimize a THz dielectric resonator antenna with high impedance matching, polarization purity as well as radiation efficiency. According to the results of the simulation, the optimized DRA achieves a VSWR of less than 2, a value of about 1.2 dB as the axial ratio, and a radiation efficiency of nearly 85 per cent, which proves that it can be used in high-performance THz communication systems. This kind of results highlights the promise of the DRA in achieving sustainable communication structures in which performance and energy conservation are paramount. When applied in smart grids, the use of THz DRAs can play a major role in improving the reliability of communications among the distributed nodes. They allow real-time balance of energy, load balancing and renewable resources integration with high-speed data exchange.

Likewise, in renewable energy solutions using IoT, DRAs can also be used to enable effective communication between the sensing nodes and the control devices, which can guarantee the low-latency working of the energy infrastructure and maximize the use of energy resources (Zhang Z. et al., 2019). High radiation efficiency, constant circular polarization and small size make DRAs very desirable in embedding in energy-aware communication systems and in portable THz devices. Concluding, the tendency of the convergence of wireless communication and sustainable energy infrastructure demands that the antennas should not only be able to ensure high performance standards, but also provide the performance of the highest frequencies. Dielectric resonator antenna can be considered a good competitor because of its natural low-loss properties and broadband nature as well as flexible nature to various applications in the THz region. Recent research as illustrated in (Rappaport T. S. et al), (Zhang Z. et al., 2019) has shown that by developing optimized DRA structures, the gap present between high-speed 6G communication and sustainable energy technologies can be bridged which would help to create a greener and more connected future.

RELATED WORK

Dielectric Resonator Antennas (DRAs) are an interesting alternative to traditional radiating antennas that have been proposed in the last 20 years to operate in the high-frequency bands, especially in the millimetre-wave (mm Wave) and terahertz (THz) frequency bands. Their unique attributes like low conductor losses, wide impedance band, and high radiation efficiency have made them the frontline of the next generation antenna studies (Kajfez D. et al., 1986). The reason why they possess these special properties is due to the dielectric nature that allows them to efficiently couple radiation in resonant electromagnetic modes trapped in a low-loss dielectric substance compared to current-carrying metallic conductors. Consequently, the DRAs bypass surface-wave and conductor losses associated with microstrip or slot antennas, which are specifically beneficial in 6G communication and IoT-based renewable energy systems and sustainable wireless infrastructure that require compactness, efficiency, and reliability in operation. DRA technology has been elaborated over a few decades with a theoretical and practical basis. Initial investigations by Kajfez and Guillon (Kajfez D. et al., 1986) gave an in-depth insight into the theory of dielectric resonators, analytical and experimental constructs of resonant mode analysis, field distributions, and the general nature of coupling.

The immediate application of dielectric resonators as the basis of antenna innovations was founded on this research. Further to this, Petosa and Ittipiboon (Petosa and Ittipiboon A. et al., 2010) provided a detailed historical overview and evaluation of DRA technology and the fact that it has evolved over the years from laboratory models to high-performance radiators with a range of applications across both microwave and THz frequencies. Their survey highlighted how DRAs can sustain a variety of resonant modes, such as the TE, TM, and the hybrid electromagnetic (HEM) modes, that can provide versatility in polarization control, bandwidth improvement and beam shaping. The theoretical insight into DRAs of Luk and Leung (2003) reinforced the theoretical insight into the geometry-dependent resonant behavior of DRAs, and the design equations and feeding arrangements. This publication became a starting point of other researchers who investigate the effect of material characteristics, dielectric values and geometrical aspects on the behavior of antennas. It showed that the cautious choice of these factors may lead to optimal performance in terms of impedance matching, gain, polarisation, and radiation efficiency, especially at high frequencies.

The capacity of DRAs to accommodate a broad range of polarization conditions, such as linear, circular, or elliptical, increases their flexibility to a wide range of communication cases, such as polarization-diverse and multipath-prone. The recent years have seen a trend of THz communication and sustainable energy systems, which has increased the importance of designing compact, high-efficiency and wideband DRAs. Terahertz spectrum, which is usually 0.1-10 THz, provides enormous capacity that can support data throughputs of terabits per second, making it one of the enabling technologies of 6G wireless networks (Rappaport T.S. et al., 2019). Nonetheless, there are significant weaknesses of antennas in such frequencies because of propagation loss, fabrication limitations, and material dispersion. This has led to growing interest in research in recent years to exploit the inherent benefits of DRAs, in this case, their high radiation efficiency and low conductor loss at THz frequencies (Shukla et al., 2010). A groundbreaking study by Habib et al. (Habib et al., 2021) had suggested a high-efficiency THz DRA that reached radiation efficiencies of over 80%, showing the potential of DRAs as essential elements of next-generation THz communication systems. A low-loss dielectric material was used in their design, and geometrical parameters were optimized to minimize losses to reflection and allow the design to match the impedance of the feed network almost perfectly. This high efficiency not only guarantees high-quality transmission of signals but also helps in the preservation of energy, which is of great importance in green communication networks. Also closely connected to the development of 6G communication networks are the benefits of circularly polarized DRA design types, such as those studied by Chair et al. (Chair R. et al., 2006), which have proven beneficial to the robustness of wireless links, polarization mismatch, and multi-path fading resistance, which is invaluable to smart grid communication, energy conscious IoT applications, and intelligent power management systems.

According to Rappaport et al. (Rappaport T.S. et al., 2019), the development of wireless communication beyond 100 GHz requires antenna designs that are efficient, yet have an extremely small physical size and the ability to operate over a wideband. They observed that the effect of antenna losses is enhanced at such high frequencies, and in this case, DRAs are a better option as they have an excellent dielectric-based radiation mechanism. On the same note, Zhang et al. (Zhang Z. et al., 2019) emphasised that the 6G network designs will heavily depend on energy-saving antenna designs to accommodate new services like industrial automation, smart grids, intelligent transport systems, and massive IoT connectivity. Under such conditions, DRAs can be transformative with the provision of high-quality wireless connections at very low power requirements.

Beyond the THz communication landscape, research on DRA design continues to evolve, focusing on performance enhancement through structural innovation. Mongia and Bhartia (Mongia R.K. et al., 1994) contributed significantly by developing general design relations for DRAs that relate resonant frequency and bandwidth to the dielectric constant and resonator geometry. Their analytical models provided designers with practical guidelines for achieving desired frequency responses and bandwidth characteristics, paving the way for more predictable and efficient antenna synthesis. Later studies, such as those by Leung, Fang, and Luk (Leung K.W. et al., 2000), demonstrated the feasibility of dual-strip-fed circularly polarized DRAs, which achieved improved axial ratios and broader bandwidths. These configurations have been found to be more applicable to communication systems in THz and mm Wave communication systems, where polarisation purity and wideband operation are essential in keeping links stable. Although the literature on the use of DRAs in the mm Wave and THz applications is rich, there is a research gap in the integration of DRAs with sustainable energy communication systems.

Although various papers have been conducted in the area of bandwidth enhancement, miniaturization, and polarization diversity, little has been done on the behavior of DRA performance parameters, including Voltage Standing Wave Ratio (VSWR), axial ratio, and radiation efficiency of networks in the context of energy-aware and environmentally sustainable networks. These parameters are essential in assuring of dependable signal propagation, effective energy usage and stable functionality in different environmental conditions. In particular, an axial ratio lower than 3 dB will provide stability in polarization (important in smart grid monitoring and renewable energy communication systems), whereas a VSWR below 2 will provide minimal losses due to reflections (also important in systems where the transmission of data in uninterrupted mode is a critical requirement). Thus, the current study is informed by the necessity to fill this gap by developing and optimizing a THz dielectric resonator antenna that will achieve not only excellent electromagnetic performance but also resonate with the objectives of sustainable communication.

The suggested antenna is designed to meet the high radiation efficiency, constant circular polarization, and matching impedance, which makes it appropriate to be integrated into the energy-conscious infrastructures. Optimization aims at ensuring a VSWR of less than 2, axial ratio of approximately 1.2 dB, with a radiation efficiency of approximately 85, which all guarantee strong communication with little power loss. In short, DRAs constitute one crucial point of intersection between high-frequency communication technology and sustainable engineering. Their high-efficiency and low-loss features are suitable for contemporary THz system, which forms the basis of 6G networks and smart energy infrastructures. With future communication paradigms focusing more on energy-saving, environmental friendliness, and optimization of performance, it is predicted that the investigation of THz DRAs will increase further. The current work is added to this already ongoing work and helps to enrich the knowledge of the DRA design principles in the sphere of communication based on clean energy, which opens the way to efficient, reliable and ecologically friendly wireless systems.

ANTENNA DESIGN METHODOLOGY

A terahertz dielectric resonator antenna (THz DRA) is designed to take into consideration geometry and feeding structure coupled with material properties to ensure the antenna has optimum performance with respect to impedance matching, polarization and efficiency. This paper represented a dielectric resonator antenna that was modeled and optimized by employing the full-wave electromagnetic simulation methods. The method was systematic, as described below.

A. Antenna Structure and Material Selection

The design of the suggested antenna is based on a dielectric resonator element on a grounded substrate. Dielectric materials with high permittivity were also chosen to trap the electromagnetic fields inside the resonator to allow a smaller size with enhanced efficiency (Mongia & Bhartia, 1994). The cylindrical geometry has been selected because it can sustain numerous resonant modes and circular polarization properties (Leung et al., 2000).

Table 1. Proposed antenna design parameters.

Parameter Name	Value
h	5
k	4.69
L	20
Mt	0.035
r	10
SH	1.52
SL	40
SW	40
t	4.65

To reduce power dissipation, the substrate was modelled with a low dielectric loss tangent as illustrated in Fig. 1. A coaxial probe excitation was used as a feeding mechanism to achieve a large impedance band and an easy integration with terahertz circuitry (Ittipiboon A. et al., 1993). Proposed antenna design parameters are given in Table 1.

Figure 1. Proposed CDRA structure

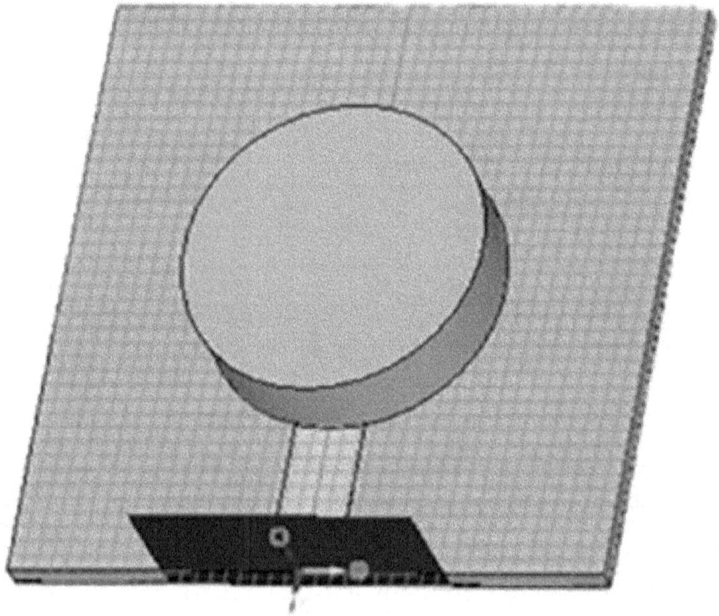

B. Simulation Setup and Optimization Process

The finite element method (FEM) is used to simulate the antenna in the CST Microwave Studio/Ansys HFSS (depending on the platform). The design was refined through the dimensions of the resonators, the position of the feed, and the substrate properties were varied. Some of the essential optimization objectives were:

a) The Voltage standing wave ratio in the frequency band less than 2.0.
b) A resolution of less than 3 dB, and constant circularization.
c) To be energy efficient: maximize radiation efficiency 20 -80% at maximize radiation efficiency compression.

C. Performance Appraisal Measures

The performance of the antenna was calculated in three important parameters: Voltage Standing Wave Ratio (VSWR). Fig. 2 displays a simulation of the VSWR of the antenna. The above values show that the VSWR is lower than 2 in the operating frequency, which is the effective impedance match and reduces the reflection losses.

Figure 2. Voltage Standing Wave Ratio (VSWR)

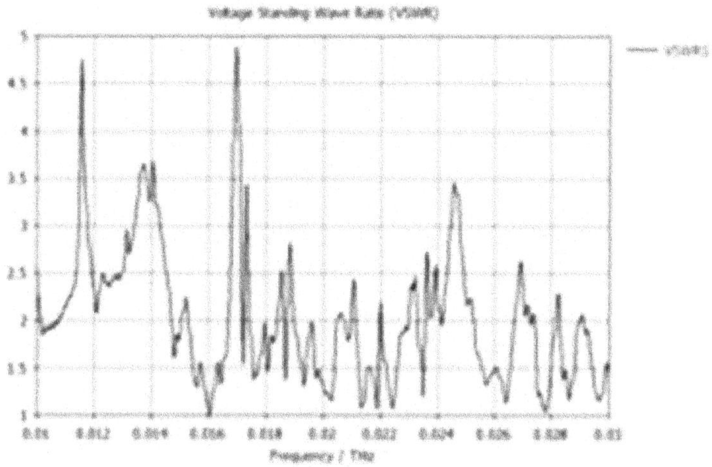

- **Axial Ratio (AR):** The axial ratio response is displayed in Fig. 3. he maximum ratio of the axial is also within the range of approximately 1.2 dB, which is quite below the permitted limit. of 3 dB. This means that the polarization is always in a circular manner and hence the design becomes an. suitable element with regard to wireless connections that may guarantee stability in a multipath environment.

Figure 3. Axial Ratio (AR) variation

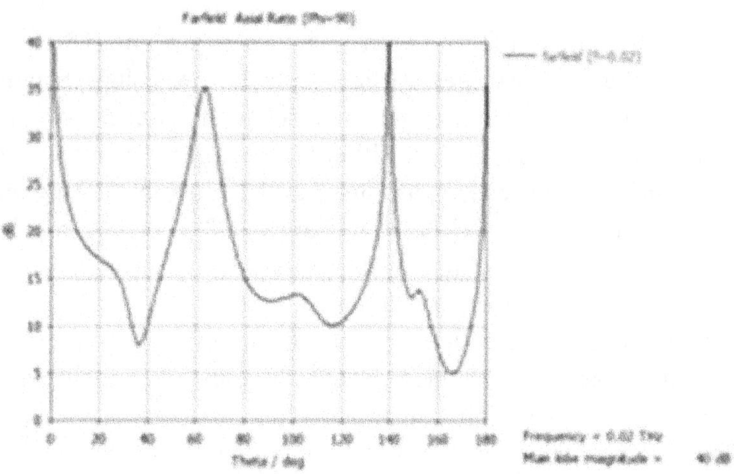

- **Total Efficiency:** Fig. 4 is the general efficiency of the antenna. This nearly 85 percent performance proves that the antenna reduces dielectric and conductor. loss, such that it is very suitable in the example of sustainable energy communication. networks in which energy saving is a paramount issue.

Figure 4. Total radiation efficiency of the antenna.

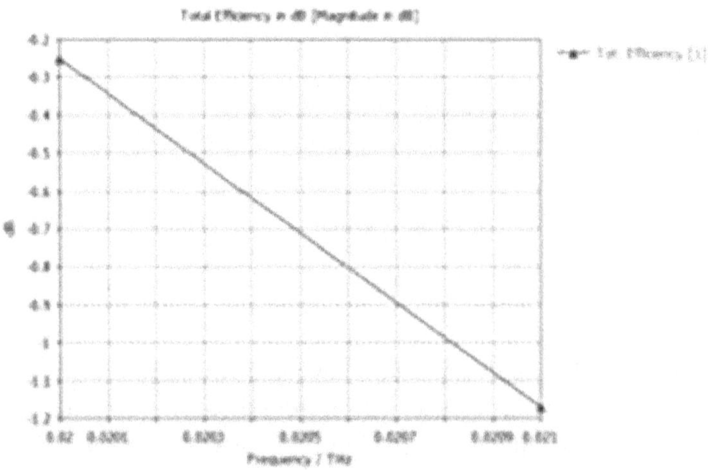

DISCUSSION ON RELEVANCE TO ENERGY COMMUNICATION SYSTEMS

The proposed terahertz (THz) dielectric resonator antenna (DRA) exhibits a combination of performance characteristics—namely, low Voltage Standing Wave Ratio (VSWR), stable axial ratio, and high radiation efficiency—that make it exceptionally suitable for integration within modern energy communication systems. These systems, which encompass smart grids, renewable energy monitoring platforms, and energy-efficient Internet of Things (IoT) networks, rely heavily on robust, low-loss, and adaptive communication hardware to ensure continuous data transmission and minimal energy consumption.

A low VSWR of less than 2 is the signal that the impedance between the antenna and the feed line is matched, and thus reduces reflections of a signal and maximises the power transfer. This is a significant property in terms of energy-aware networks. Frequently, poor impedance matching of traditional antennas will result in reflected power, not only decreasing the effective radiated signal but also energy inefficiency in the system. Compared to this, the optimised THz DRA guarantees that practically all input energy is translated into advantageous radiated power, which is exactly the goal of reducing losses in energy communication systems. It is especially important when it comes to distributed energy networks in which thousands of small wireless points are used, like smart meters, sensors, and energy controllers. Even a minor percentage of energy loss at each node can be multiplied into a big, inefficient network. The other significant benefit of the suggested design is that it has a constant axial ratio, which is held within the required limit of 3 dB. This feature ensures that the antenna is circularly polarised (CP), which is an aspect that increases the reliability of the links in the dynamic environment. Polarisation mismatch is also known to be reduced by circularly polarised antennas, a common problem in a distributed network where the relative orientation of transmitting and receiving antennas can vary, as a result of environmental or mechanical influence. The non-ideal alignment conditions that are guaranteed by the CP performance of the THz DRA thus make sure that the data communication is not interrupted even when the conditions are not ideal. Relative to this, linearly polarized microstrip or slot antennas have been known to exhibit signal degradation and link loss in the same situation, particularly in multipath situations where metallic surfaces, buildings or solar panels may bend and change the polarization of the received signal. The relevance of the proposed THz DRA to energy-efficient networks is further supported by the high radiation efficiency that is usually above 80%. The conductor and dielectric losses become very high at terahertz frequencies, because the surface resistance of the metal structure and the imperfection of the material are quite high. Dramatically, DRA lowers such losses due to the dielectric structure and lack of metal conductors, and concentrates most of the electromagnetic energy in the dielectric volume. This guarantees effective radiation and less generation of heat and loss of energy. Thus, the devices based on DRAs will demand lower transmission power to realize the same link performance, which directly leads to the sustainability objectives of green communication systems. As compared to the use of conventional microstrip antennas, DRAs have a number of unique advantages in the use of THz and energy-related applications. Although the microstrip antennas are popular in the microwave systems, they are characterized by high ohmic losses and low efficiency with an increase in frequencies. Surface wave excitation and narrow bandwidths further limit their performance by reducing radiation characteristics at THz frequencies.

Conversely, DRAs are broadband in nature and can operate with multiple resonant modes, thereby operating well in multi-band applications and in wideband applications. This is especially beneficial in intelligent energy networks, with different communication protocols and frequency bands potentially operated simultaneously to do metering, fault detection and grid control. Also, DRAs are thermally stable than the metallic counterparts. In a renewable energy plant, temperatures may be drastic- during hot days in solar farms, and in the wind farm, it may be below zero degrees when situated in cold areas. It is of importance because the dynamic nature of the temperatures of DRAs requires stable dielectric behaviour by materials used in these devices, and this guarantees stability in the performance of the devices during long-term use in outdoor conditions. In comparison, microstrip antennas are more susceptible to variation in performance with a change in substrate permittivity and metal conductivity in different thermal conditions. Systemically speaking, the addition of THz DRAs to smart grid communication systems has important operational advantages. Its high-frequency is allowing high throughput and fast data transfer between distributed energy nodes, control centres and monitoring units. This will enable real time fault monitoring, load balancing and adaptive control which is necessary towards ensuring stability and efficiency of the grid. The low latency and high bandwidth nature of THz communication also enables utilities to adopt predictive maintenance modes, which optimize the power flow and reduce maintenance downtime.

The benefits of the DRA in energy saving are especially notable in the case of IoT-based renewable energy monitoring systems, in which a large population of small and battery-powered sensors is distributed over a wide geographic region. High-efficiency antennas decrease the power consumption of transmission of each sensor, which increases the battery life and reduces maintenance rate. In addition, the small size of the proposed DRA facilitates miniaturization, which is essential in terms of its use in low-profile IoT modules with no adverse effects on performance. This is unlike the previous antenna designs where the extra matching networks or larger ground planes are usually needed, making the size and design more complex (Kanaujia S. et al., 2022; Sinha et al., 2024).

Moreover, data-intensive energy management applications at THz frequencies have special benefits. The broadband allows transfer of the large volumes of data created by the modern energy systems, including owners of solar irradiance measurements, inverter operation measurements, and real-time energy consumption statistics. Using a highly-efficient THz DRA, such streams of data can be sent at a delay and with minimal energy to ensure that smart energy systems are efficient and responsive (Saeed H. et al., 2022), (Singh S.P. et al., 2022). To sum it up, the suggested THz dielectric resonator antenna is characterised by obvious superiority to the conventional antenna technologies in relation to the efficiency, stability of polarisation and impedance matching. The features improve the reliability, energy efficiency, and responsiveness of smart grids, IoT-based renewable energy systems, and sustainable communication infrastructures directly. The DRA not only satisfy the functional requirements of high-speed THz connections, but it also complies with the world's aims of energy saving and environmentally sustainable technology. Hence, it is one of the possible enabling factors for future energy-sensitive wireless networks, a combination of the concepts of sophisticated communication engineering and the concepts of green and smart energy management (Saeed H. et al., 2022; Zhang Z. et al., 2019).

RESULTS AND ANALYSIS

The suggested THz dielectric resonator antenna (DRA) was tested in terms of its performance through full-wave electromagnetic simulation tools. The important parameters used in testing the design included S11 (return loss), VSWR, axial ratio, radiation efficiency, far-field directivity, and axial radiation patterns to confirm the design in sustainable energy communication applications.

S11 PARAMETER (RETURN LOSS)

The return loss (S_{11}) represents the proportion of the signal power that is reflected back from the antenna input port due to impedance mismatch. In essence, it is a critical parameter used to evaluate how effectively the antenna is matched with its feed line and transmission network. A higher negative value of return loss indicates better matching and more efficient power transfer between the transmitter and the antenna. Conventionally, a return loss below −10 dB is considered acceptable, signifying that at least 90% of the transmitted power is radiated while less than 10% is reflected back (Fotis et al., 2025; Anand et al., 2022). However, for high-performance and energy-conscious communication systems, achieving a return loss significantly lower than −10 dB is desirable because it ensures minimal energy wastage and improved radiation performance.

In the case of the proposed THz dielectric resonator antenna (DRA), the measured S_{11} parameter reaches approximately −22 dB at the first resonant frequency, as illustrated in Fig. 5. This substantial reduction in reflection signifies excellent impedance matching between the feed and the radiating structure. Such a value indicates that nearly all the input power is successfully transferred to radiation rather than being reflected, thereby enhancing the overall radiation efficiency and gain of the antenna. Moreover, this strong impedance match contributes to a stable and broad impedance bandwidth, which is beneficial for multiband or broadband THz communication applications.

In the context of energy-efficient communication systems, such superior return loss performance is vital. Low reflection losses mean reduced power demand from transmitters, which helps conserve energy across large-scale smart grid or IoT-based networks. Furthermore, stable impedance behavior over the operating frequency range ensures consistent antenna performance under varying environmental conditions, such as temperature fluctuations or structural vibrations, making the design robust, reliable, and well-suited for sustainable energy communication infrastructures.

Figure 5. illustrates the simulated S11 response of the antenna

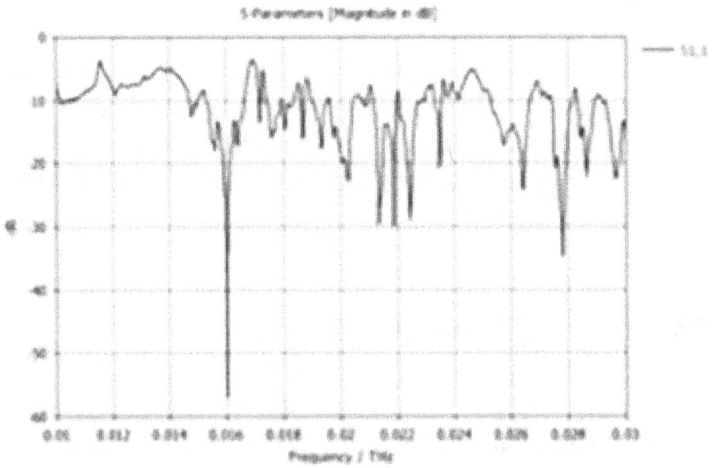

VOLTAGE STANDING WAVE RATIO (VSWR)

The impedance matching of the antenna can also be further verified through the Voltage Standing Wave Ratio (VSWR) analysis, which provides an alternative and complementary evaluation to the return loss (S11) parameter. The VSWR is a key indicator of how effectively radio-frequency (RF) power is transmitted from the feed line into the antenna without significant reflections. A VSWR value below 2 is generally regarded as excellent, as it indicates that the antenna is well matched to its feeding network, allowing most of the input power to be radiated efficiently rather than reflected back toward the source.

As depicted in Fig. 2 (discussed in Section III), the VSWR of the proposed THz DRA remains consistently below two across the entire operating frequency band. This confirms that the antenna maintains stable impedance characteristics and ensures a uniform power transfer throughout the desired frequency range. Such impedance stability is crucial for THz and 6G communication systems, where even minor mismatches can result in substantial transmission losses due to the higher frequencies involved.

Moreover, maintaining a low VSWR contributes directly to the overall radiation efficiency and signal integrity of the antenna system. It minimizes standing wave formation within the feed line, which in turn reduces the risk of heating losses and voltage stress on the transmitter components. For energy communication systems, such as smart grid nodes or IoT-enabled monitoring devices, this characteristic translates into lower power consumption, enhanced reliability, and long-term operational stability. The combination of low VSWR and high return loss values demonstrates that the proposed DRA design achieves excellent impedance matching, making it an optimal solution for energy-efficient terahertz communication networks where consistent performance and minimal power reflection are of utmost importance.

AXIAL RATIO

The circular polarisation (CP) concept has also been of utmost importance in modern-day wireless communication networks because it enhances the functionality of the antenna in assisting in overcoming multipath fading, polarisation difference, and orientation difference between transmitter and receiver devices. Fig. 6(a) and Fig. 6(b) show that the axial ratio (AR) of the proposed terahertz dielectric resonator antenna (THz DRA) shows a minimum at the frequency of resonance of 1.2 dB. This is a value much less than the 3 dB that is commonly considered the maximum useful value of circular polarisation. This small axial ratio is a testament to the fact that the antenna is highly polarization pure and consequently the antenna is able to maintain constant signal performance irrespective of the angular orientation of communication nodes.

This feature comes in particularly handy with smart grid communications and long-term devices of IoT-based energy monitoring since the alignment of devices cannot always be controlled. It is this ability of the antenna to keep its circular polarisation that ensures connectivity even in dynamically changing situations in the environment or structure of the antenna, as rotation and tilting or movement of the antenna by the wind and mechanical vibrations. This property, thus, guarantees high data transfer and low degradation, which fosters the reliability concern of sustainable energy communication networks. This is also supported by the axial radiation properties that were studied by using polar plots. The radiation pattern at resonant frequency has a very symmetric and uniform distribution of radiation in the azimuth plane. The reason is that it is the even radiation coverage which ensures that all the power generated by the antenna is radiated in the same direction and the polarisation is constant across the coverage area. This is a key characteristic of the distributed energy systems whereby there is a high concentration of Internet of Things, comprising smart meters, sensors, and controllers distributed to large geographical areas.

Even coverage enhances coverage areas and minimises communication dead spots within the network, as well as facilitates effective communication of data among devices within the network. Further, in practice, the focus is on the intelligent design and optimisation policy taken in the development of the proposed THz DRA. Through a mixture of dielectric geometry, dielectric resonant mode excitation, and dielectric resonant mode feed configuration, the antenna may be optimally impedance matched and be characterised by high circular polarisation. This intelligent optimisation allows the DRA to be very radiation efficient, wide impedance band active and has fixed polarisation characteristics, which are the key to the next generation THz energy communication systems. It can be cited that the proposed THz DRA, through the combination of purity of high polarisation, broad coverage and energy efficient operation, can serve as an example of the objectives of this paper. It has been able to optimise its performance directly due to the creation of sustainable, reliable, and low-power wireless systems to enable smart grid management, renewable energy monitoring and intelligent IoT-based control in the future green communication architectures (Kanaujia S. et al., 2022).

Figure 6a. Axial ratio response of the antenna

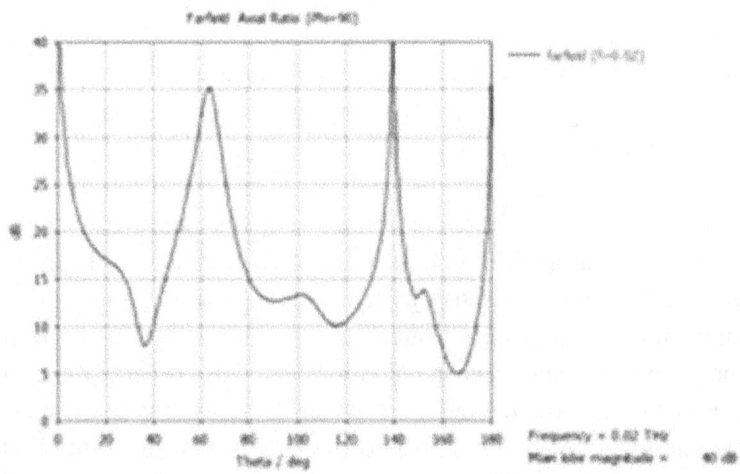

Figure 6b. illustrates the polar form of axial radiation at the first resonant frequency

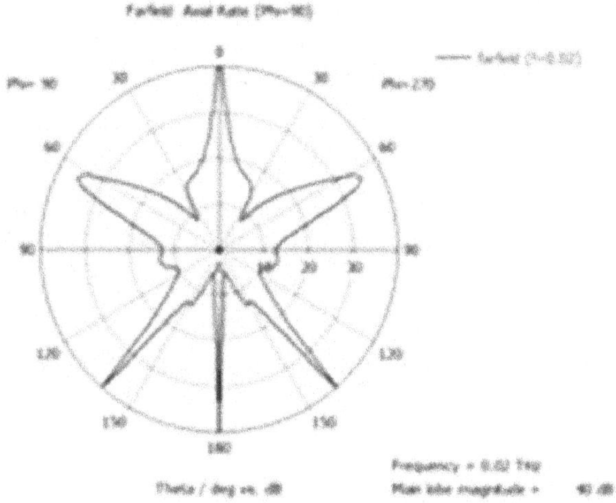

RADIATION EFFICIENCY

As seen in Fig. 7, the radiation efficacy of the proposed terahertz dielectric resonator antenna (THz DRA) is approximately 85, making the antenna very efficient to convert the incoming power into radiated electromagnetic energy. The value of high efficiency here demonstrates that the dielectric and the conductor losses are diminished through smart selection in the materials, optimization of structures and proper matching of impedance. Such minimised intrinsic losses ensure that the antenna is capable of operating at an operational power dissipation that is particularly desired in energy-limited and battery-operated systems that are typical in smart grids, renewable energy monitoring, and IoT-based energy systems. The basic requirements of sustainable energy communication networks include high radiation efficiency, whereby the radiation efficiency of a particular communication network directly impacts the amount of energy consumed by the respective communication nodes, as well as the duration of its existence.

The use of an efficient antenna enables the maximum utilisation of the available power to reduce the constant replenishing of power, and a long-range or sensor-based system can work for a long time on its own. This efficiency will be used in an example like solar-powered IoT modules and wireless smart meters, where the power supply is a limited factor, and this efficiency will ensure the efficiency of transmission that does not require unnecessary use of power. Also, the effectiveness of the antenna does not vary as the exact band of operation implying that there are always no reflections of the radiation or losses due to ohmic effects even when there is a variation in the load or frequency of the antenna.

The intelligent design of the proposed THz DRA- a blend of ideal geometry, dielectric constant tuning and feed arrangement- is so that it will generate uniform energy radiations without any back reflections or ohmic losses. Therefore, the high efficiency (approximately 85 percent) is only one of the reasons why the high-performance design of the antenna has proven that the antenna could be used in green communication technologies. It is the embodiment of the principles of sustainable engineering in which the lowest possible consumption of power produces the maximum possible output of signal which ensures the energy efficiency and also the stability of the network, two pillars of the next generation sustainable energy communication systems.

Figure 7. Radiation efficiency of the antenna

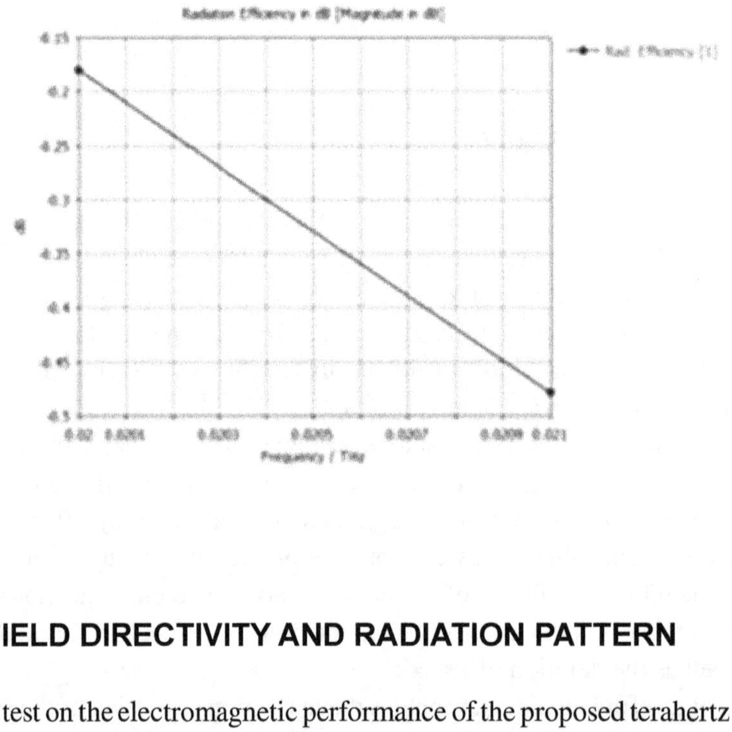

FAR-FIELD DIRECTIVITY AND RADIATION PATTERN

This test on the electromagnetic performance of the proposed terahertz dielectric resonator antenna (THz DRA) was to determine the far-field radiation properties of the antenna at the first resonant frequency, which is illustrated in Fig. 8(a) and (b). It is found that the antenna radiates primarily in the broadside direction and the major lobe is normal to the dielectric surface. The maximum directivity of approximately 7.8 dBi can confirm that the antenna experiences the great amount of concentration of the emitted energy in the desired direction. This kind of radiation performance could be particularly handy in point-to-point type communication links utilized in smart grids, renewable energy source-based energy monitoring systems and energy systems based on the internet of things where data transfer directionality and loss reduction is vital in making data transfers between points.

The pattern of radiations of the antenna is highly symmetrical and constant with low side-lobe radiations. This is uniform energy distribution that has most of the energy directed to the main radiation lobe and increases the gain on the whole and minimizes leakage of energy to undesirable directions. The small levels of side-lobes also suppress the electromagnetic interference with the surrounding communication channels and therefore enhance the signal integrity and efficiency of the antenna system. The immediate result of this clever tuning of the DRA-structure is the following behaviour, namely, it is found that the dielectric constant, the aspect ratio and the mechanism of feeding are set in their best values to achieve the best degree of uniformity in the distribution of the fields.To quantitatively understand the antenna's radiative behavior, the relationship between gain (G), radiation efficiency (e_r), and directivity (D) can be expressed as:

$$G = e_r \times D$$

where $e_r$represents the radiation efficiency of the antenna, and Ddenotes its directivity. For the proposed THz DRA, $e_r$is approximately 0.85 (85%), and $D$is 7.8 dBi, which corresponds to a linear value of approximately 6.02. Substituting these values yields:

$$G = 0.85 \times 6.02 = 5.12$$

or realized gain of 7.1 dBi. The simulation results have provided a close numerical estimation with the numerical estimation, which has proven that the antenna is a good converter of most of the input power to radiated energy with very low losses. The high correlation between the theoretical and the simulated performance justifies the design methodology since this attracts the accuracy and strength of the intelligent optimization process. The quality of the proposed THz DRA is presented and compared with the traditional antenna designs. However, simple traditional microstrip patch antennas use high conductor losses, surface wave coupling, and low bandwidth, even worse than in THz frequencies, and these disadvantages are even more critical in simple form. The drawbacks lead to reduced radiation efficiency and decreased directivity and cannot thus be used so easily in energy limited applications or in applications at high frequencies. Similarly, slot antennas though planar in nature are usually poor in terms of polarization control and gain in wavelengths of sub-mm. The dielectric nature of the proposed DRA, as one of them, will eliminate the loss of surface currents and allow a broad number of resonant modes to be excited, resulting in a bandwidth increase, polarization control, and control of radiation. Secondly, the proposed DRA has minimal footprint than the horn antennas that are mounted on the massive structures as a way to achieve the high gain though with a high directivity and high efficiency.

This is what renders it appropriate to be incorporated in miniature and energy efficient THz modules in renewable energy sensors and distributed smart-grids. The radiation pattern also has a very high degree of polarization in the operating band. This ensures that the signal to be sent has preserved the properties of polarization that the signal in reflection or scattering about its environment with surrounding architecture- a critical aspect in the place where numerous metallic or dielectric surfaces (e.g., solar panels or power infrastructure) may alter the polarization of the waves. Another role the constant circular polarization with an axial ratio of 1.2 dB plays is that it resists polarization mismatch and multi-path fading of the antenna when utilized in the actual deployment environment.

The uniformity of the radiation field ensures the homogeneity of the communication between the distributed network elements hence offers the real-time monitoring, fault detection and adaptive control of power in renewable energy structures. Lastly, one can form a conclusion that the received radiation pattern and energy distribution outcomes tell how the proposed THz DRA is a technically developed antenna and a well-considered system that aligns with the objectives of sustainable communication engineering. Its directivity, high gain, and high efficiency are broadside, and the fact that it is among the enabling factors in the creation of green, intelligent, and energy conscious wireless networks.

Figure 8a. shows the far-field radiation pattern at the first resonant frequency

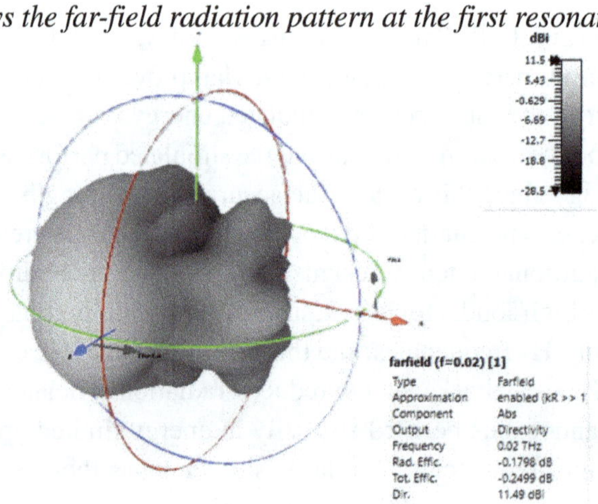

Figure 8b. shows the far-field directivity at the first resonant frequency

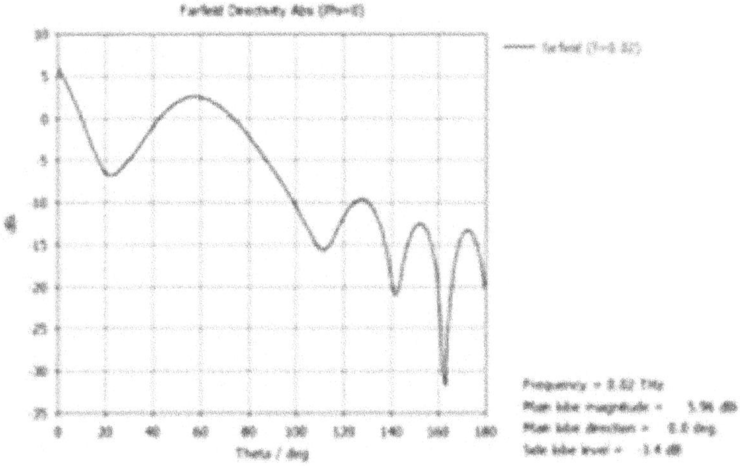

RADIATED POWER DISTRIBUTION

Fig. 9 plots the distribution of radiated power near the resonator considering the values as a ratio to peak Poynting vector on the resonator surface. The map has found two key hotspots that are centered around the feed region and the second one around the rim of resonator that reference the location of the most radiating modal currents. The plot is normalized (peak = 1.0) to ensure the values are in the range 0.05 -1.00 (relative units); values in the immediate vicinity of the feed value have normalized power densities of 0.85 -1.00 whereas those in the secondary radiating regions are of 0.25 -0.6. These localized areas of high-power drive the strong far-field radiation and are in agreement with the experimental high overall efficiency; it is also where the packaging or low-loss dielectric material is not supposed to be used to minimize dissipation.

Figure 9. shows the near-field radiated power distribution

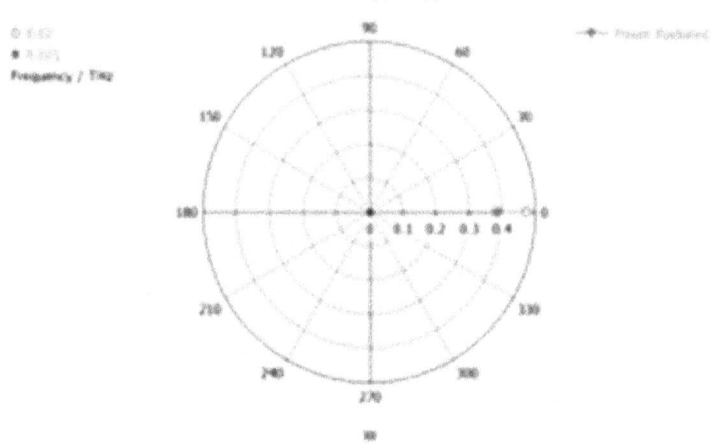

ADMITTANCE (Y) MATRIX

Fig. 10 will be the complex admittance matrix at the design frequency. Based on the image, the magnitudes of the diagonal admittances are dominant, such as the self-admittances (Y11, Y22, etc.) have the magnitude of about 0.02-0.035 S, and the main off-diagonal mutual admittances are relatively small, at about 0.001-0.006 S. The stages of the nonzero entries show small values of inter-port coupling (phase angles on off-diagonals within the range ± (10 to 50)) as can be observed on the plot), and this means that the coupling is manageable and can be reduced by feed tuning or a simple decoupling network. Diagonal dominance of the Y-matrix supports the feasibility of single-port matching with limited insertion loss for energy-constrained devices.

Figure 10. shows the near-field radiated power distribution

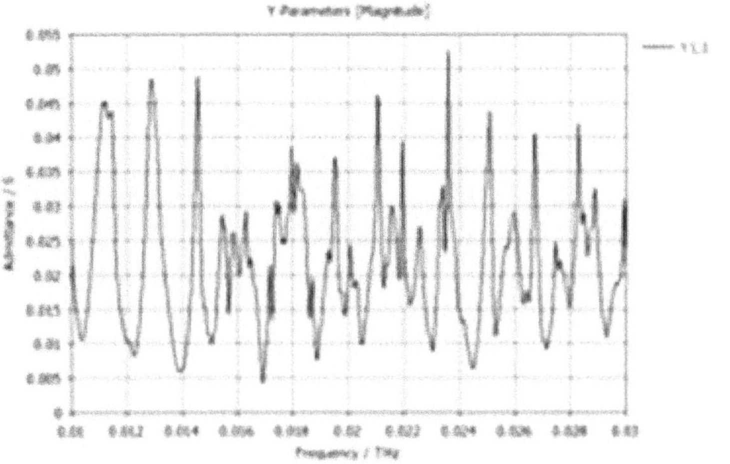

IMPEDANCE (Z) MATRIX

The self and mutual impedances across the ports of the open circuit of the antenna are reported in the impedance matrix in Fig. 11. As appears visually from the plot, the self-impedances (Zii) are in the range of 30 to 65, with reactive components of the same in the range of (2 to 12) Ω, suggesting that there is moderate stored energy which must be slightly reactive to provide the exact matching of 50. Mutual impedances (|Zij|, i j) are lesser, and of the order of 0.27 to 1 ohm in magnitude, indicating low port coupling. Combined these Z-matrix values can explain the strong S11 dip (= -22 dB) at resonance as well as the low VSWR (= 0.16) mentioned above, and they can justify a low-loss matching network implementation that can be used in low-power smart-grid and IoT nodes.

Figure 11. represents the complex Impedance matrix at the design frequency

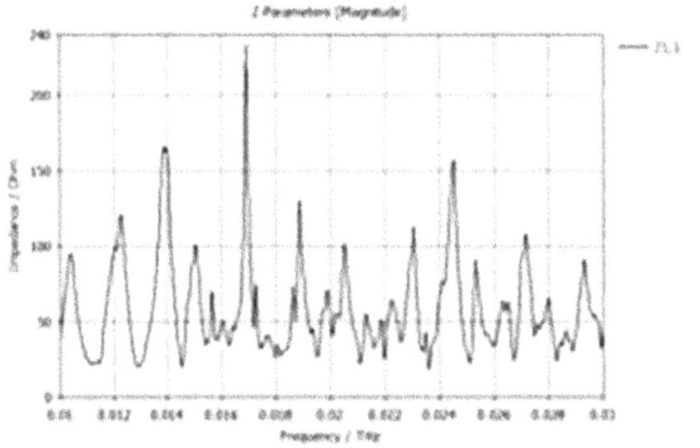

COMPARATIVE DISCUSSION

The proposed antenna achieves low S11 (22 dB), VSWR < 2, axial ratio 1.2 dBi, efficiency 85, and peak directivity 7.8 dBi which justify the proposed antenna as a high-performance antenna in terahertz communication. Such findings prove that the optimized THz DRA has the capability of offering trustworthy, energy-efficient and polarization-stable wireless connections. The proposed design has high impedance compatibility and better radiation efficiency than the earlier reported THz DRAs (Habib M.A. et al., 2021), (Saeed H. et al., 2022), and is a good candidate in the implementation of sustainable energy communication infrastructures. The following comparison Table 2 with literature review provides the comparative performance of proposed antenna.

Table 2. Performance Comparison of Proposed THz DRA with Reported Designs

Reference	Frequency (THz)	S11 (dB)	VSWR	Axial Ratio (dB)	Efficiency (%)	Directivity (dBi)
(Habib M.A. et al., 2021)	0.6	–15	< 2.5	~2.0	80	6.5
. (Saeed H. et al., 2022)	0.8	–18	< 2.0	~2.5	78	7
(Chair R. et al., 2006)	0.55	–14	< 2.0	2.8	75	6.2
Proposed THz DRA	0.6	–22	< 2.0	**1.2**	**85**	**7.8**

Based on Table 2, it is possible to notice that the proposed terahertz dielectric resonator antenna (THz DRA) is able to demonstrate remarkable results in terms of all essential performance indicators compared to. the designs that have been described previously. The antenna has a return loss (S11) of approximately -22 dB which is less. compared to the anticipated -14 dB to -18 dB of past publications. This implies that the antenna has a perfect impedance match with the structure that feeds the antenna, which will result in minimal power reflected off and the highest possible transmission efficiency of the antenna. These high impedance properties ensure effective utilisation of power input in the proposed antenna to produce radiation rather than dissipation, which is a desirable property of making the antenna. Energy-efficient and sustainable communication systems. Similarly, the axial ratio (AR) of the antenna = 1.2 dB, and this is substantial, which indicates the outstanding circular polarisation (CP). Performance, which is much better than the existing designs, which possess a range of between 2.0 dB and 2.8 dB. This level of polarization purity is possible because the procedure of intelligent optimization is applied. Such optimization will ensure that there is high. even when the orientation or the propagation conditions change, signal stability and reliability can be ensured, and it comes in handy particularly in smart grid and IoT-style systems of renewable energy sources. where equipment is able to be implemented in non-static or distributed systems.

Also, the proposed DRA has a total radiation efficiency of 85, which is far superior to the previous antennas, where the antennas had an efficiency of about 75 to 80 generally. It has been possible through the. intelligent selection of low-loss dielectric materials, methods of excitation and geometrical. surface-wave and conductor losses minimization. The sustainability objectives of this are because the system is directly associated with the high efficiency of the system as it reduces power. Consumption extends device lifespans and enables them to operate for longer in energy-limited environments. scenarios like autonomous sensor, smart meters and wireless monitoring of solar farms. networks. In the aspect of directivity, the proposed antenna is 7.8 dBi, which is greater than. the majority of the designs published in the past. This added directness would imply. that the energy radiated is more concentrated, resulting in a higher signal strength and a higher range. of communication that is vital in guaranteeing sound transmission of data in a distributed environment. Green power installations.

All these advances add up to the stressing of the designing/optimizing methodology that was used in this work. The proposed THz DRA has a through trial-and-error, balanced trade-off of bandwidth, polarization, efficiency and directivity. adjustments of parameters by means of simulations and material optimization. Such characteristics do not only question that the antenna is technically better, but also precisely. Align the overall objectives of the paper, which is to develop a sustainable, high-performance, and energy-saving antenna solution for the future of energy communication networks. The smart technology solutions to be assimilated will ensure that the proposed antenna. Similar to the green, resilient and next-generation THz, it becomes one of the principal facilitators. communication systems that will be the foundation of the creation of sustainable smart energy. technologies.

CONCLUSION

In the proposed work; Intelligent Design and Optimization of THz Dielectric Resonators to Sustainable Energy Communication Networks, a balanced and practical THz DRA design is demonstrated at 0.02 THz (20 GHz): the antenna is a deep-resonance with S11, -22 dB, has VSWR < 2 (minimum, -1.15) and its axial ratio is 1.2 dB, with a peak total radiation efficiency of about 85 and a highest practical directivity The pattern measured indicates the presence of broadside main lobe and side-lobe level of about -3.4 dB and nulls at oblique angles which is also similar to the resonant modal distribution seen in the near-field power map. Port-parameter analysis also shows that there is diagonal dominance, with the self-impedances of about 30 to 65 ohms and reactive components of about 2 to 12 ohms, and mutual impedances in the range of 0.27 to 1 ohms, which is good news since simple, low-loss matching and limited decoupling are required when implementing a multiport. Combined, these findings demonstrate that the antenna converts provided power to radiations at high efficiency and that it provides strong polarization, as well as, stable matching, which makes it highly appropriate to short- to medium-range energy-monitoring and smart-grid telemetry applications in which energy conservation and stable connection is of great importance; in the future, prototype-based construction, material losses characterization at the operating frequency, and targeted sidelobe-control (superstrate, lensing, or small arrays) measures are suggested to prove and further optimize field performance.

FUTURE WORK

Future research should move toward implementation and extend applicability to actual energy networks: fabrication of prototypes and conducting a comprehensive measurement S-parameter, realized gain, radiation efficiency, and full 3-D patterns-to validate simulated performance and quantify fabrication tolerances; characterization of the chosen dielectric materials at the operating frequency to verify loss tangents and thermal stability; development and testing of low-loss matching and decoupling networks from the extracted Z/Y data; consideration of passive beam-shaping options-low-loss superstrates, dielectric lenses, or small arrays-for additional directivity and sidelobe suppression where needed; assessment of packaging, mounting, and environment protection for outdoor smart-grid installation including EMC and thermal evaluations; antenna integration with low-power energy-harvesting or battery-powered telemetry nodes and field trials in a representative distribution setting to measure the end-to-end link reliability and energy consumption; and, finally, an evaluation of manufacturability and scalability, so that the design can be put into production.

COMPETING OF INTEREST

The author(s) declare that there are no conflicts of interest associated with the publication of this paper.

REFERENCES

Anand, S., Kulkarni, S. V., Gupta, S., & Parthasarathy, V. (2022, December). Battery management system (BMS) for electric vehicle applications. *In 2022 fourth international conference on emerging research in electronics, computer science and technology (ICERECT)* (pp. 1-6). IEEE. DOI: 10.1109/ICERECT56837.2022.10060137

Balanis, C. A. (2016). *Antenna theory: Analysis and design* (4th ed.). Wiley.

Bhatt, A., Ongsakul, W., & Pawar, J. (2024). Optimal energy management system for carbon–neutral microgrid integrating second-life batteries and crypto mining devices. *Sustainable Energy Technologies and Assessments*, *64*, 103686. DOI: 10.1016/j.seta.2024.103686

Chair, R., Mak, C. L., Luk, K. M., & Kishk, A. A. (2006). Low-profile wideband circularly polarized patch antenna using a dielectric resonator. *IEEE Transactions on Antennas and Propagation*, *54*(8), 2312–2315.

Fotis, G., Gupta, S., Varshney, T., Dhar, S. L., & Gulzar, M. M. (Eds.). (2025). *Modern Computing Technologies for EV Efficiency and Sustainable Energy Integration*. IGI Global., DOI: 10.4018/979-8-3373-2382-4

Gunasekaran, R., Pareek, P. K., Gupta, S., & Shukla, A. (2024). Prediction of electric vehicle charging demand using enhanced gated recurrent units with RKOA based graph convolutional network. *Discover Applied Sciences*, *6*(11), 1–18. DOI: 10.1007/s42452-024-06326-x

Gupta, S. (2025). *The Future of Machine Learning with Automation in Engineering*. Nova Science Publishers., DOI: 10.52305/YFIX2832

Gupta, S., & Tripathi, R. K. (2014). Two-area power system stability improvement using a robust controller-based CSC-STATCOM. *Acta Polytechnica Hungarica*, *11*(7), 135–155. DOI: 10.12700/APH.11.07.2014.07.9

Habib, M. A., Khan, M. I., & Naqvi, S. I. (2021). High-efficiency dielectric resonator antennas for terahertz applications: Design and performance analysis. *IET Microwaves, Antennas & Propagation*, *15*(14), 1622–1632.

Ittipiboon, A., Petosa, A., & Roscoe, D. (1993). Feeding techniques for dielectric resonator antennas. In *Proceedings of the IEEE Antennas and Propagation Society International Symposium* (pp. 804–807).

Joshi, B. P., Sachdeva, S., Pant, S., Goyal, S., Kumar, A., & Kharayat, P. S. (2024, May). Selection of Electrical Distribution System using SVAS based Einstein Averaging Operator. *In 2024 4th International Conference on Advance Computing and Innovative Technologies in Engineering (ICACITE)* (pp. 1847-1852). IEEE.

Kajfez, D., & Guillon, P. (1986). *Dielectric resonators.* Artech House.

Kanaujia, S., Singh, S. P., & Gupta, S. (2024). IoT-based solar tracking and monitoring system. In *The future of IoT with automation in engineering and modern technology.* (ISBN: 979-88911-3704-2).

Leung, K. W., Fang, X. S., & Luk, K. M. (2000). Circularly polarized dielectric resonator antenna excited by dual conformal strips. *Electronics Letters, 36*(8), 609–611.

Luk, K. M., & Leung, K. W. (2003). *Dielectric resonator antennas.* Research Studies Press.

Malik, S., Joshi, B. P., Bisht, V., Kharayat, P. S., & Goyal, S. (2024, May). Ambiguous Fuzzy Einstein Ordered Averaging Operator: Application to the classification of Power Generation methods. *In 2024 4th International Conference on Advance Computing and Innovative Technologies in Engineering (ICACITE)* (pp. 1853-1857). IEEE.

Meena, G., Singh, V. P., Dixit, A., Mathur, A., & Bhatt, A. (2024, April). Steady-state models of ac microgrid for load flow solutions. *In 2024 1st International Conference on Innovative Sustainable Technologies for Energy, Mechatronics, and Smart Systems (ISTEMS)* (pp. 1-6). IEEE.

Mongia, R. K., & Bhartia, P. (1994). Dielectric resonator antennas—A review and general design relations for resonant frequency and bandwidth. *International Journal of Microwave and Millimeter-Wave Computer-Aided Engineering, 4*(3), 230–247. DOI: 10.1002/mmce.4570040304

Nishad, D. K., Tiwari, A. N., Khalid, S., & Gupta, S. (2024). Power quality solutions for rail transport using AI-based unified power quality conditioners. *Discover Applied Sciences, 6*(12), 651. DOI: 10.1007/s42452-024-06372-5

Petosa, A., & Ittipiboon, A. (2010). Dielectric resonator antennas: A historical review and the current state of the art. *IEEE Antennas & Propagation Magazine, 52*(5), 91–116. DOI: 10.1109/MAP.2010.5687510

Rappaport, T. S., Xing, Y., Kanhere, O., Ju, S., Madanayake, A., Mandal, S., Alkhateeb, A., & Trichopoulos, G. C. (2019). Wireless communications and applications above 100 GHz: Opportunities and challenges for 6G and beyond. *IEEE Access : Practical Innovations, Open Solutions, 7*, 78729–78757. DOI: 10.1109/ACCESS.2019.2921522

Saeed, H., Zubair, M., & Alomainy, A. (2022). Terahertz antennas for 6G communication systems: Challenges, design strategies, and applications. *IEEE Access : Practical Innovations, Open Solutions, 10*, 375–390.

Semwal, S., Kant, S., Bandyopadhyay, K., Asim, M., Garg, R., & Goyal, S. (2024, October). Implicit Control of DC-DC Converter based DC Motor for the Charging of Electric Vehicle using Fuzzy Logic Controller. *In 2024 5th IEEE Global Conference for Advancement in Technology (GCAT)* (pp. 1-6). IEEE.

Shankar, R., Chaudhary, B. P., Mehraj, H., Gupta, S., Arslan, F., Srivastava, A. K., & Goswami, C. (2023). Impact of node mobility on the DL based uplink and downlink MIMO-NOMA network. *International Journal of Information Technology : an Official Journal of Bharati Vidyapeeth's Institute of Computer Applications and Management, 15*(6), 3391–3404. DOI: 10.1007/s41870-023-01362-z

Shukla, R. D., Tripathi, R. K., & Gupta, S. (2010, November). Power electronics applications in wind energy conversion system: A review. *In 2010 International Conference on Power, Control and Embedded Systems* (pp. 1-6). IEEE. DOI: 10.1109/ICPCES.2010.5698663

Singh, S. P., Kumar, M., Kumar, A., & Saxena, R. (2022). UWB antenna for ITS application. In A. Dhawan, V. S. Tripathi, K. V. Arya, & K. Naik (Eds.), *Recent trends in electronics and communication (VCAS 2020)* (Lecture Notes in Electrical Engineering, Vol. 777). Springer. DOI: 10.1007/978-981-16-2761-3_102

Sinha, R., Sinha, K. K., Patel, M., Gupta, S., & Priya, S. (2024, July). Detection of leukemia disease using convolutional neural network. *In 2024 5th International Conference on Image Processing and Capsule Networks (ICIPCN)* (pp. 451-456). IEEE. DOI: 10.1109/ICIPCN63822.2024.00080

Tyagi, M., Gupta, S., & Khan, M. I. (2024, May). Reduction of voltage instability in complex power networks through the utilization of D-STATCOM. *In 2024 4th International Conference on Advance Computing and Innovative Technologies in Engineering (ICACITE)* (pp. 1662-1666). IEEE. DOI: 10.1109/ICACITE60783.2024.10616856

Zhang, Z., Xiao, Y., Ma, Z., Xiao, M., Ding, Z., Lei, X., Karagiannidis, G. K., & Fan, P. (2019). 6G wireless networks: Vision, requirements, architecture, and key technologies. *IEEE Vehicular Technology Magazine, 14*(3), 28–41. DOI: 10.1109/MVT.2019.2921208

Chapter 2

Application of Artificial Intelligence to Wind Energy for the Control of Variable Speed Wind Turbine

Navdeep Singh
https://orcid.org/0000-0002-8745-806X
EED, MMEC, Maharishi Markandeshwar University, Mullana, India

Bhanu Pratap
National Institute of Technology, Kurukshetra, India

Nitin Sundriyal
https://orcid.org/0000-0001-9771-7398
EED, Graphic Era University, India

Surinder Chauhan
https://orcid.org/0000-0001-8576-8147
University of Petroleum and Energy Studies, Dehradun, India

Prashant Singh
https://orcid.org/0000-0002-5715-1120
VTT Technical Research Center of Finland, Espoo, Finland

ABSTRACT

Motivated by growing worries about urgent environmental issues such as global warming and the rapid depletion of fossil fuel supplies, people all over the world

DOI: 10.4018/979-8-3373-7041-5.ch002

are turning to cleaner, more sustainable alternatives. Among various energy saving technologies, wind energy conversion system (WECS) is among the most prominent and rapidly evolving technologies and playing a critical role in renewable power generation. During generation variable speed wind turbine (VSWT) present numerous issues including wind variability, pitch regulation, and rotor speed fluctuation. To meet rising demand and reduce the susceptibility of VSWT, an effective control system is required which assures the efficient power production. Recently, researchers are focusing on advances in the field of artificial intelligence (AI) which improves both sustainability and effectiveness of electricity production in WECS. Accordingly, this chapter presents an extensive review of existing literature on the application of Artificial Intelligence (AI) based control strategies for the operating zones of VSWT in WECS.

INTRODUCTION

Today, with constantly expanding demand for electrical power and growing worries about environmental deterioration, clean and pollution-free energy sources are more important than ever. Conventional energy supplies such as coal, oil, and natural gas are not only diminishing at an alarming rate due to unsustainable usage, but they also play a substantial role in air pollution, greenhouse gas emissions, and global climate change. These problems underline the critical importance of transitioning to renewable and sustainable power generation (Moshtaghi et al., 2025; Kundu et al., 2025). Wind and solar energy has emerged as one of the most promising and commonly embraced renewable energy sources, amongst hydro, biomass, and geothermal (Li et al., 2022; Sundriyal et al., 2025). The primary benefits of wind power include its economic feasibility, environmental friendliness, and ability to create electricity without emitting hazardous emissions. Unlike fossil fuels, wind is an infinite natural resource that may be used as long as atmospheric conditions persist (Singh et al., 2018).

Wind energy conversion systems (WECS) are typically distinguished by enormous commercial wind farms, which may include dozens or even hundreds of wind turbines strategically installed onshore and offshore regions. These turbines operate together to turn the kinetic energy of moving air into mechanical energy, which is then transformed to electrical energy by generators (Masenge & Mwasilu, 2024). With continual breakthroughs in turbine design, efficiency, and grid integration, wind energy is playing an important part in the global transition to sustainable power generation and is likely to dominate the renewable energy market in the next decades.

As per the Renewables Global Status Report 2025, in 2024, renewable electricity capacity climbed by 18%, reaching a record 741 GW. Solar PV was the key driver, accounting for 602 GW and 81% of the overall capacity increase. Out of which wind energy added 117 GW globally (Galain & Adib, 2025). In continuation of that, the electricity sector remained the driving force for renewable energy adoption. During the year, several countries produced electricity from renewable sources, while 15 countries effectively integrated more than 30% variable renewable electricity (Galain & Adib, 2025) with a hybrid combination of solar and wind power, as depicted in Fig. 1.

Figure 1. Countries with a large share of Hybrid generation in their power systems

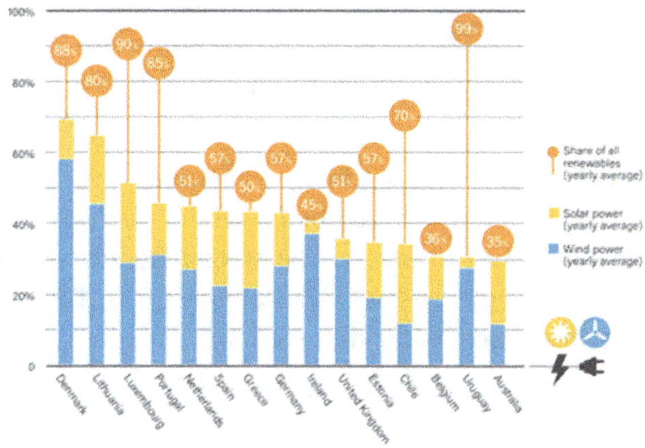

Source: Galain and Adib (2025)

Wind energy is going to be a major contributor to the aims of governments worldwide to become net-zero carbon over the next years, as indicated by the 2023 Global Wind Report, which predicts 680 GW of extra wind energy potential by 2028 (Hassan et al., 2026). As of March 31, 2024, the predicted prospect towards energy output from renewable resources has reached 21,09,655 MW. Wind has the most potential for energy generation (@ 150 meters), with 11,63,856 MW (55.17%), followed by solar energy at 7,48,990 MW (35.50%) and large hydro at 1,33,410 MW (6.32%) as shown in Fig. 2 (Singh et al., 2025).

Figure 2. Estimated Potential of Renewable Power in India

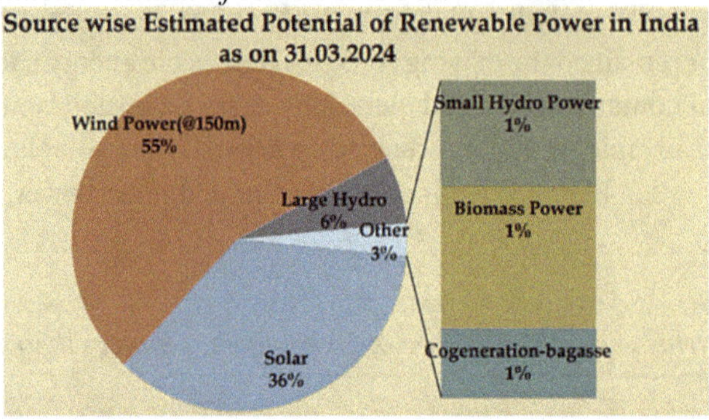

Source: Singh et al. (2025)

From Fig.2, Wind power is India's primary renewable energy source, accounting for an estimated 1,163,856 MW (55% of total potential). India is continually working toward sustainable and clean energy. The MNRE intends to reach 500 GW of alternative fuels based energy generation capacity by 2030. The classifications of WT's are well described below in the next section.

WIND TURBINES CLASSIFICATIONS

WTs are basically classified into two types as per the axis of the rotation of their rotors/orientation and as per the speed (Shukla et al., 2010): As per their orientation: (i) HAWT and (ii) VAWT. A HAWT features a rotor propeller that is parallel to the level of ground and the wind direction. These types of turbines have a low cut-in wind speed and a large Cp. A Yaw mechanism is also required to spin the entire nacelle and rotor to keep the turbine facing squarely into the wind (Awasthi et al., 2024). However, the design of HAWT is more sophisticated and expensive as it requires a gearbox and an electric generator at the top of the tower (Gupta, 2025). Although it has several benefits, HAWT has several limitations, such as requiring a wide land area, causing aerodynamic and mechanical noise, a threat to animals, particularly birds and bats, and a high capital cost. Whereas in VAWT the rotor shaft is positioned vertically, perpendicular to the ground and is capable to capture wind from any direction (Rawajfeh & Gomaa, 2023). As a result, these turbines do not require the yawing control system to align them in the direction of the wind. It has a simpler and less expensive installation cost as gearbox and generator are located at the bottom of the tower. The comparative analysis between the two is well depicted in Table 1 (Rawajfeh & Gomaa, 2023).

Table 1. Comparison between HAWT and VAWT

Sr. No.	*HAWT*	*VAWT*
1	*It rotates horizontally.*	*Rotation is Vertical*
2	*Axis of Rotation: parallel to the wind*	*Axis of Rotation: perpendicular to the wind*
3	*Location of Gearbox, generator: top of the tower*	*Location of Gearbox, generator: at the ground*
4	*Yaw mechanism is required*	*Yaw mechanism is not required*
5	*The design and installation is complicated.*	*The design and installation is easy*
6	*Demands a vast area for blade operations.*	*Demands a less area for blade operations.*
7	*Cause a significant difficulty for birds.*	*Cause less disturbance to birds*
8	*High Cost*	*Less Expensive*

Both turbines are well suited for the power production as per the usage. Although HAWT is more efficient as in comparison to VAWT is it is well suitable for turbulent and changeable wind conditions. Similarly as discussed above as per their axis of rotation, turbines are also categorized as per their speed (Singh et al., 2016) to: (i) CSWT and (ii) VSWT.

CSWTs have a fixed rotational speed regardless of wind velocity, which limits their performance. Because the rotor speed cannot react to changing wind conditions, these turbines frequently exhibit poorer aerodynamic efficiency, resulting in lower power generation (Singh et al., 2016). Furthermore, the flexibility in operation decreases mechanical stress on the drive train, resulting in less component wear, which increases the WT's operational lifespan. As a result, variable-speed technology has become the favored option in modern wind energy systems, resulting in increased energy production, system reliability, and overall reduce the mechanical stress on the turbines. The overall mechanical modelling of VSWT covering all the aspects of the models is well presented in the next section.

Mechanical modelling of VSWT

The mechanical modelling of VSWT mainly comprises of different subsystems such as aerodynamic subsystem, rotor subsystem, pitch angle subsystem, electrical subsystem, and tower dynamics (Singh et al., 2018; Fotis et al., 2025).

Aerodynamic subsystem: The power absorbed from the rotor of VSWT is expressed as (Fadili & Boumhidi, 2025):

$$P_r = \frac{1}{2}\rho\pi R_r^{\,2} v^3 C_p(\lambda,\beta) \tag{1}$$

where the power coefficient $C_p(\lambda,\beta)$ is the function of tip speed ratio λ and blade pitch angle β expressed as (Singh et al., 2018)

$$C_p(\lambda,\beta)= \{0.5 - 0.0167(\beta - 2)\}\sin\frac{\pi(\lambda + 0.1)}{18 - 0.3(\beta - 2)} - 0.00184(\lambda - 3)(\beta - 2) \quad (2)$$

Similarly, the aerodynamic torque T_a from the available wind is given below:

$$T_a = \frac{P_r}{\Omega_r} = \frac{\pi\rho R_r^{\,3} v^2 C_p(\lambda,\beta)}{2\lambda} \quad (3)$$

where tip-speed ratio λ is given as:

$$\lambda = \frac{\Omega_r R_r}{v} \quad (4)$$

where, P_r is the power captured by the rotor, Ω_r is the rotor speed, R_r is the radius of the rotor v is the wind speed.

Rotor subsystem: The overall dynamics of the drive train is represented in the context of turbine, generator and of shaft speed as (Singh et al., 2017):

$$J_r\dot{\Omega}_r = T_a - T_{ls} - K_r\Omega_r \quad (5)$$

$$J_g\dot{\Omega}_g = T_{hs} - K_g\Omega_g - T_{em} \quad (6)$$

$$T_{ls} = B_{ls}(\theta_r - \theta_{ls}) + K_{ls}(\Omega_r - \Omega_{ls}) \quad (7)$$

Drive train proportion is defined as:

$$n_g = \frac{T_{ls}}{T_{hs}} = \frac{\Omega_g}{\Omega_{ls}} = \frac{\theta_g}{\theta_{ls}} \text{ or } T_{ls} = n_g T_{hs} \quad (8)$$

Pitch Angle Subsystem: The dynamic equation of the pitch subsystem is expressed as:

$$\dot{\beta} = -\frac{1}{\tau}\beta + \frac{1}{\tau}\beta_{opt} \quad (9)$$

where β_{opt} is the desired angle and τ is the time constant depend on the pitch actuator.

Electrical Subsystem: The dynamics of the generator subsystem is given below

$$\dot{T}_g = -\frac{1}{\tau_t}T_g + \frac{1}{\tau_t}T_{gref} \quad (10)$$

where T_g is the actual generated torque, T_{gref} is the reference torque and τ_t is the time constant (Jain et al., 2015). Finally, the output power is given as $P_e = T_g \Omega_g$.

Tower Dynamics: The motion of the tower has a considerable impact on actual power production. Internal vibrations, motions, and the tower's flexibility can all affect the aerodynamic performance, which in turn influences energy conversion efficiency. The modeling of a tower is represented as

$$M_t \ddot{\xi} + D_t \dot{\xi} + K_t \xi = F_t \tag{11}$$

where M_t, D_t and K_t are the mass, damping, and spring constant of tower. The dynamical two-mass model of VSWT covering the aerodynamic and mechanical aspects is well depicted in Fig. 3 and its simulation parameters are given in Table 2.

Figure 3. Two- Mass Dynamical model of VSWT

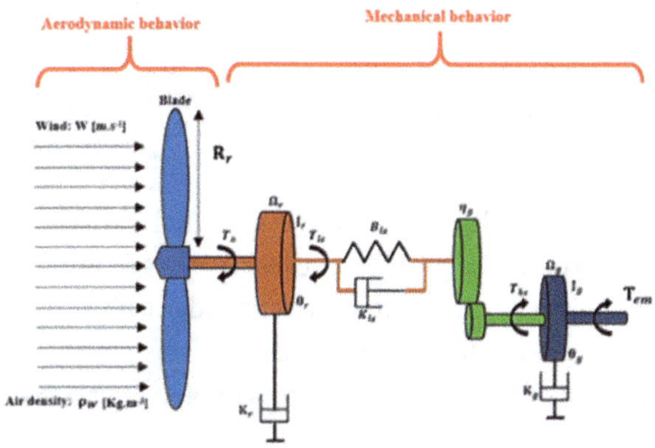

Source: Singh et al. (2018) and Fadili and Boumhidi (2025)

A VSWT complicated dynamical modeling necessitates thorough mathematical descriptions of its numerous components in order to effectively capture and anticipate system behavior. These interactions must be accurately modeled using differential equations, aerodynamic models, and electrical system formulations (Kumar et al., 2017).

Table 2. Simulation Parameters of Two- Mass Dynamical model of VSWT (Fadili & Boumhidi, 2025)

Parameters	Symbols	Values
Rotor inertia	J_r	325×10^3 kg m^2
Generator inertia	J_g	34.4 kg m^2
Rotor external damping	K_r	27.36 N m/rad×s
Generator external damping	K_g	2×10^{-1}N m/rad×s
Shaft Stiffness coefficient	B_{ls}	2.691×10^5 N m/rad
Shaft damping coefficient	K_{ls}	9500 N m/rad/×s
Ratio of gearbox	n_g	43.165
Air density	ρ	1.29 kg/m^3
Rotor Diameter	R_r	43.3 m
Output Power	P_e	3 KW
Mass constant of tower	M_t	4.2278×10^5 kg
Damping constant of tower	D_t	2.0213×10^3
Spring constant of tower	K_t	1.6547×10^6 N m/rad
Tip speed ratio &Co-efficient of Performance	λ_{opt} and Cp_{opt}	8.1484 and 0.4745

A WT's runs under a wide range of environmental variables, including fluctuating wind speeds, turbulence, and load demands, so precise modeling is critical. Researchers and engineers can better understand turbine dynamics, anticipate potential instabilities, and enhance performance across many operating zones by providing exact mathematical descriptions, resulting in increased dependability and efficiency in real-world applications.

Operating Zones of VSWT

A VSWT operates under three crucial wind speeds and four operating zones, which govern its power-producing behavior (Singh et al., 2019a).

- In Zone I, there is no power generation as the wind speed is too low.
- Whereas in Zone II, the cut-in wind speed is the lowest wind velocity required for the turbine to overcome system inertia and begin producing usable electrical power.
- Similarly, in Zone III at the rated wind speed, the turbine generated its optimum output, or rated power capacity.
- Finally, in the last Zone IV, the cut-out wind speed indicates the highest safety limit for operation and the turbine automatically shuts down. The four operating zones with its varying wind speed is shown in Fig. 4 below.

Figure 4. Operating Zones with different wind speeds

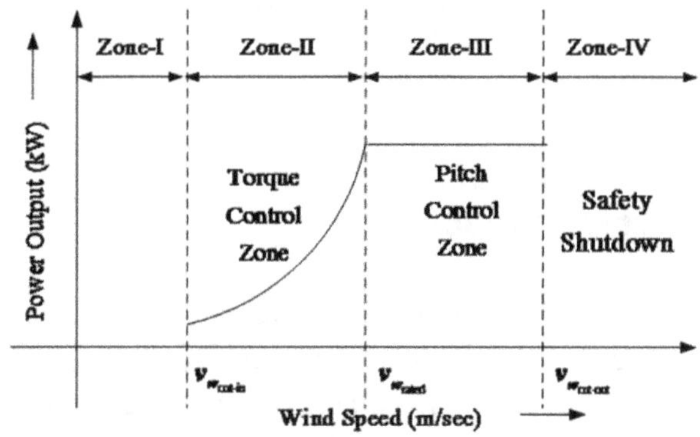

Source: Singh et al. (2019a)

From Fig. 4, it seems that VSWT can adapt to an extensive range of nature circumstances while maintaining high performance and dependability.

The majority of current AI applications in WECS have focused on system installation, problem detection, operational monitoring, predictive and corrective maintenance in recent years, 2022, 2025 (Marquez & Gonzalo, 2022; Rao et al., 2023). In accordance with this trend, a comprehensive analysis is published in 2025 on VSWT condition monitoring systems and maintenance management strategies (Sun & Han, 2025).

LITERATURE REVIEW

Based on nonlinear dynamical modeling and the unique operating zones of VSWT, very few literature reviews have been done towards the various control strategies of VSWT using artificial intelligence (AI). This chapter focuses on reviewing the use of AI-based control strategies in WECS. The study mainly focuses on several intelligent and adaptive control systems developed for torque control in Zone II (TCZ) and pitch control in Zone III (PCZ), both of which play crucial roles in controlling turbine performance under changing wind conditions. Furthermore, the study investigates AI-driven approaches for MPPT, which ensures harnessing of optimal energy from available wind resources, as well as for immeasurable state, nonlinearity, and forecasting, which are critical for improving the reliability, stability, and predictive maintenance of WECS. Collectively, these domains show how AI techniques are rapidly being used in WT's control to improve efficiency, adaptability, and operational reliability. A systematic literature survey has been conducted from 2020 to 2025 for the VSWT control; however, few research articles have been considered from previous years to match the scope of the chapter. The survey is characterized into three categories: linear control, nonlinear control, and hybrid control. The applications of AI-based linear control for VSWT are presented in the next section.

Linear Controllers with Adaptive and Intelligent Control Approach

Currently, the bulk of WT's systems use traditional control strategies such as PID or simplified PI control. The PI control method is popular due to its basic structure and simple execution, making it suitable for practical applications of WECS. However, simplicity frequently comes at the expense of performance constraints in terms of overshoot and poor dynamic response, particularly when wind speed swings fast or there is considerable turbulence.

In contrast, the PID controller allows for greater tuning flexibility and can increase dynamic performance as compared to PI control. WT's operate in highly nonlinear and time-varying environments due to constant changes in wind speed and atmospheric conditions, therefore fixed-parameter PID controllers frequently fail to adapt efficiently to these variations. As a result, it becomes difficult to ensure output power stability, which may result in greater mechanical stress, decreased efficiency, and a worsening of the overall reliability of the WECS.

As the hydraulic pitch system is highly nonlinear and susceptible to time-varying disturbances, typical fixed-parameter control approaches are frequently insufficient. So to deal with it, a real-time fuzzy adaptive PID control for PCZ is applied in (Sheng & Jun, 2013) to the VSWT, combining the standard PID structure along with FLC to dynamically tune the gain of controller. This versatility makes it particularly good at dealing with the nonlinearities and uncertainties that are typical arise in WT's operations. By incorporating fuzzy logic into PID control, the system can better handle variations, make smoother pitch angle adjustments, reduce overshoot, and improve the overall stability and efficiency of VSWT operation.

An implementation of a neuro-fuzzy-based PID controller has been utilized for PCZ of a HAWT. VSWT has inherent issues due to the nonlinearity and irregularity of wind speed, which leads to substantial swings in power production (Goyal et al., 2022). The primary goal of control is to keep the VSWT output steady, even in the face of extremely changing and uncertain wind speed conditions. To counteract these problems the neuro-fuzzy PID controller combines NN adaptive learning approach with the conceptual and linguistic rule-driven structure of FLC. This fusion allows the controller to adjust system specifications in real-time to account for wind fluctuations and operational unpredictability. The proposed controller improves the resilience of the electric power managing mechanism. Furthermore, by efficiently decreasing stress and minimizing variations in power, this control procedure not only increases productivity but also extends the life and dependability of the VSWT.

To meet these issues, a RBFNN-based FOPID controller is constructed for the PCZ. The primary function is to minimize tensions and enhance the longevity of the VSWT system. In this process, an ideal database is initially established and then used for training the RBFNN (Asgharnia et al., 2020). After training, the NN adaptively select the most suitable control parameters in real time, enhancing the adaptability of the system to chaotic and dynamic-varying operating conditions. One of the most salient advantages of this technology is that the RBFNN is able to predict the gains of the PID/FOPID controller even if it requires precise measurement of wind speed, without additional sensors and lower overall sophistication. The efficacy of the proposed controller has been assessed based on simulation with the FAST simulator, and the results indicate a monumental improvement in performance, with reduced load and more stable VSWT execution compared to traditional methods

A type-2 FOPID controller was used to control for both the zones PCZ for pitch angle and TCZ for MPPT of a VSWT (Oussama et al., 2022). The designed control effectively manages the uncertainties and nonlinearities involved in WECS. Simulation experiments proved that controller realized better parameters leading to improved stability, quicker dynamic response, and more effective energy harvesting under fluctuating wind speed.

A fuzzy-PID controller was developed for PCZ of a small-scale VSWT. The control method integrates the robustness and simplicity of PID control with fuzzy systems being to deal with nonlinearities generated by the changes in wind speed. The efficacy of the given approach was established using simulation on a small WT's equipped with a PMSG (Ngo et al., 2020). The findings indicate that the applied strategy effectively controls the pitch angle, leading to improved power control, stability, and reliable performance in changing wind conditions.

A LQR with an NN has also been employed to control the torque of a VSWT to attain MPPT (Li et al., 2024; Sharma et al., 2024). The introduced control structure exploits the NN model's prediction ability to properly replicate the VSWT nonlinearity and enhances the LQR's performance. The control is particularly designed for TCZ to forecast and optimize the rotor speed and blade pitch angle in real time. This predictive regulation improves the entire VSWT system's stability, robustness, and adaptability. Following that for optimal power generation a PID-assisted LQR control with NN approach has utilized for TCZ to the VSWT (Wen et al., 2023). The hybrid method couples the LQR optimal control abilities in managing system disturbances, and optimizing control effort. The results indicate that the technique based on LQR performs better compared to the conventional PID controller dynamically performance, especially in terms of controlling rotor speed of VSWT.

For solving the tuning issue of the PI control, a LQR-based FLC for PCZ is utilized for VSWT. The goal is to maintain the system's rated power production at a constant level, even under fluctuating wind speed conditions (Jeon & Paek, 2021). The comparative findings are extremely effective at reducing power fluctuations, with a reduction of up to 38.93% as compared to the classical approach which results in smoother power delivery, and increases the structural strength of the VSWT. A fuzzy PID-based PSO adaptive approach for pitch regulation has been developed in (Shao et al., 2022). In this control, the PID parameters are optimized using the PSO algorithm, and additional fuzzy rules are constructed to dynamically alter its parameters to deal with VSWT uncertainty. From results, it is seen that the applied strategy increases the system's robustness and improves its effectiveness in reducing different perturbations.

To tackle the challenge of power swings of VSWT, a fuzzy-based H∞ control strategy for the PCZ has been developed (Jeon et al., 2023). This combination merges the reliability of H-infinity control with the versatility of fuzzy logic to improve system stability and power conservation. The merging of these control approaches accomplishes greater performance in actual WT's operations. Experiments were conducted to a WT model with a capacity of 40 W to check the effectiveness of control applied. The experimental findings show that the suggested fuzzy-based H∞ controller may significantly decrease power fluctuations, providing greater reliability and effective operation than traditional methods.

To enhance the performance of VSWT, an ANN with H∞ control for the PCZ is used. The suggested method improves VSWT dynamic performance, overall stability and energy production availability (Mohamed et al., 2025). ANN possesses an real-time estimation of nonlinearities, while H∞ control provides a robustness to the VSWT dynamics. Incorporating these capabilities, the ANN-based H∞ controller provides resilient operation against uncertainties, and improves the life of VSWT.

The combination of linear and adaptive controllers provides credible and reliable results, yet the unexpected and highly changeable nature of wind speed presents substantial obstacles, as wind speed does not follow a regular or clearly foreseeable pattern, so the combined approach cannot capture the entire spectrum of VSWT dynamic fluctuations. This uncertainty adds a high level of nonlinearity and complexity to the modeling of VSWT, making it difficult to achieve constant, precise performance and efficient energy conversion. Hence, it is quite a difficult challenge to formulate a controller for the VSWT system to deal with the highly nonlinearity and provide the global stability to the overall system. Some of the most frequently utilized nonlinear control tools are sliding mode control, quasi sliding mode control, backstepping control and dynamic surface control. So to deal with the nonlinearity, robust nonlinear controllers are necessary for smooth and effective functioning of

VSWT, despite of parametric uncertainty. The next subsection briefs the crucial role of nonlinear controllers based on AI for VSWT.

Nonlinear Controllers with Adaptive and Intelligent Control Approach

To tackle with the nonlinearity of VSWT, the most prominent traditional control strategy is the robust control approach so-called sliding mode control technique (SMC). SMC has several advantages such as insensitivity to variable fluctuations, outside influence tolerance capacity, and quick dynamic response (Osman & Alsokhiry, 2022). Other significant features of SMC are the simplicity of its implementation and the order reduction of the closed loop systems.

SMC design consists of two step process. In the first step, a suitable sliding surface also known as switching surface is designed so that the system states are driven from any initial state to reach that switching manifold (Osman & Alsokhiry, 2022). Then according to that sliding surface, a suitable control law is derived which keeps the states of system attracted towards the switching function and forced toward the sliding surface to stay on it. The control phases before the state trajectory reaching the sliding surface is called the reaching phase. Once the plant states reach the sliding surface, they hold on it and continuously sliding along the sliding surface which is known as sliding phase. The whole process is known as sliding mode. Due to its attractive features of robustness it has been widely employed in VSWT systems to cope with variable variance (Rajendran et al., 2024).

SMC has its roots in relay control, so one of its most interesting aspects is the discontinuous nature of the control action whose major job is to transition between two entirely distinct patterns around some specified surface such that a new type of system motion called sliding mode exists around that surface (Rajendran et al., 2024). Furthermore, this control approach has the advantages of the development and execution simplicity, as well as the possibility of lowering the ordering of the closed loop system.

However, the major drawback of SMC is the chattering that is inherent with the discontinuity of the control law. Usually it occurs due to the negligence of unmodeled dynamics during the modelling of the plant. Unmodeled dynamics come from actuators and sensors which are usually ignored when designing the controller. For plants with SMC, rapid switching can excite these unmodeled dynamics leading to chattering phenomena (Hassan et al., 2025).

To overcome the chattering problem, a SOSM technique is an efficient tool. SOSM is also known as higher order sliding mode HOSM or super-twisting SOSM (Hassan et al., 2025). It is a technique that uses higher order sliding surface which has the effect of integrating the discontinuity from the control law and making the input of continuous nature. This operational characteristic leads to reduce the chattering effect. Indeed, the attractive features of HOSMs are chattering-free behavior, robustness with respect to unmodeled dynamics (turbine and generator) and external disturbances and reducing the stress acting on the drive train.

To solve the control problems of the highly nonlinear systems with parametric variations, Backstepping control (BSC) approach is considered as a powerful tool to design the controllers (Roy et al., 2023; Gupta & Tripathi, 2014). It is a systematic and recursive design control approach for the class of nonlinear system which requires pure strict-feedback form. It recursively selects the state variables as the virtual control inputs for lower dimension subsystems of the overall system. A virtual controller is obtained from the every Backstepping step which is expressed in terms of virtual control designs from the earlier designed stages (Roy et al., 2023). At the end of the procedure, a suitable stable control law can be obtained by combining the Lyapunov functions associated with each designing stages (Anand et al., 2022; Johri et al., 2025). Therefore, the BSC technique maintains the robustness properties with respect to the uncertainties (Echiheb et al., 2024). The benefit of using BSC is that it has the ability to prevent cancelations of valuable nonlinearities and follows the goals of equilibrium and tracking. To handle the systems with unknown parameters Backstepping design can be further extended to the adaptive Backstepping design. The BSC proves to be an effective tool for nonlinear system, but there is a substantial drawback of computational burden. The BSC design approach contains large number of explosion of terms in design procedure. Also, certain functions remain linear in the unknown parameters, and require a tedious analysis to calculate them, (Zhenhua et al., 2025). To overcome the complexity arising in BSC approach, a dynamic surface control (DSC) algorithm is used to avoid the explosion of terms.

The Dynamic Surface Control (DSC) is an advanced version of integral Backstepping control approach through the use of dynamic low pass filters. It is similar to backstepping consist of multiple sliding surface with the addition of '(r–1)' numbers of low pass filters, where 'r' represents the relative degree of the output to be controlled (Song et al., 2024). The DSC is composed of two blocks: multiple sliding surface (MSS) block and low-pass filter (LPF) block. The MSS block builds several sliding surfaces and determines the impact state values, relying on both the state details and the filtered input. Then, all the forcing state values pass through the LPF block and the corresponding filtered values go back into the MSS. After several numbers of iterations, the control input is calculated and fed into the system. Several studies for VSWT have been reported here of nonlinear controllers with adaptive and intelligent control approach.An adaptive nonlinear control methodology for VSWT has been designed by applying RBFNN. The methodology has been successfully used for both zones TCZ and the PCZ (Jafarnejadsani et al., 2013). The main intention of the proposed algorithm is to implement MPPT under changing wind conditions, ensuring that the turbine harvests the maximum amount of energy from the wind in the TCZ. In addition, the PCZ control approach aims at varying the pitch angle of the blade in an effort to maintain the output power of the turbine constant, even at high wind velocities. The technology enhances the strength and efficiency of the system through adaptive nonlinear control together with RBFNN's learning ability in enabling the turbine to perform optimally across a broad range of wind velocities.

To tackle nonlinear dynamics, variability of wind speed, and grid demands, a cascaded (SMC +Fuzzy) nonlinear control method has been utilized (Abdeddaim & Betka, 2013) for optimal performance of a grid-connected VSWT with a DFIG. The controller is implemented for TCZ primarily to derive an MPPT scheme based on FLC to enable the system to extract maximum aerodynamic energy, irrespective of uncertain or changing wind profiles. Whereas on the other end a strong II-order SMC control method is employed to control stator active and reactive power components with high accuracy, leading to smooth power flow, minimizing the chattering effects. The combination of FLC with II-order SMC creates an efficient, stable, and dependable framework for improving the overall performance of DFIG-based WECS in real grid-network situations.

By adjusting the pitch of the VSWT, an adaptive nonlinear SMC together with BSC is proposed for PCZ in (Yin et al., 2015) to ensure smooth tracking of pitch angle. As observed from the result, the suggested control technique greatly lowers changes in generator output power and turbine mechanical loads, boosting energy capture consistency, overall system reliability, and the operational lifespan of WT's components. Similarly, an observer-based adaptive BSC technique has been designed in (Bossoufi et al., 2015) for DFIG-based VSWTs operating for TCZ. The primary goal of this control approach is to accomplish MPPT. The observer is utilized to estimate immeasurable system states and disturbances, and the BSC technique assures increase in the DFIG's dynamic performance, robustness against uncertainties, and maximizes wind energy capture under shifting wind speed situations. In (Beniysa et al., 2021), the authors suggested an adaptive BSC for TCZ of VSWT system powered by a PMSG. The primary purpose of this controller is to provide robust and reliable control of the turbine's rotor speed while also adjusting the generator's mechanical power production. To address modeling imperfections and system nonlinearities, the control design includes an ANN which helps in assessing and mitigates the impacts of unknown dynamics and unpredictable variations in the WECS. The applied controller improves system stability, tracking performance, by using adaptive bonding of (BSC) and ANN-based compensation, even under uncertain or fluctuating operating conditions.

A higher order adaptive SMC based control algorithm for PCZ is implemented (Zhang & Plestan, 2021) for the regulation of output power of floating offshore VSWT. The adaption of gain values makes the SMC robust to achieve the objectives for pitch control regulation. The fundamental control objective is to manage the electrical power output at the rated value, which allows maximum energy acquisition while protecting the generator and associated equipment from overload. Secondly, the controller aims to diminish platform pitch motion, or the cyclic tilting of the floating wind turbine platform generated by wind and wave interactions. Eliminating these vibrations is critical for preserving structural integrity, increasing turbine reliability, and boosting energy conversion efficiency.

To enhance the quality of the control, also an adaptive observer based ST-SMC has used in (Ma et al., 2024) for PCZ to the floating offshore 5 MW VSWT. The control strategies use the adaptive gains for stabilizing the output power of the generator, ensuring that the electrical power supplied is stable and close to the desired reference. This stability is critical for maintaining grid compliance and increasing total energy conversion efficiency. Finally, the control technique helps to reduce the motion of the floating platform, namely pitch and heave oscillations caused by wind and wave disturbances which lead to reduces tower loads. In (Dong et al., 2022), the authors suggested an adaptive fractional-order fault-tolerant SMC technique for WECS using a DFIG. The controller is developed for TCZ to provide MPPT by managing the DFIG electromagnetic torque and provides a greater freedom in structuring the control dynamics of VSWT. Furthermore, the fault-tolerant provides consistent operation even when actuator or sensor failures occur, and ensures high tracking accuracy for the optimal torque reference, efficient wind energy extraction. A RBFNN based ST-SMC strategy was developed in (Didier et al., 2025) for the PCZ of a floating offshore VSWT. The primary goal is to attain the MPPT area for maximize aerodynamic power extraction. As the wind speed surpasses the rated value the control task shifts to keep the speed of rotor at the same rated value. The combination of RBFNN with ST-SMC improves system robustness by compensating uncertainty and eliminating chattering which overall results in maximizing energy acquisition while also assuring safe, stable operation in the harsh offshore environment.

A SMC based DSC technique was used in (Keighobadi et al., 2022) to regulate both the TCZ and PCZ of a floating VSWT. The DSC framework aids in eliminating the explosion of complexity problem that frequently occurs in BSC, whereas SMC ensures significant robustness to system variability and external perturbation. A RBFNN is used to increase control performance even more, particularly when modeling uncertainties and intermittent disturbances exist. The RBFNN is used to estimate unpredictable nonlinear movements and compensate for inaccuracies in real time, of torque and pitch control. This combination method offers higher adaptability, reduced chattering effects, and increased overall stability of the floating VSWT system and harnessing smooth power under fluctuating behaviour of wind speed.

A robust nonlinear SMO for a two-mass VSWT model has also developed in (Singh & Pratap, 2025) to estimate the states and frictional effects occurring in WT's dynamics. The suggested method's goal is to forecast the immeasurable states, which can then be employed in the robust control design process for VSWT. The next sub-section discusses the function of AI with adaptive and hybrid controllers in VSWT.

Hybrid Controllers

A fuzzy SMC based on a RBFN is applied in (Lin et al., 2011) to accomplish MPPT in turbine pitch control systems. In this approach, the fuzzy SMC is used for TCZ and PCZ to ensure for the optimal power production under fluctuating winds. Whereas, RBFNN is used to estimate nonlinear WT dynamics and regulates and limits the pitch angle to minimize severe stress on the structure and preserve system stability implementing uniform control responses. The results reveal that the implemented strategy allows for efficient power capture while assuring safe and reliable functioning of the WT. A unique control technique for PCZ based on ANN has applied in (Dahbi et al., 2016) to maximize wind power utilization. This intelligent control system enables the PMSG to run at its best rotational velocity under varying wind conditions and guarantees the system captures as much energy as possible from the wind. Under high wind speeds above the rated speed, the pitch angle control automatically deflects the blade position to lower aerodynamic efficiency, avoiding surplus power production and electrical overloading. In the same manner, for the control of output power, a recurrent NN fuzzy based pitch control of 5 MW VSWT has been used in (Asghar et al., 2024). The job of the controller is to restrict the power in PCZ while maintaining the turbine unsubjected to mechanical stress.

An ANFIS centered control scheme has been applied in (Abdalla et al., 2025) to control a hybrid (wind + solar + fuel) power generation system. The work is focused on the optimization of conventional controllers through the integration of NN with FLC. The findings indicated that application of the ANFIS-based control technique not only led to substantial power reduction, but also system stability and reliability were improved, which made it a viable substitute for current smart-grid solutions.

A VSWT pitch control system was designed for PCZ through NN and FLC primarily in order to maintain constant output power. In this technique, the NN is responsible for approximating and predicting the current and expected wind speeds,(Garcia & Santos, 2022). The predicted wind speeds are utilized in the regulation process and provided as an input to the FLC, for computes the required blade pitch angle adjustments. The FLC update the angle of the VSWT in real time, ensuring the rated output is fixed under varying wind conditions. The hybrid approach results in robustness and flexibility, leading to enhanced power regulation, and enhanced safety of VSWT components against impact caused by wind swings.

An ANFIS control method was utilized for enhancing the efficiency of VSWT power generation. In this hybrid method, NN is utilized for forecasting the coefficient of performance (Cp), of VSWT (Petkovic et al., 2013) that subsequently aids in attaining MPPT under changing wind speeds. Moreover, FLC employs this data for PCZ to control the blade pitch angle for optimizing the aerodynamic performance of the WT. This regulated adaptation allows achieving maximum amount of energy from the wind while maintaining consistency in power output. Additionally results indicate that ANFIS yields stable and intelligent solution for achieving efficient MPPT in present WECS. Also, a control strategy on the basis of ANFIS is proposed for TCZ in order to achieve MPPT for a 3.5 kW VSWT. Here, NN is employed to forecast the continuously varying wind speed conditions and the FLC, based on this data, detects and follows the most efficient rotor speed of the VSWT, ensuring maximum wind resource power absorption (Hermassi et al., 2024). Based on computational simulations, the proposed NN-FLC hybrid control system works better than traditional MPPT methods in terms of responsiveness, tracking precision, and robustness. Additionally, it also offers a rational and effective solution, to ecologically friendly and cheap renewable power generation.

A PSO based ANFIS pitch angle regulation technique has been proposed for a fluctuating speed DFIG basedWECS. In this method, PSO is used to adjust the parameters of the ANFIS controller, in dynamic wind situations (Sahoo, 2024). The goal of control method in PCZ is to adjust the pitch angle of the rotating blades so that the speed of the generator remains constant. Furthermore, the combination of PSO with ANFIS enables the controller to resulting in increased WECS robustness and dependability under fluctuating wind speeds. A nonlinear adaptive observer-based QSMC technique for TCZ and PCZ has been proposed in (Singh et al., 2019b) for VSWT. The goal of this control is to precisely forecast the VSWT's unobserved states, which are unavailable due to system nonlinearities, and uncertainties. After the approximation of states, an adaptive control law is design using a quasi SMC technique. The gain of the control is tuned or updated with RBFNN which improves the durability of the control system. The adaptive quasi-SMC technique reduces chattering while providing excellent tracking performance, including a quick convergence of state estimates, accurate approximations of unknown dynamics, and effective disturbance rejection. A QSMC method is used for VSWT primarily to extract the most power. The control approach is intended for TCZ, and its goal is to achieve the best possible rotor speed for efficient power generation (Pratap et al., 2018). Finally, the adopted methodology is compared to traditional SMC to assess its effectiveness which provides robust results in terms of tracking and accuracy.

An adaptive ATFC control is used to predict aero torque with the goal of achieving the MPPT for VSWT (Qi et al., 2025). The approach is used for TCZ which modifies the MPPT and allows the VSWT to better adapt to quick fluctuations in aerodynamic loads. Comparative results with traditional control methods shows that the augmented control strategy improved the output and decreased torque variations.

Similarly, an adaptive control scheme for both operating regions, of VSWT has been developed and demonstrated through a real-time case study in (Bashetty et al., 2020). The task of this control is to regulate the rotor speed and pitch angle of the blades, enabling optimum energy capture and structural protection in the PCZ by modifying the pitch angle to restrict aerodynamic torque. The real-time validation emphasizes the adaptability and resilience of the technique, demonstrating its efficacy to preserve VSWT stability under realistic environment.

To handle VSWT nonlinearities, an adaptive hybrid (DNN+MPC) management system is implemented for TCZ and PCZ (Xie et al., 2023). The hybrid technique aims to improve WECS performance by upgrading VSWT dynamics in real-time using DNN, including estimations of state and real time fluctuations of wind speed. The suggested method demonstrates significant adjustability and durability, successfully coping with uncertainties and unforeseen defects and resulting in better outcomes than commonly-used MPC.

An ANFIS has been applied to the TCZ and PCZ of VSWT to improve running effectiveness. The primary goal of this strategy is to keep the pitch angle in ideal range while ensuring that the generator torque stays within functioning limits (Sahoo et al., 2019). The ANFIS manages speed variation and uncertainties linked to wind speed changes by integrating NN learning capabilities with fuzzy logic reasoning flexibility. A comparison with standard control systems, (PI and Fuzzy PID), demonstrates that the ANFIS approach outperforms them in terms of accuracy, quickness, and adaptability which results in lower stress on VSWT and higher energy capture efficiency, thus demonstrating the usefulness of the suggested control process.

OUTCOME

The majority of extant research focuses on integration of AI to the pitch control, as pitch regulation also has a great impact on extraction of power and load reduction of VSWT. Torque control has gotten relatively little coverage in the literature. Although, there are few studies addressing torque control, but they are often limited in scope as the applications of AI tend to emphasize them for online monitoring and diagnostic methods rather than regulating the operating regions. As a result, there remains a significant research gap in developing AI-driven approaches specifically tailored for torque control, highlighting an opportunity for further investigation and innovation in this area. Finally, the extensive survey has been redrafted in the form of summary as presented in tabular form with Table 3. The sequential evolution shown in the table will serve as a reference source for new scholars in the research area to continue forward in the future decade.

Table 3. Summary of recent AI applications to VSWT

Year	*Reference*	*Type of Study*	*Control Scheme with AI*	*Applicability and Operation Zone*	*Software Used*
Linear +Adaptive Controllers					
2013	*(Sheng & Jun, 2013)*	*Numerical & real-time*	*Fuzzy Adaptive + PID*	*Predict uncertainty and is applicable to Zone III (PCZ)*	*MATLAB/Simulink*
2022	*(Goyal et al., 2022)*	*Numerical & Experimental*	*PID with Neuro Fuzzy + ANN*	*To estimate nonlinearity and to keep rotor speed fixed for Zone III (PCZ)*	*MATLAB/Simulink*
2020	*(Asgharnia et al., 2020)*	*Numerical & Comparative*	*FOPID + RBFNN*	*To decrease load on VSWT structure and fixed power for zone III (PCZ)*	*Fast Simulator/5 MW HAWT*
2022	*(Oussama et al., 2022)*	*Numerical & Experimental*	*FOPID+ Type-2 FLC*	*To attain MPPT and holding rotor speed fixed Zone II & Zone III (TCZ and PCZ)*	*MATLAB/ 10 kW-WT*
2020	*(Ngo et al., 2020)*	*Comparative*	*PID+FLC*	*Blade Pitch is regulate Zone III (PCZ)*	*MATLAB/Simulink*
2024	*(Li et al., 2024)*	*Comparative & Experimental*	*LQR+NN*	*Forecast the aerodynamic behaviour for optimal power extraction Zone II (TCZ)*	*MATLAB/Simulink/FAST Simulator*
2023	*(Wen et al., 2023)*	*Numerical & Comparative*	*LQR+PID+NN*	*Update LQR parameters to attain MPPT Zone II (TCZ)*	*MATLAB*

continued on following page

Table 3. Continued

Year	*Reference*	*Type of Study*	*Control Scheme with AI*	*Applicability and Operation Zone*	*Software Used*
2021	*(Jeon & Paek, 2021)*	*Comparative & Experimental*	*LQR+FLC*	*To improve performance of VSWT in Zone III (PCZ)*	*MATLAB/Simulink/ NREL 5 MW WT*
2022	*(Shao et al., 2022)*	*Comparative & Simulation Analysis*	*PID+PSO+FLC*	*PSO initially optimizes PID parameters, further adaptively adjusted by the FLC. Zone III (PCZ)*	*Simulink/ 2 MW WT*
2023	*(Jeon et al., 2023)*	*Comparative & Experimental*	*H∞ +FLC*	*Improve VSWT durability Zone III (PCZ)*	*MATLAB/40W WT*
2025	*(Mohamed et al., 2025)*	*Comparative & Simulation Analysis*	*H∞ +ANN*	*To maintain Output power fixed Zone III (PCZ)*	*Simulink*
Non-Linear +Adaptive Controllers					
2013	*(Jafarnejadsani et al., 2013)*	*Numerical & Experimental*	*Adaptive + RBFNN*	*Regulate rotor speed for optimal generation. Zone II (TCZ) and Zone III (PCZ)*	*Simulink//5MW WT Simulator*
2013	*(Abdeddaim & Betka, 2013)*	*Numerical & Comparative Analyses*	*Higher order SMC+FLC*	*To achieve MPPT Zone II (TCZ)*	*MATLAB/Simulink*
2015	*(Yin et al., 2015)*	*Numerical & Comparative Analysis*	*Adaptive + SMC + BSC*	*Regulate the blade pitch angle, Zone III (PCZ)*	*LabView/750kw WT Simulator*
2015	*(Bossoufi et al., 2015)*	*Comparative & Experimental*	*Adaptive + BSC*	*Estimate VSWT states for MPPT Zone II TCZ*	*MATLAB/Simulink/Sim power Systems/ 200 W WT*
2021	*(Beniysa et al., 2021)*	*Comparative Analysis*	*Adaptive + BSC + ANN*	*Regulate rotor speed for constant generator output. Zone II TCZ*	*MATLAB*

continued on following page

Table 3. Continued

Year	*Reference*	*Type of Study*	*Control Scheme with AI*	*Applicability and Operation Zone*	*Software Used*
2021	*(Zhang & Plestan, 2021)*	*Simulation & Comparative*	*Adaptive + SMC*	*Hold the power at rated value Zone III (PCZ)*	*MATLAB/Simulink/FAST WT simulator/ 5MW OC3-Hywind*
2024	*(Ma et al., 2024)*	*Numerical, comparative & Experimental Analyses*	*Adaptive+ ST-SMC*	*Production output constant Zone III (PCZ)*	*MATLAB/Simulink/ FAST WT simulator/NREL-5MW*
2022	*(Dong et al., 2022)*	*Numerical Analysis*	*Adaptive +SMC*	*Predict unknown disturbances of VSWT to keep output stable Zone II (TCZ)*	*MATLAB/Simulink*
2025	*(Didier et al., 2025)*	*Numerical, comparative & Experimental Analyses*	*RBFNN+ST-SMC*	*estimate nonlinearities to make VSWT robust, Zone III (PCZ)*	*MATLAB/Simulink/HOMER/Open FAST WT Simulator/NREL*
2022	*(Keighobadi et al., 2022)*	*Numerical & comparative Analyses*	*SMC+DSC+RBFNN*	*Estimate Uncertainties and disturbances Zone II (TCZ) and Zone III (PCZ)*	*MATLAB*
2025	*(Singh & Pratap, 2025)*	*Numerical Analyses*	*SMC+SMO*	*Estimation of VSWT immeasurable states for future control of operating regions.*	*MATLAB*
Hybrid (Adaptive + Intelligent) Controllers					
2011	*(Lin et al., 2011)*	*Numerical & Experimental Analyses*	*Fuzzy + SMC+RBFNN*	*estimate upper bound of uncertainties, MPPT, Zone II (TCZ) and Zone III (PCZ)*	*MATLAB/dSPACE*

continued on following page

Table 3. Continued

Year	*Reference*	*Type of Study*	*Control Scheme with AI*	*Applicability and Operation Zone*	*Software Used*
2016	*(Dahbi et al., 2016)*	*Numerical Analyses*	*ANN*	*operate at an appropriate speed to get MPPT, Zone III (PCZ)*	*MATLAB/Simulink*
2024	*(Asghar et al., 2024)*	*Comparative & Experimental Analyses*	*ANFIS (NN+FLC)*	*Limit the generator power fixed Zone III (PCZ)*	*MATLAB/Simulink/5 MW WT*
2025	*(Abdalla et al., 2025)*	*Comparative Analyses*	*ANFIS (NN+FLC)*	*Optimize the performance of Hybrid RES*	*MATLAB/Simulink/Sim Power Systems*
2022	*(Garcia & Santos, 2022)*	*Comparative & Experimental Analyses*	*FLC+NN*	*Predict the current and future wind speed, for attaining the max power, Zone III (PCZ)*	*MATLAB/Simulink*
2013	*(Petkovic et al., 2013)*	*Analytical analysis*	*ANFIS (NN+FLC)*	*Predict Cp to attain MPPT and further used for Zone III (PCZ)*	*MATLAB/Simulink*
2024	*(Hermassi et al., 2024)*	*Numerical & comparative Analyses*	*ANFIS (NN+FLC)*	*Estimate fluctuating wind speed and nonlinearity to attain MPPT for Zone II (TCZ)*	*MATLAB/Simulink/3.5kw WT*
2024	*(Sahoo, 2024)*	*Numerical & comparative Analyses*	*PSO+ANFIS*	*Maintain the generator speed fixed in Zone III (PCZ)*	*MATLAB/Simulink*
2019	*(Singh et al., 2019b)*	*Numerical & comparative Analyses*	*QSMC + Adaptive*	*Regulate the rotor speed for both Zone II (TCZ) and Zone III (PCZ)*	*MATLAB/Simulink*

continued on following page

Table 3. Continued

Year	*Reference*	*Type of Study*	*Control Scheme with AI*	*Applicability and Operation Zone*	*Software Used*
2018	*(Pratap et al., 2018)*	*Numerical & comparative Analyses*	*QSMC + Adaptive*	*Measure optimal speed of VSWT to reach MPPT in Zone II (TCZ)*	*MATLAB*
2025	*(Qi et al., 2025)*	*Numerical & comparative Analyses*	*AFTC*	*Kalman observer is used to estimate aerodynamic torque to enhance the MPPT in Zone II (TCZ)*	*Matlab/Simulink/NREL 5 MW WT*
2020	*(Bashetty et al., 2020)*	*Numerical & Case Study Analyses*	*Adaptive*	*To increase the power extraction of VSWT for both Zone II (TCZ) and Zone III (PCZ)*	*Matlab/Simulink/NREL 2 MW and 5 MW WT*
2023	*(Xie et al., 2023)*	*Numerical & comparative Analyses*	*Adaptive NN+MPC*	*Approximate Updated VSWT dynamics to regulate rotor speed in both Zone II (TCZ) and Zone III (PCZ)*	*NREL TurbSim/5MW WT*
2019	*(Sahoo et al., 2019)*	*Numerical & comparative Assessment*	*Adaptive NN+FLC*	*Reduce stress on VSWT blades*	*MATLAB/Simulink*

CONCLUSION

Improving WT's performance is a popular research topic nowadays, which has led to several research publications on various topics related to VSWT operations. This chapter attempted to analyze various applications of AI control strategies for the operating regions of VSWT to attain optimal energy in correspondence to state and nonlinearity estimation. Efficient and accurate controllers are necessary for VSWT to function at the nonlinear behaviour of wind speed. After reading and analyzing all of the publications, it's evident that controlling rotor speed in various zones holds a significant importance for improved performance in terms of MPPT and the life span of VSWT. Despite the foregoing, the literature review makes no claim to be comprehensive, and the absence of any work is inadvertent. Future research will concentrate on cyber-attacks while developing control mechanisms for VSWT to maintain data security.

ACKNOWLEDGEMENT

The authors gratefully thank Sardar Patel Police Library, Police Lines, Moginand, Panchkula, Haryana, India, for providing the space to complete this chapter work.

REFERENCES

Abdalla, S. A., Abdullah, S. S., & Kassem, A. M. (2025). An adaptive frame and intelligent control approach for an autonomous hybrid renewable energy technology consisting of PV, wind, and fuel cell innovation. *Alexandria Engineering Journal, 114*, 279–291. DOI: 10.1016/j.aej.2024.11.048

Abdeddaim, S., & Betka, A. (2013). Optimal tracking and robust power control of the DFIG wind turbine. *International Journal of Electrical Power & Energy Systems, 49*, 234–242. DOI: 10.1016/j.ijepes.2012.12.014

Anand, S., Kulkarni, S. V., Gupta, S., & Parthasarathy, V. (2022, December). Battery management system (BMS) for electric vehicle applications. *In 2022 fourth international conference on emerging research in electronics, computer science and technology* (ICERECT) (pp. 1-6). IEEE. DOI: 10.1109/ICERECT56837.2022.10060137

Asghar, A. B., Ehsan, R., Naveed, K., Ammar, E. A. A., Ejsmont, K., & Nejman, M. (2024). Recurrent neural network for pitch control of variable-speed wind turbine. *Science Progress, 107*(2), 00368504241243160. Advance online publication. DOI: 10.1177/00368504241243160

Asgharnia, A., Jamali, A., Shahnazi, R., & Maheri, A. (2020). Load mitigation of a class of 5-MW wind turbine with RBF neural network-based fractional-order PID controller. *ISA Transactions, 96*, 272–286. DOI: 10.1016/j.isatra.2019.07.006

Awasthi, B., Gupta, S., Kumar, S., & Tripathi, A. (2024, January). Business management and planning using queuing model with heterogeneous customers for electrical applications. *In 2024 2nd International Conference on Intelligent Data Communication Technologies and Internet of Things (IDCIoT)* (pp. 279-283). IEEE. DOI: 10.1109/IDCIoT59759.2024.10467673

Bashetty, S., Guillamon, J. I., Mutnuri, S. S., & Ozcelik, S. (2020). Design of a robust adaptive controller for the pitch and torque control of wind turbines. *Energies, 13*(5), 1195. Advance online publication. DOI: 10.3390/en13051195

Beniysa, M., El Idrissi, A. E. J., Bouajaj, A., Britel, M. R., & Ariwa, E. (2021). Neural network adaptive backstepping control via uncertainty compensation for PMSG-based variable-speed wind turbine: Controller design and stability analysis. *Wind Engineering*, *46*(2), 439–458. DOI: 10.1177/0309524X211031269

Bossoufi, B., Karim, M., Lagrioui, A., Taoussi, M., & Derouich, A. (2015). Observer backstepping control of DFIG-generators for wind turbines variable-speed: FPGA-based implementation. *Renewable Energy*, *81*, 903–917. DOI: 10.1016/j.renene.2015.04.013

Dahbi, A., Said, N. N., & Said, M. S. N. (2016). A novel combined MPPT-pitch angle control for wide range variable speed wind turbine based on neural network. *International Journal of Hydrogen Energy*, *41*(22), 9427–9442. DOI: 10.1016/j.ijhydene.2016.03.105

Didier, F., Basbas, H., Larioumlil, D., Laghrouche, S., & Depernet, D. (2025). Neural network-based super-twisting control for floating wind turbines: Design and real-time validation. *Ocean Engineering*, *338*, 121973. Advance online publication. DOI: 10.1016/j.oceaneng.2025.121973

Dong, Y., Wang, J., Ding, S., & Li, W. (2022). Adaptive fractional-order fault-tolerant sliding mode control scheme of DFIG wind energy conversion system. *Proceedings of the Institution of Mechanical Engineers. Part I, Journal of Systems and Control Engineering*, *237*(1), 15–25. DOI: 10.1177/09596518221123101

Echiheb, F., Elkafazi, I., Bossoufi, B., Bhiri, B. E., Almalki, M. M., & Alghamdi, T. A. H. (2024). Nonlinear robust sliding mode–backstepping hybrid control for WECS: Theoretical design and experimental evaluation. *Heliyon*, *10*(5). Advance online publication. DOI: 10.1016/j.heliyon.2024.e31767

Fadili, Y. E., & Boumhidi, I. (2025). Enhanced nonlinear sliding mode control technique for wind power generation systems application: Theoretical design and comparative study. *e-Prime Advances in Electrical Engineering. Electronics and Energy*, *11*, 100937. Advance online publication. DOI: 10.1016/j.prime.2025.100937

Fotis, G., Gupta, S., Varshney, T., Dhar, S. L., & Gulzar, M. M. (Eds.). (2025). *Modern Computing Technologies for EV Efficiency and Sustainable Energy Integration*. IGI Global., DOI: 10.4018/979-8-3373-2382-4

Galain, R. M., & Adib, R. (2025). Renewables 2025 global status report: *Global overview*. https://www.ren21.net/renewables-2025-global-status-report-global-overview

Garcia, J. E. S., & Santos, M. (2022). Deep learning and fuzzy logic to implement a hybrid wind turbine pitch control. *Neural Computing & Applications*, *34*(13), 10503–10517. DOI: 10.1007/s00521-021-06323-w

Goyal, S., Deolia, V. K., & Agrawal, S. (2022). An advanced neuro-fuzzy tuned PID controller for pitch control of horizontal axis wind turbines. *ECTI Transactions on Electrical Engineering, Electronics, and Communications*, *20*(2), 296–305. Advance online publication. DOI: 10.37936/ecti-eec.2022202.246911

Gupta, S. (2025). *The Future of Machine Learning with Automation in Engineering*. Nova Science Publishers., DOI: 10.52305/YFIX2832

Gupta, S., & Tripathi, R. K. (2014). Two-area power system stability improvement using a robust controller-based CSC-STATCOM. *Acta Polytechnica Hungarica*, *11*(7), 135–155. DOI: 10.12700/APH.11.07.2014.07.9

Hassan, A., Ahmad, G., Shafiullah, M., Islam, A., & Shafiul Alam, M. (2025). Review of the intelligent frameworks for pitch angle control in wind turbines. *IEEE Access : Practical Innovations, Open Solutions*, *13*, 29864–29885. DOI: 10.1109/ACCESS.2025.3540367

Hassan, A. S., Chandarana, N., Groh, R., & Macquart, T. (2026). Multi-rotor wind turbines: A review of modern research efforts and challenges. *Renewable and Sustainable Energy Reviews,226*(A). DOI: 10.1016/j.rser.2025.116252

Hermassi, M., Krim, S., Kraiem, Y., Hajjaji, M. A., Mimouni, M. F., & Mtibaa, A. (2024). Wind speed estimation and maximum power point tracking using neuro-fuzzy systems for variable-speed wind generator. *Wind Engineering*, *48*(6), 1088–1104. DOI: 10.1177/0309524X241247231

Jafarnejadsani, H., Pieper, J., & Ehlers, J. (2013). Adaptive control of a variable-speed variable-pitch wind turbine using radial-basis function neural network. *IEEE Transactions on Control Systems Technology*, *21*(6), 2264–2272. DOI: 10.1109/TCST.2012.2237518

Jain, A., Schildbach, G., Fagiano, L., & Morari, M. (2015). On the design and tuning of linear model predictive control for wind turbines. *Renewable Energy*, *80*, 664–673. DOI: 10.1016/j.renene.2015.02.057

Jeon, T., & Paek, I. (2021). Design and verification of the LQR controller based on fuzzy logic for large wind turbine. *Energies*, *14*(1), 230. Advance online publication. DOI: 10.3390/en14010230

Jeon, T., Song, Y., & Paek, I. (2023). A study on H∞-fuzzy controller for a non-linear wind turbine with uncertainty. *Applied Sciences (Basel, Switzerland), 13*(21), 11930. Advance online publication. DOI: 10.3390/app132111930

Johri, S., Selvakumar, P., Satyanarayana, P., Das, A., Gupta, S., & Bhattacharya, S. (2025). Urbanization and Energy: The Nexus of Global Challenges. In Geopolitical Landscapes of Renewable Energy and Urban Growth (pp. 453-478). *IGI Global Scientific Publishing*. DOI: 10.4018/979-8-3693-8814-3.ch016

Keighobadi, J., Ansar, H. M. K., & Naseradinmousavi, P. (2022). Adaptive neural dynamic surface control for uniform energy exploitation of floating wind turbine. *Applied Energy, 316*, 119132. Advance online publication. DOI: 10.1016/j.apenergy.2022.119132

Kumar, V., Singh, N., & Pratap, B. (2017). Adaptive pitch control design for variable speed wind turbine using Chebyshev neural network. *In 2017 International Conference on Information, Communication, Instrumentation and Control (ICICIC)*, 1–6, IEEE. DOI: DOI: 10.1109/ICOMICON.2017.8279166

Kundu, S., Sundriyal, N., Singh, N., Singh, S., & Ramirez, J. M. (2025). *Control strategy for a single-phase grid-connected PV system with bidirectional EV charging support. Modern computing technologies for EV efficiency and sustainable energy integration*. IGI Global Scientific Publishing., DOI: 10.4018/979-8-3373-2382-4.ch007

Li, J., Wang, Y., Lin, S., & Zhao, X. (2022). Nonlinear modelling and adaptive control of smart rotor wind turbines. *Renewable Energy, 186*, 677–690. DOI: 10.1016/j.renene.2022.01.020

Li, T., Liu, X., Lin, Z., Yang, J., & Ioannou, A. (2024). A linear quadratic regulator with integral action of wind turbine based on aerodynamics forecasting for variable power production. *Renewable Energy, 223*, 119605. Advance online publication. DOI: 10.1016/j.renene.2023.119605

Lin, W. M., Hong, C. M., Ou, T. C., & Chiu, T. M. (2011). Hybrid intelligent control of PMSG wind generation system using pitch angle control with RBFN. *Energy Conversion and Management, 52*(2), 1244–1251. DOI: 10.1016/j.enconman.2010.09.020

Ma, R., Siaw, F. L., Thio, T. H. G., & Yang, W. (2024). New adaptive super-twisting extended-state observer-based sliding mode scheme with application to FOWT pitch control. *Journal of Marine Science and Engineering, 12*(6), 902. Advance online publication. DOI: 10.3390/jmse12060902

Marquez, F. P. G., & Gonzalo, A. P. (2022). A comprehensive review of artificial intelligence and wind energy. *Archives of Computational Methods in Engineering*, *29*(5), 2935–2958. DOI: 10.1007/s11831-021-09678-4

Masenge, I. H., & Mwasilu, F. A. (2024). Modeling and control dynamics of microgrid with renewable energy systems. In *Modeling and control dynamics in microgrid systems with renewable energy resources* (pp. 199–242). Academic Press., DOI: 10.1016/B978-0-323-90989-1.00015-4

Mohamed, M. A. E. H., Kamel, S., & Meymand, H. Z. (2025). Intelligent integration of ANN and H-infinity control for optimal enhanced performance of a wind generation unit linked to a power system. *Optimal Control Applications & Methods*, *46*(1), 79–93. DOI: 10.1002/oca.3199

Moshtaghi, P., Hajialigol, N., & Rafiei, B. (2025). A comprehensive review of artificial intelligence applications in wind energy power generation. *Sustainable Futures : An Applied Journal of Technology, Environment and Society*, *9*, 100638. Advance online publication. DOI: 10.1016/j.sftr.2025.100638

Ngo, Q. V., Yi, C., & Nguyen, T. T. (2020). The fuzzy-PID based-pitch angle controller for small-scale wind turbine. *International Journal of Power Electronics and Drive System*, *11*(1), 135–142. DOI: 10.11591/ijpeds.v11.i1.pp135-142

Osman, A. M., & Alsokhiry, F. (2022). Sliding mode control for grid integration of wind power system based on direct drive PMSG. *IEEE Access : Practical Innovations, Open Solutions*, *10*, 26567–26579. DOI: 10.1109/ACCESS.2022.3157311

Oussama, M., Abdelghani, C., & Lakhdar, C. (2022). Efficiency and robustness of type-2 fractional fuzzy PID design using salps swarm algorithm for a wind turbine control under uncertainty. *ISA Transactions*, *125*, 72–84. DOI: 10.1016/j.isatra.2021.06.016

Petkovic, D., Cojbasic, Z., & Nikolic, V. (2013). Adaptive neuro-fuzzy approach for wind turbine power coefficient estimation. *Renewable & Sustainable Energy Reviews*, *28*, 191–195. DOI: 10.1016/j.rser.2013.07.049

Pratap, B., Singh, N., & Kumar, V. (2018). Robust control of variable speed wind turbine using quasi-sliding mode approach. *Procedia Computer Science*, *125*, 398–404. DOI: 10.1016/j.procs.2017.12.052

Qi, L., Zhao, M., Wu, S., Zhang, X., Meng, P., Zhao, Y., & Deng, W. (2025). Adaptive torque feed-forward control for wind turbine MPPT considering predicted wind speed characteristics. *International Journal of Electrical Power & Energy Systems*, *167*, 110598. Advance online publication. DOI: 10.1016/j.ijepes.2025.110598

Rajendran, S., Jena, D., Diaz, M., & Rodriguez, J. (2024). Design of modified complementary terminal sliding mode controller for wind turbine at region II using a two-mass model. *Results in Engineering, 24*, 103026. Advance online publication. DOI: 10.1016/j.rineng.2024.103026

Rao, M. S., Saraswathy, S., Neela, V., Arun, M. R., Gupta, S., & Taqui, S. N. (2023, November). Efficient Model Training in Decentralized Systems with Federated Learning. *In 2023 International Conference on Communication, Security and Artificial Intelligence (ICCSAI)* (pp. 900-904). IEEE. DOI: 10.1109/ICCSAI59793.2023.10421469

Rawajfeh, M. A. A., & Gomaa, M. R. (2023). Comparison between horizontal and vertical axis wind turbine. *International Journal of Applied Power Engineering, 12*(1), 13–23. DOI: 10.11591/ijape.v12.i1.pp13-23

Roy, T. K., Ghosh, S. K., & Saha, S. (2023). Robust backstepping global integral terminal sliding mode controller to enhance dynamic stability of hybrid AC/DC microgrids. *Protection and Control of Modern Power Systems, 8*(1), 1–13. DOI: 10.1186/s41601-023-00281-2

Sahoo, S. (2024). Enhancing pitch angle control in variable speed doubly fed induction generator based wind energy conversion system using advanced adaptive neuro-fuzzy approach with particle swarm optimization. *Wind Engineering, 49*(2), 444–461. DOI: 10.1177/0309524X241275145

Sahoo, S., Subudhi, B., & Panda, G. (2019). Torque and pitch angle control of a wind turbine using multiple adaptive neuro-fuzzy control. *Wind Engineering, 44*(2), 125–141. DOI: 10.1177/0309524X19849825

Shao, Y., Liu, J., Huang, J., Hu, L., Guo, L., & Fang, Y. (2022). The implementation of fuzzy PSO-PID adaptive controller in pitch regulation for wind turbines suppressing multi-factor disturbances. *Frontiers in Energy Research, 9*, 828281. Advance online publication. DOI: 10.3389/fenrg.2021.828281

Sharma, R. K., Saran, V. B., Kanaujia, S., & Gupta, S. (2024, May). Evaluation of mppt controller performance using fuzzy logic design with wireless sensor node. *In 2024 International Conference on Smart Systems for applications in Electrical Sciences (ICSSES)* (pp. 1-6). IEEE. DOI: 10.1109/ICSSES62373.2024.10561365

Sheng, H. W., & Jun, D. Y. (2013). Research on new pitch control algorithm of wind power generators. *Power Electronics, 47*(2), 53–54.

Shukla, R. D., Tripathi, R. K., & Gupta, S. (2010, November). Power electronics applications in wind energy conversion system: A review. *In 2010 International Conference on Power, Control and Embedded Systems* (pp. 1-6). IEEE. DOI: 10.1109/ICPCES.2010.5698663

Singh, D., Singh, A., Chowdhury, I. R., Gupta, R., Ram, P., & Gupta, N. (2025). Government of India, Ministry of Statistics and Programme Implementation, National Statistics Office. *Energy statistics India 2025*. https://www.mospi.gov.in

Singh, N., & Pratap, B. (2025). Robust observer design for variable speed wind turbine using a two-mass model. *In Proceedings of the International Conference on Systems, Control and Automation (ICSCA 2023) Lecture Notes in Electrical Engineering, 1260*, 171–180, Springer. DOI: 10.1007/978-981-97-7384-8_15

Singh, N., Pratap, B., & Swarup, A. (2016). Design of robust control for wind turbine using quantitative feedback theory. *IFAC-PapersOnLine, 49*(1), 718–723. DOI: 10.1016/j.ifacol.2016.03.141

Singh, N., Pratap, B., & Swarup, A. (2017). Discrete-time robust control of variable speed wind turbine using quantitative feedback theory. *In 2017 IEEE India Council International Conference (INDICON)*, 1–6, IEEE. DOI: DOI: 10.1109/INDICON.2017.8487583

Singh, N., Pratap, B., & Swarup, A. (2018). Robust control design of variable speed wind turbine using quantitative feedback theory. *Proceedings of the Institution of Mechanical Engineers. Part A, Journal of Power and Energy, 232*(6), 691–705. DOI: 10.1177/0957650917751463

Singh, N., Pratap, B., & Swarup, A. (2019). Design of dynamic surface controller for robust performance of variable speed wind turbine. *IET Renewable Power Generation, 13*(9), 1511–1521. DOI: 10.1049/iet-rpg.2018.5922

Singh, N., Pratap, B., & Swarup, A. (2019). Nonlinear robust observer based adaptive control design for variable speed wind turbine. *Journal of Engineering Research, 7*(3), 258–285. DOI: 10.1016/S2307-1877(25)00618-2

Song, J., Zhang, H., Zhang, J., Hu, P., & Feng, J. (2024). Observer-based adaptive fuzzy backstepping control of pitch angle of wind turbine. *2024 IEEE 2nd International Conference on Electrical, Automation and Computer Engineering (ICEACE)*, 887–892. DOI: 10.1109/ICEACE63551.2024.10898733

Sun, Y., & Han, W. (2025). A review of enhancing wind power with AI: Applications, economic implications, and green innovations. *Digital Economy and Sustainable Development, 3*(11), 11. Advance online publication. DOI: 10.1007/s44265-025-00059-4

Sundriyal, N., Kundu, S., Gupta, S., Bhatt, A., & Singh, N. (2025). *Hybrid EV Charging Systems: A Study Report on Integration of Fuel Cell with EV for Clean Solution. Hybrid Renewable Power Infrastructure for Sustainable Electric Vehicle Development. Studies in Infrastructure and Control*. Springer., DOI: 10.1007/978-981-95-1323-9_9

Wen, H., Liu, Y., Lee, S. M., & Park, J. H. (2023). Optimal power regulating for wind turbine with LQR and disturbance accommodation pitch control. *9th International Conference on Control, Automation and Robotics (ICCAR)*, 262–267. DOI: DOI: 10.1109/ICCAR57134.2023.10151773

Xie, J., Dong, H., & Zhao, X. (2023). Data-driven torque and pitch control of wind turbines via reinforcement learning. *Renewable Energy*, *215*, 118893. Advance online publication. DOI: 10.1016/j.renene.2023.06.014

Yin, X. X., Lin, Y. G., Li, W., Liu, H. W., & Gu, Y. J. (2015). Adaptive sliding mode back-stepping pitch angle control of a variable-displacement pump controlled pitch system for wind turbines. *ISA Transactions*, *58*, 629–634. DOI: 10.1016/j.isatra.2015.07.006

Zhang, C., & Plestan, F. (2021). Adaptive sliding mode control of floating offshore wind turbine equipped by permanent magnet synchronous generator. *Wind Energy (Chichester, England)*, *24*(7), 754–769. DOI: 10.1002/we.2601

Zhenhua, L., Yaming, G., & Qiang, L. (2025). Asymptotic tracking stability control method for grid-connected wind turbine systems based on quantitative trajectory constraints. *IEEE Access : Practical Innovations, Open Solutions*, *13*, 32248–32259. DOI: 10.1109/ACCESS.2025.3542774

Chapter 3

Renewable Energy Integration in Industrial Operations With Intelligent Systems

Sandeep Gupta
https://orcid.org/0000-0002-3734-3723
Graphic Era University, Dehradun, India

Tarun Varshney
https://orcid.org/0000-0002-4047-7351
SSES, Sharda University, Greater Noida, India

Nikita Rawat
Graphic Era University, Dehradun, India

Tianyi Han
https://orcid.org/0000-0001-5908-1851
Tsinghua University, Beijing, China

Deeptanshu Dwivedi
United University, Prayagraj, India

ABSTRACT

The increasing global importance of sustainable and "green" approaches has transformed renewable energy integration into a priority for multiple sectors, including the industrial sector. Industrial processes and activities that are typically energy-intensive and pose both challenges and opportunities for the use of on-site Renewable Energy (RE) sources such as solar, wind, geothermal, and biomass.

DOI: 10.4018/979-8-3373-7041-5.ch003

In this chapter, we introduce the theme of incorporating renewable energy into industrial processes by analyzing the technical, economic, and regulatory aspects. An extensive examination of the current renewable energy literature, encompassing various technological solutions, challenges to integration, and case studies on successful applications, is presented. Furthermore, the role of EMS, smart grids, clean energy systems, and energy storage in the optimal utilization of renewable energy is also discussed in this chapter.

INTRODUCTION

Importance of Sustainability in Industrial Operations

The industrial sector represents a significant proportion of global energy demand and environmental impact, accounting for some 38% of global energy consumption and 23% of global CO_2 emissions (IEA, 2020; Sudha et al., 2024). Heavy industrial sectors like manufacturing, processing of food, chemical manufacturing, and others highly rely on fossil fuels (coal, oil, natural gas, and so on) for energy (Husin & Zaki, 2021). Non-renewable energy sources can damage the climate, degrade the environment, pollute the air, and deplete resources.

In recent years, there has been a growing awareness of the need to see industry as more sustainable. It is partly due to regulatory pressure and partly as a reaction to the urgent need to respond to climate change. Jones (2017) explained that there are various ways of achieving industrial sustainability. One is the consideration of renewable energy usage. Some examples of industrial sustainability include energy use, waste, recycling, and efficiency. These goals can be achieved by the industries through the use of sustainable energy systems that ensure a reduced carbon footprint, while also minimising impacts on operating costs and the national sustainability goals. As renewable technologies get better and cheaper, they can help industries stay in the game while preserving the environment. Sustainable corporation development is a win-win opportunity for businesses because they help the world meet its sustainability goals while building a stronger future for themselves.

Role of Renewable Energy in Achieving Net-Zero Targets

Countries across the globe are promising to attain net-zero emissions by mid-century, as displayed in Figure 1 (Zobaa & Bansal, 2011). If these goals are to be achieved, the biggest energy-consuming industries have to significantly alter how they power themselves. In other words, net-zero indicates that greenhouse gas emissions made by us in the air are neutralised (by taking out the same amount) to keep the climate stable. The phrases "climate change" and "global warming" are not just mere words; they form the foundation of climate action globally. Industries are dependent on huge consumption of energy, making them absolutely essential in achieving those goals (Awasthi et al., 2024; Gupta & Tripathi, 2014. It is renewable energy, which is the long-term replacement of fossil fuels, the revolution is all about. One route to dramatically reducing their greenhouse gas emissions and the industry will be to move on from the use of coal, oil and natural gas to cleaner energy sources, like wind, solar power or bioenergy (Taibi et al., 2012). Through the integration of renewable technology with energy storage and smart grid systems, industrial processes have better access to a flexible and reliable energy supply (Semwal et al., 2024; Shukla et al., 2010). It also enables organisations to acquire energy at a cheaper rate.

Figure 1. Types of Renewable Energy Sources and Their Industrial Applications

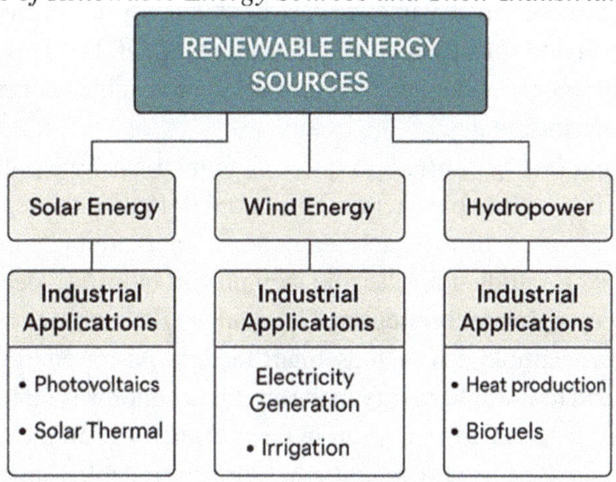

Purpose and Scope of Chapter

This chapter provides more detailed information about how businesses generate electricity and their efforts to transition to large-scale renewable energy use. It looks at more than just the technical side of things. It also examines the economic and political factors influencing this shift. The goal is to determine what companies need to do to successfully transition to cleaner energy, what challenges they may face, and what new opportunities they can capitalise on along the way. This chapter talks about the technical details, pros, and cons of each system. The chapter is divided as follows:

Integration barriers: The main impediments to implementation on the industry's part include renewable energy integration, namely intermittency, grid integration challenges, storage, penetration and cost-effectiveness (Fotis et al., 2025; Lu, 2016).

Technical solutions: Technical solutions pertaining to the power grid increase the stability of the grid and address the problems of renewable energy being intermittent (Gupta et al., 2010).

Case studies: Case studies showcasing how different sectors have employed renewable energy solutions, the techniques they employed, the results they obtained and the lessons learnt (Thandlam et al., 2024; Rao et al., 2023). Through examples, useful lessons can be gained through the use of renewable energy tools in business situations.

Economic/financial implications: The economic features associated with the installation of renewable energy involve cost-benefit analysis, return on investment (roi), and tax credit and subsidy facilities (Sharma et al., 2024). This chapter examines the financial impact on a business from using renewable energy and how this can lead to long-term savings.

Policy and regulatory context: A quick examination of the policy context for the industrial shift to renewable energy. This part will look at the renewable energy government incentives, subsidies, and regulations. International commitments and standards will also be studied for the role they play in industrial decarbonization.

Trends and emerging technologies: The chapter will conclude with a brief look at trends and new technologies such as smart factory, AI for energy optimisation, green hydrogen and renewable energy in the circular economy (Gupta, 2025; Meena et al., 2024; Sinha et al., 2024). Progress in these capacities could change industrial energy systems and support a lower-carbon, more sustainable future.

The goal of this chapter is to provide technical and planning aspects of renewable energy integration into industrial processes, as well as guidance for the industrial sector for a first step or to complete engagement in renewable energy. By interpreting the present findings, stakeholders in the industry, policymakers, and energy managers can make informed decisions on investing in renewables to align with global sustainability goals and operational efficiency.

Structure of the Chapter

Renewable energy offers an alternative to conventional energy forms that is both sustainable and affordable for various industries. This chapter investigates the function, prospects, and conversion and storage systems of industrial applications. In addition to financial and economic challenges, it addresses the incorporation of renewable energy, with intermittency, grid integration, and storage issues among them. This chapter also covers technical solutions such as energy management systems, smart grids and storage technologies. This report also looks at the financial and economic ramifications of renewable energy investment, along with cost-benefit and return-on-investment analysis. The future of the renewable energy industry and recommendations for businesses that plan to adopt renewable energy.

Because of this organisation, the book covers all major areas of Renewable generation and integration. " After reading this chapter, the reader will have a strong understanding of the technological, economic, and policy aspects of the diffusion of renewable energy in the industrial sector and get practical ideas on how industries could move smoothly towards a sustainable, low-carbon future.

RENEWABLE ENERGY SOURCES FOR INDUSTRIAL OPERATIONS

Renewable energy sources, such as solar, wind, biomass, geothermal, and hydropower, are being explored for their compatibility with industrial activities, including the sugarcane industry. Understanding their technical aspects, advantages, and disadvantages is crucial for efficient harnessing of power.

Solar Energy

Solar energy is becoming an increasingly flexible and scalable energy option that can be utilised for various industrial applications. While the sun's energy can be tapped using photovoltaic (PV) cells, it can also be tapped through concentrated solar power (CSP).

- **PV Systems:** These systems convert sunlight directly into electric power by means of semiconductor materials. These are widely used for industrial rooftops, open ground, and industrial complexes. The technology can be most effectively utilised in sectors with high energy usage, which require a considerable amount of electricity. PV plants are easy to maintain at a low cost and are also easy to install, especially in areas with intense sunlight (Tang et al., 2021).
- **CSP:** This is another example, and it concentrates sunlight from a large area onto a small area (like a storage medium or a heat exchanger), and subsequently, the collected heat is used to produce steam, for example, which can be used to drive turbines. CSP, which is more suitable for large-scale industrial processing requiring base load power, such as factories with even energy

requirements. Compared to PV power plants, CSP can be combined with thermal storage to provide electricity in cloudy or dim light conditions.

Benefits:

- Abundant and renewable resource
- Long-term savings in energy costs are possible
- Scalability for broader industry sizes
- Saves some of the expenditure on foreign energy supplements

Limitations:

- Intermittence (sun energy is only available during the day)
- Needs a considerable amount of land for large-scale systems
- Relatively high up-front cost (but grants and incentives may be available)

Applications:

Solar power has been conducive to industries such as textiles, food processing, and electronics production. The advantages of such large-scale solar installations increasingly require little introduction: companies like Tesla and Google have significantly reduced their utility bills and carbon footprints by covering industrial buildings with solar panels.

Wind Energy

Wind power is another efficient renewable energy source that has been increasingly adopted in industrial applications. Wind energy is captured by turbines, which transform wind's kinetic energy into electrical power. Wind power is particularly suitable in regions with strong winds, such as offshore and high-altitude sites.

- **Onshore Wind:** Onshore wind farms are the most common type of wind power generation, based onshore. These turbines are usually installed at well-suited, windy locations. Onshore wind is the best option for power-intensive industries to meet their large-scale requirement of wind energy.
- **Offshore Wind:** Wind turbines may be placed out at sea where wind blows more powerfully and more steadily. As chemical plants and steel mills require a lot of power, offshore wind farms are becoming increasingly popular.

Benefits:

- High wind potential in wind areas
- Wind farms can produce substantial amounts of electricity
- Low costs of operation and maintenance after the turbines have been installed
- Reduces reliance on fossil fuels

Limitations:

- Intermittency and dependence on wind-resource availability
- High capital for the setup of a wind farm.
- Effect on wildlife and hydrological / land use restrictions.

Applications:

The use of wind energy is best when its peak hours correspond with demand. In this regard, the industry is an ideal case. For instance, companies making aluminium and steel, which consume large amounts of electricity, can potentially take advantage of wind farms. Wind power helps to boost the demand for renewable energy in various parts of the world, such as Europe and North America.

Biomass Energy

Organic materials like wood shavings, ash, agricultural remains, municipal waste, and other waste are used to derive biomass energy. Biomass is often directly burnt to generate heat or electricity or converted into biofuels. The technology is suitable for various scales, ranging from industrial plants to power plants.

- Biofuels, such as bioethanol, biodiesel, and biogas. Biofuels are particularly relevant in industries that use oil as a liquid fuel for powering engines and heating applications. Bioethanol and biodiesel are widely utilised in agriculture, transport, and heavy machinery.
- **Biogas:** Biogas is obtained through the fermentation of the organic matter. It is usually burned to generate heat or produce electricity for processes such as food or waste processing. Biogas is also used in cogeneration (CHP) plants, which are an efficient way to utilise biogas for both electricity and heat at an industrial facility.
- **Combustion:** Biomass can be combusted directly to produce heat. Biomass is also used as a heating fuel in some industrial facilities, and utilising it is significantly cheaper than other fuels, especially in industrial applications (e.g., paper or craft mills).

Benefits:

- Keeps waste out of landfills and turns organic waste into beneficial energy
- Closed-loop of hazardous materials can be combined with current industrial processes
- Furnishes a source of renewable power for energy-consuming businesses such as industrial users that need energy and heat continuously
- Boon to rural economies: Uses locally-mined biomass.

Limitations:

- Take-off feedstock availability may vary and may be linked to agricultural cycles
- Land and water are extensive for biomass cultivation
- Large initial investment costs for biogas and biomass conversion facilities

Applications:

Biomass is widely utilised in various sectors, including paper and pulp, food and beverages, and agriculture. For example, numerous paper mills utilise biomass waste to generate heat and power, thereby significantly reducing their reliance on fossil fuels. Biogas is increasingly making its way into the food processing sector, where organic waste can be converted into a valuable energy source.

Geothermal Energy

Geothermal power is based on tapping into the heat from the earth's interior to generate electricity or offer direct heating. Geothermal power plants typically harness steam from beneath the surface to power turbines connected to generators, which in turn generate electricity. Geothermal heat pumps provide space and water heating in industrial settings.

- **Geothermal Power Plants:** For industries situated in regions rich in geothermal activity (e.g., close to tectonic plate boundaries, such as Iceland, New Zealand, and some parts of the United States). Geothermal energy is dispatchable and provides baseload power, making it a reliable long-term renewable source for industrial processes that require constant power.
- **Geothermal Heat Pumps:** Systems that heat and cool industrial buildings by moving heat between the building and the ground. Geothermal heat pumps are highly energy-efficient and offer a sustainable solution for maintaining comfortable temperatures in industrial facilities.

Benefits:

- Predictable and dependable supply of energy (baseload power)
- Clean and environmentally-friendly power generation with low emissions and no impact on the environment
- Saves money on heating and cooling energies

Limitations:

- Geography constraints (only near geothermal reservoirs)
- High up-front capital cost for drilling and infrastructure
- The hazard of future reservoir depletion if not operated in a sustainable manner

Applications:

Industries that utilise geothermal power for heating or cooling include the food processing and chemical industries, as well as the timber drying process.[290] The Cities of Klamath Falls, Oregon, and Boise, Idaho, use geothermal water to heat homes and businesses. Companies in geothermally active regions around the world are increasingly implementing this technology to meet their own power needs.

Hydropower (Where Applicable)

Hydropower harnesses the energy of flowing water to generate electricity. Hydropower is generally associated with large dams, but small units are also developed along surface water bodies, such as the canals found in valleys. Hydropower can be categorised into two major types: large hydropower systems and Small (run-of-river) hydropower projects.

Benefits:

- Reliable, renewable energy source
- Long life and low operating costs
- Can produce large amounts of power, particularly in regions with water access.

Limitations:

- Geographically restricted to regions of flowing water
- Environmental issues concerning the construction of dams (loss of habitat, fish migration)

Applications:

Hydro power plants constructed between 1000 kW and 10000 kW, having a small capacity, are employed to utilise energy and are known as Small Hydro. Industries can set up a small-hydro power plant to fulfil their electric needs. Hydropower can be an excellent alternative in regions where solar and wind power are frequently used.

The renewable sources that are the focus of this industry are discussed in this chapter. The place influences the selection of the most appropriate renewable energy technology for the energy demand arising from the industrial activity and its nature.

CHALLENGES IN THE INTEGRATION OF RENEWABLE ENERGY INTO INDUSTRY

Although renewable energy can be beneficial in terms of sustainability, cost savings, compliance, and other benefits, it will involve challenges in the industrial sector. These barriers can be classified as technical, economic and policy barriers. According to Farret and Simões (2017) and Tyagi et al. (2024), understanding these challenges is important for industries trying to incorporate renewable energy applications. In order to deal with issues of integration, this chapter outlines the key barriers.

Technical Barriers

This DE A 7 226 513 recognises and identifies the main technical difficulties in energy integrating renewables into industrial processes, i.e. intermittency, energy storage, grid integration, and matching supply with demand. To achieve reliability, efficiency, and scalability, renewable energy systems must be designed creatively and thoughtfully to meet these challenges.

- **Renewable Intermittency:** Solar and wind energy sources are unstable because of their inherent nature. Sunlight is present only in the day, and wind is limited by ideal speed and strength. As a result of this intermittency, on-site energy supply does not coincide with the industry demand period, especially in the case of energy-intensive industries requiring electricity in a constant and continuous manner (Anand et al., 2022). For instance, steel mills operate 24 hours a day, and if the power supply were to be intermittent, production could be stopped.
- **Energy Storage Solutions:** Storage Devices that store excess energy whenever there is a peak in generation, whether it be from the sun or the wind, in order to overcome the intermittency effect. Batteries, pumped hydro, compressed air energy storage and other technologies can store energy for later use. Lithium-ion and sodium-sulfur batteries are two of the best options.

Despite their potential, energy storage technologies are still hampered by multiple challenges, including costs, scaling issues, and short life.

- **Grid Integration:** It is a technical challenge to integrate renewables into the grid, too. In many cases, industrial facilities are connected to the grid, and the grid does not necessarily have the infrastructure to manage the effects of decentralised renewable generation (Onyinyechukwu et al., 2024). For example, with solar panels on the rooftops of factories, power is locally produced, which prevents losses in transmission. However, grid technologies like smart grid and demand-side response are helping to manage distributed generation and grid stability, but those are not enough. Our experts believe that customers who deploy volt/VAR optimisation technologies become less likely to experience voltage fluctuations, loss of power or damage to equipment.
- **Flexibility and Robustness of the System:** Industrial processes need energy systems that can be modified in response to changes in load requirements. The supply of renewable energy sources, particularly solar and wind, is not constant. Therefore, it is necessary to match the peak demand of industrial activity. In many industries, a sufficient solar and wind energy supply isn't enough for baseload power. Mixed systems integrating renewable energy sources and conventional backup systems (for instance, natural gas or biomass) are being investigated to ensure more flexibility and reliability.

Solutions:

- Energy management systems (EMS) could be developed to track and control energy usage and to avoid disturbance of industrial processes due to the integration of renewable energy.
- Hybrid renewable systems (such as solar and wind with other renewable energy) with backup capacity can help bring stability to industrial processes.
- Smart grids, which allow real-time monitoring of the energy produced and bidirectional communication between users and generators, can be a solution that contributes to better integration of grids.

Economic and Financial Barriers

Besides its long-term economic and ecological benefits, the initial investment required for the use of renewable energy sources is a significant obstacle to implementing the technology, especially for Small and Medium-sized Enterprises (SMEs) in the industry. The financial difficulties of implementing renewable energy can be divided into the following aspects:

- **Large Capital Expenditure:** A massive amount of investment is needed for the implementation of renewable resources for power generation; for example, solar panels, wind farms, etc. This includes the cost of the equipment, as well as its installation and other necessary infrastructure (such as connection to the grid or energy storage) required to take power from it. The high upfront cost of renewable energy devices in many cases still provides future savings. However, this cost may inhibit businesses from investing in new equipment. In particular, the upfront cost of renewable energy devices is often compared to the low cost of fossil fuels.
- **Length Continues to Pay Back:** The returns from renewable energy investments may take a few years to pay off. Based on energy use, location and the size of the alternative energy system in use, the payback is estimated to be between five and 15 years. In sectors with limited financing capacity, the long payback time could make renewables less conducive than traditional energy.
- **Financial Risk and Uncertainty:** Investment in renewable energy projects is often deterred by financial risk, which includes volatility in energy prices and uncertainty in future government incentives. The financial viability of the project may be called into question when the subsidies or tax breaks that support investment in renewable energy change.
- **Economically Feasible Small-Scale:** The process of utilising renewable energy on a large scale can provide economies of scale. However, these economies of scale may not be very feasible when it comes to small or medium-sized industrial processes. Implementing renewable energy technologies can create a circumstance where renewable energy can be integrated so far that little or no economic rationale applies unless government incentives are strong or financing access reduces the up-front cost.

Solutions:

- Governments and banks could reduce the costs of renewable energy projects by directly offering them money in the form of grants, subsidies, and low-interest loans.
- Use financing, like power purchase agreements (PPAs) or energy-as-a-service, to allow businesses to use renewable energy without having to pay for the installation up front. They can simply pay according to the energy they use.
- EPCM uses financial modelling and cost-benefit analysis tools to support industries in their long-term savings and return on investment understanding. In turn, that will help stakeholders to justify investment in renewable energy sooner.

Regulatory and Policy Barriers

Regulatory and policy environment: The speed and scale of integrating renewable energy into industrial processes are largely influenced by decision-makers. The usage of renewable energy technologies can be aided or hindered by government policy. Regulatory obstacles are mainly:

- **Non-standardisation:** The renewable integration rules being unstandardised, especially in some parts of the globe, creates confusion for the industry. Standards in such areas will help integrate renewable energy technologies in the infrastructure while ensuring safety and performance specifications. Without clear laws, companies will not commit to investing in renewables, as they could end up breaking the laws and getting shut down.
- **Policy uncertainty:** Uncertain and shifting policies, alongside new trade and payment mechanisms, have created major risks. Government policies such as consumer tax credits, rebates, and renewable energy standards are responsible for this. Nonetheless, there will be risks associated with the new policies, as a change in politics could lead to the alteration or cancellation of subsidies. If companies are unsure whether the policies that favour renewables will stay in place, they may avoid investing in them.
- **Interconnection Standards:** Industrial facilities wishing to feed in renewable energy onto the grid may face regulatory issues about interconnection standards, namely, the rules and regulations concerning plugging renewable energy systems into the national grid. These regulations are mostly elaborate, differ from country to country and are very time-consuming and expensive to obtain.

The renewable energy sources are great alternatives to traditional energy sources. The government and regulatory bodies play an important role in promoting them and streamlining their standardisation.

TECHNOLOGY SOLUTIONS FOR INTEGRATION OF RENEWABLE ENERGY

Technology development is necessary to adapt key industrial sectors to new wind and solar applications in order to improve the performance, flexibility, and reliability of energy systems. This includes using cutting-edge technology to mitigate intermittency, as well as energy storage and grid integration, allowing for deeper penetration of niche energy resources.

Energy Management Systems (EMS)

An energy management system (EMS) is software that allows a plant to monitor, control and optimise energy usage (Vu et al., 2020). Renewable energy sources can be integrated into the system by them. Critical components of industries enable the use of energy efficiently, reduce costs and the emission of carbon and GHG. As shown in Figure 2, EMS measure energy use in real time. Tracks and monitors the utilisation of energy with the help of EMS. As it helps to find the wastage and inefficiencies, businesses can modify processes as per the demand.

They assist in demand response and load forecasting, allowing industries to adjust their consumption according to changing supply conditions. The EMS can also forecast load, aiding in scheduling loads for preferred usage times. EMS is also used to determine the optimal operating point for hybrid systems to ensure smooth operation using renewable energy (Igogo et al. 2021; Kundu et al. 2024).

Benefits:

- Improving the energy efficiency of air conditioning and reducing operational costs
- Maximal integration of renewable energy (RES) under the limitations posed by the flexibility of a power plant
- Reinforces energy systems for reliability and stability
- Facilitates meeting the criteria for economy and environmental protection

Figure 2. The role of energy management systems in industrial renewable energy integration

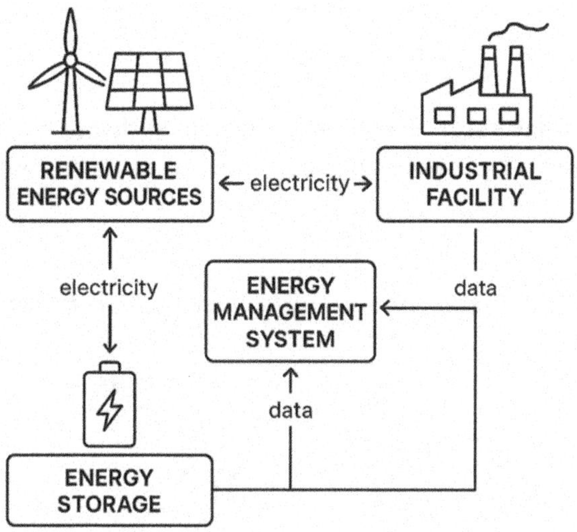

Smart Grids

A smart grid is a modernised electrical grid system that uses digital technology to monitor and manage the transmission of electricity from all power generation sources, thereby meeting the varying electricity demands of end users, as shown in Figure 3. Unlike legacy grids, which are often one-dimensional, smart grids enable two-way communication between energy-generating sources (such as solar panels) and energy-consuming sources. Smart grids are the backbone for integrating renewable energy into the system, especially given the flexibility they provide for managing intermittent resources such as wind and solar (Mishra & Singh, 2023; Kundu et al., 2022).

- **Distribution of Energy in Real Time:** Smart grids enable real-time distribution of energy, thereby maximising the use of renewable energy resources. They can quickly divert electricity to areas of high demand, use excess renewable energy to backfeed into a local facility, and reduce loss of electricity during transmission.
- **Grid Stability & Frequency Control:** The smart sensors, as well as automated technology built into the smart grids, will ensure proper regulation of frequency. Further, it will help eliminate the occurrence of power outages and the fluctuation of voltage. This component proved to be all the more essential

in the case of the integration of renewable energy sources. This is because renewables have the potential to create issues with grid stability. A smart grid is capable of drawing power from reserves or from distributed renewable energy sources in response to changes to maintain frequency on the grid.

Utility systems have also become more efficient, thanks to greater renewable generation. Energy losses in distribution systems are reduced, and industries enjoy greater flexibility.

Figure 3. Renewable Energy Integration in the Industrial Grid (Conceptual Diagram)

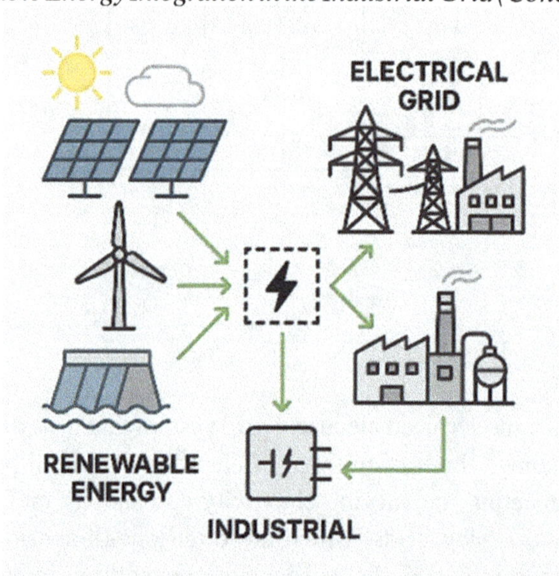

Energy Storage Systems (ESS)

Energy Storage Systems (ESS) modern technologies will play an important role in improving the reliability and consistency of renewable energy sources like solar and wind. Energy storage facilities can balance high production with periods of low output. It can also help in cases of high demand for renewable energy sources. Various methods exist with one or more advantages/disadvantages.

- **Battery Storage:** Battery storage systems are one of the most popular types of ESS, and they store power using chemical processes. The lithium-ion battery is one of the most popular industrial rechargeable batteries due to its high energy density, excellent rate capability and long cycle life. Batteries can also be paired with solar or wind systems to store excess power for later use, thereby reducing the need for backup power from fossil fuels such as gas or coal.

- **Pumped Hydro Storage (PHS):** Pumped hydro storage is a mature way of storing energy and utilises the gravitational potential energy. During off-peak or high-rain renewables, any excess energy is used to pump water up to a higher reservoir. In peak usage times, water is then released to spin turbines that generate power. Pumped hydro is a highly efficient technology and can store enormous amounts of energy. Still, it needs to be located in a geographically suitable location (with a source of water and sufficient height difference).
- **CAES (Compressed Air Energy Storage):** CAES, or compressed air energy storage, stores energy in the form of compressed air, which is stored in underground caverns or large tanks. The compressed air is then released and used to power turbines that produce electricity. CAES is a nascent technology capable of providing bulk storage for industrial use cases where there is no room for pumped hydro.
- **Sensible-thermal storage:** Sensible thermal storage systems store heat that can be utilised directly for industrial process heat or for thermodynamic conversion into electricity. Thermal energy stored by common molten salt and ice storage formats is sterile. EAMUs are especially beneficial for businesses that use large amounts of heat, such as the cement industry.

Benefits:

- Assures continuity and stability of the energy supply on the grid from variable renewable sources
- Lessens the need for fossil fuel-based backup power
- Improves grid stability and flexibility
- Provides a permanent solution to energy storage, specifically at a large-scale industrial level

Microgrids

A microgrid is a local energy system that can work independently or in connection with the central grid. Thus, it provides energy security and independence in industrial applications. This combines solar, wind, and bioenergy with other renewable sources to ensure supply stability. The energy independence and resilience that microgrids offer lead to a decrease in dependence on the grid, as well as enable backup power in the event of failure. They also create economic value by offering energy commodities at a lower cost from sources of renewable energy.

Smart Sensors and IoT for Energy Management

Smart sensors and the Internet of Things (IoT) are being used in many industries today to reduce energy use. These objects can provide real-time retrieval of energy consumption, equipment operation and environmental data, allowing energy management through enhancing energy efficiency.

- **Energy Monitoring:** IoT sensors can monitor energy consumption in specific areas of a plant to identify the areas where energy loss occurs. Therefore, businesses can devise plans to minimise their energy usage and give preference to utilising renewable sources.
- **Predictive Maintenance:** Smart sensors in your equipment can warn you when they're about to break or are running inefficiently. Hence, implementing predictive maintenance becomes simple. This avoids losing energy and ensures that renewable energy systems, such as solar panels and windmills, get maximum efficiency.
- **Impact on Real-Time Data Analytics:** The IoT Systems can enable cloud real-time data analytics with prompt changes for maximum use of energy. For example, a renewable energy system may perform poorly during cloudy or low wind periods, which may result in the system reducing energy utilisation and relying more on stored energy or grid energy.

Benefits:

- Monitoring in real time and optimising consumption of energy (Gupta et al., 2023; Sinha et al., 2023).
- Reduced operational costs as a result of forecasting maintenance and better resource allocation.
- Make better use of renewable energy in an industry
- Enhanced operational efficiency and sustainability

Wind turbines or solar panels' installation is but a facet of renewable energy integration in the industry; another is the addition of technology to enhance the reliability, flexibility and efficiency of energy systems (Gupta et al., 2015). There are several important components that enable the smooth incorporation of renewable energy. Energy management and monitoring systems; smart grids; energy storage systems; microgrids; and IoT. By deploying these next-generation technologies, industrial organisations can resolve irregularities in grid performance, improve energy efficiency and, thus, reduce dependency on fossil fuels with a focus on sustainability in the future.

CASES ANALYSIS OF RENEWABLE ENERGY INTEGRATION

Insights into best practices in mainstreaming renewables for industry adoption. We hope that these factories implementing these solutions can learn from one another and apply some lessons learned from successful implementation at earlier sites. It's now time to look at other industries and see how they are integrating renewable energy into their operations – lessons can be learnt across sectors.

Case Study 1: Solar Energy for the Textile Industry (India)

In India, the textile industry is one of the largest and most energy-intensive industries and accounts for a significant share of industrial energy use in the country. A textile manufacturer based in Gujarat, India, was looking to reduce its dependence on conventional power and lower carbon emissions through a solar project.

- **Project Description:** The manufacturer installed a rooftop solar photovoltaic (PV) system on the roofs of its textile manufacturing site to generate a significant portion of the company's electricity usage during daylight hours to reduce dependence on the grid.
- **Project:** At the manufacturing site, a total system of 1 MW solar PV system was installed, which will meet 20% of the facility's energy needs. The textile maker collaborated with a local solar company to develop a system at the site that could integrate with the existing power grid and meet the power demand of its energy-intensive daytime operations. It further incorporated an energy management system (EMS) with the project for monitoring and managing energy usage.

Results:

- The solar system contributed approximately 20-25% to the facility's overall energy requirements.
- The organisation cut energy spending by 15% annually and saw a return on investment in five years.
- The programme even saves 1,000 tons of CO_2 per year.

- The firm became known as committed to sustainability, and this improved its image in the locality and also overseas.

Lessons Learned:

- **Scalable:** One of the benefits of solar energy is that it is easily scalable. This is especially true in those areas of the world that are sunny for most of the day. Rooftop installation is highly suitable for industries that have a large roof area.
- **Energy Management:** The use of EMS to control the load requirements in conjunction with solar power and the grid can help some industries realise their full potential from renewables invested in their Illume Ability.

Case Study 2: Wind energy in steel production (Germany)

Wind Germany has been a world leader in adopting renewable energy, especially wind power. A steel producer in the north of the nation leveraged the beneficial German wind profile to add wind power to its energy mix.

- **Project description:** This steel plant, with a high and consistent demand for power, was developed around an onshore wind farm to produce electricity for the steel site. The wind farm and the plant were connected to the plant's grid, and the plant was both an electricity producer and an electricity consumer.
- **Actuation:** A 5 MW wind turbine farm was applied on an adjacent agricultural land by the company to offer about 50% of its electricity demand. The wind farm was linked to the plant's energy management system so the power produced by the rotors could be efficiently "fed" to the plant.

Results:

- The wind farm cut down the amount of power the plant drew from the national grid, meeting half the factory's energy demands, and made substantial savings in energy cost.
- The business realised a CO2 reduction by some 12,000 tons per year;
- The project has enabled the company to meet its sustainability targets and strict EU emissions standards.

Lessons Learned:

- **Long-term Investment:** wind energy projects need a large initial outlay, but also provide financial benefits over the long term, and can help the industry reduce its energy bill over the long term of the project.
- **Energy Independence:** For industries such as the steel industry, which are high energy consumers, investing in renewable energy options like wind can offer increased security of supply and lower the risk from escalating energy prices.

Case Example 3: Wood/Non-wood Energy in Paper Production (USA)

The paper industry consumes a significant amount of electricity and heat, making it a power-intensive sector. To reduce its reliance on coal and natural gas, a paper mill in the Midwest of the United States turned to biotechnology, renewable energy, and other strategies.

- **Overview of the Project:** The pulp and paper mill utilised biomass waste (wood chips, sawdust, paper sludge, etc.) to generate biogas and Fuller's earth-based biomass pellets for utilising the same thermal and electric power generation inside the mill.
- **Use:** A CHP system fueled by biomass that generates heat and electricity was run by the business. The system utilises leftover heat from the process to warm the mill, thereby minimising the usage of fossil fuels.

Outcomes:

- The mill's energy consumption has dramatically fallen after the biomass project takes 70% of its energy requirements.
- The business cut its grid-powered energy consumption and reduced annual CO_2 emissions by up to 20,000 tonnes.
- The company achieved pollution-free waste management, which turned waste into a quality energy source and a component of the circular economy.

Knowledge Gained:

- **Waste to Energy:** Businesses that generate organic waste during the course of production may find biomass to be a potential renewable source of energy. Generating energy from waste helps save on disposal costs and avoids the discharge of pollutants to the air and water.

Case study 4: Geothermal Energy in Food Processing (Iceland)

Iceland's rich geothermal resources make geothermal energy a practical option in Iceland. A company in Reykjavik that uses a great deal of energy uses the geothermal energy of Iceland for heating purposes.

- **Project Summary:** A food processing company installed a geothermal heating system to provide space and water heating at the facility. The nearby geothermal field of the plant provides steam for the heating requirement of the plant, which is needed at a low cost.
- **Implementation:** The firm utilised geothermal energy from the earth with the help of a ground-source heat-pump system. The heat pump system was designed to work along with the existing boiler, so the site always had hot water.

Results:

- The company's heating requirements are met exclusively by the geothermal system that cuts heating costs by 40% annually.
- The company's reliance on foreign fossil fuels and carbon footprint was drastically reduced.
- The initiative is helping the company keep its energy costs competitive while enhancing its green profile.

Lessons Learned:

- Geothermal: Industries located in regions with considerable geothermal resources (like Iceland) can use this energy source, which is well established and effective in heating and cooling.
- Savings: Geothermal has the potential to save the chemical and food industries a lot of money.

Case 5: Hybrid Renewable Energy in the Cement Plant (Mexico)

A cement plant in Mexico has been outfitted with a hybrid renewable energy system utilising solar and wind energy. The plant's dependency on fossil fuels at peak times is reduced by a 5 MW solar PV installation and a 10 MW wind turbine farm that powers the plant with extra energy. The EMS of the plant interfaces for maximum utilisation of green power. The facility's demand for internal energy is now being met 60% of the time, and it has resulted in a drop of about 25% in power costs against grid costs. The plant saves 15,000 metric tonnes of CO_2 each year, reducing energy costs and limiting exposure to changing energy prices. The hybrid system provides flexibility and stability at all times of changing energy demand.

The cases below demonstrate that industries can function entirely on renewable energy (100%), and this can be advantageous to many sectors. Feedstuffs and producers across the globe are achieving considerable savings through on-site renewable energy sources, such as solar, wind, biomass and geothermal. Recent case studies contain useful lessons for overcoming technical, economic, and regulatory challenges. These case studies can serve as roadmaps for other sectors.

The successful application of renewable-energy technology in so many varied enterprises underscores the importance of thoughtful consideration and careful assessment of energy demand in relation to available RE resources and technology that can use them. This means that renewable energy can also be used at any industrial plant, allowing said firm to contribute to global sustainable development while increasing its competitiveness and sustainability in the long run, both financially and non-financially as well.

ECONOMIC CONSIDERATIONS AND COST-BENEFIT ANALYSIS

When it comes to integrating RE in the industrial processes, there is a high degree of economic factors shown in Figure 4. Though the long-term economic and environmental benefits of using renewable energy technologies (RETs) are higher, investment costs can pose an important barrier from a financial as well as industry viewpoint. Renewable energy is the focus of this unit of study, which examines the capital, cost-benefit analysis, financial modelling, ROI, and payback periods concerned with a case study energy option.

Figure 4. Cost-Benefit Analysis of Renewable Energy Integration in Industrial Settings

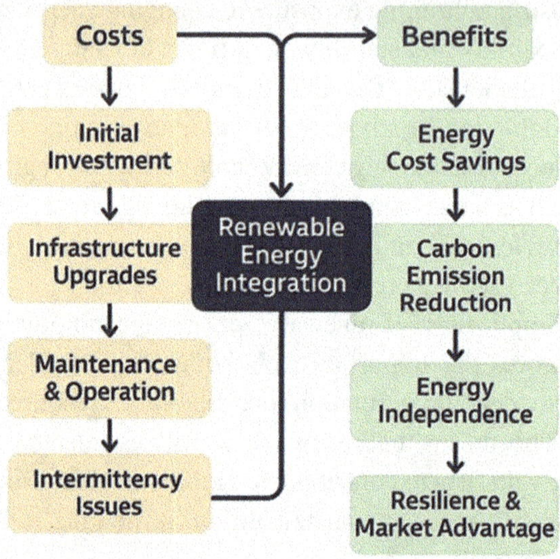

Initial Capital Investment and Financial Outlay

Costs of Renewable Energy Technology: Energy options such as solar panels, wind turbines and biomass boilers all require high-end capital equipment and can be costly to an industry, more so to small and medium-scale industries. As mentioned earlier, the initial investment also includes various other things, such as:

- **Equipment Capital Cost:** The expense of acquiring and installing renewable energy systems varies with technology; equipment capital cost is defined by IRENA. For illustration, the installation cost of solar photovoltaic (PV) ranges from $1,000 to $3,000 per installed kilowatt (kW), and a wind turbine averages over $1.5 million per installed megawatt (MW). Biomass and geothermal will be costly to build out, just like solar and wind, as these are infrastructure, equipment, and site preparation-heavy.
- **Installation and Infrastructure Costs:** An industry will have to include the installation cost over and above the equipment cost of the RE systems. The installation cost includes land preparation, grid connection, and energy storage systems (where required). The overall expenditure of the project rises because of these expenses.
- **Interconnection cost:** The link-up of renewable energy systems and the national grid or industrial micro-grid may have another cost, which is the cost

incurred in upgrading the system, smart grid integration, energy management system (EMS), etc.

- **Funding Costs:** Industries often need to borrow funds for renewable energy projects due to funding costs. The cost of loans or other financial transactions (i.e. interest rates, loan terms, and any other relevant costs) can be an important part of the total project cost. Government grants, tax breaks and other subsidies may sometimes offset the associated cost, but this does not always equate to paying the true cost of investment.

Challenges:

- One of those prohibitive barriers to entry can be high up-front costs, particularly for those in budget-strapped industries.
- Sectors that cannot access affordable finance may be unable to adopt renewable energy systems.

Cost-Benefit Analysis (CBA) for Renewable Energy Integration

CBJWs enable firms to conduct a comprehensive cost-benefit analysis (CBA) for the introduction of renewable energy. When the project is economically feasible over the long term, both direct and indirect costs, as well as projected returns, must be calculated according to the CBA.

Costs:

- Initial Capital outlay: As mentioned above, renewable energy services typically require substantial amounts of investment in the near term.
- Costs of maintenance: Although renewable energy systems generally have lower maintenance costs than non-renewable energy systems, it is essential to have regular check-ups. The maintenance of wind turbines and biomass systems requires special attention.
- Integration and Storage: Grid integration-related costs, such as infrastructure updates, could add to overall expenditures. Moreover, the use of energy storage devices (e.g., batteries or thermal storage) results in an increase in total investment (Igbinenikaro et al., 2024).

Benefits:

- Low Cost of Power: A key argument for the integration of renewables is the lower long-term cost of energy. In this way, industries can reduce or avoid the need for grid electricity during peak demand periods when electricity prices are at their highest.

- Incentives & Subsidies: Government incentives (grants, tax benefits, feed-in tariffs (FiTs)) can de-risk renewables projects to a great extent. They differ in each region and by technology, but they act to bring the economics of renewable energy projects closer to reality.
- Decreases in Carbon Taxes: Various governments charge carbon taxes for using fossil fuels by industries. Industries will achieve long-term savings by reducing or eliminating their carbon tax liability through the use of renewable energy.
- Environmental benefits and leadership: which are difficult to measure, but adopting a renewable energy program can help to reduce a company's carbon footprint and help contribute towards combating climate change; this can improve the corporate social footprint of a company, as well as develop its brand image, and ultimately result in attracting socially responsible customers and investors.

Example of CBA

A business puts in a 1 MW solar PV system that costs $1 million to install (with an annual 10% maintenance cost). The 1.4 MWdc system is expected to save $150,000 per year. Since the company receives a 30% tax credit, the upfront cost is reduced to $700,000. It should be noted that over a 20-year interval, the company will save $3 million in energy costs, resulting in a positive NPV and 10% IRR.

Challenges:

- Measuring the softer returns, like brand value or customer loyalty, isn't easy.
- It is often challenging to forecast very long-term energy prices, which leads to additional financial analysis uncertainty.

Financial Models for Renewable Energy Projects

The viability of a renewable energy project from an economic standpoint is often checked with the help of different models. These resources help to determine the costs, savings, ROI and financial risks of any renewable energy implementation. The most popular models used include:

Discounted Cash Flow (DCF) Analysis: Through discounted cash flow (DCF) analysis, the time value of money is considered (i.e., future cash flows or savings are discounted to present value). Calculating the NPV of the project in question is useful, as it provides an estimation of costs and benefits over time.

Levelized Cost of Energy (LCOE): The comparison of the projected total cost of electricity generation from a given renewable energy generation system to the corresponding total cost of electricity generation over its life cycle is derived as Levelized Cost of Energy (LCOE). It includes various costs, including set-up, maintenance, financing, and energy return costs, to find the cost of 1 unit of energy (usually measured in Kwh). A reduction in LCOE makes the renewable energy system more economically feasible.

Payback Period Analysis: The payback period analysis model we have already met. It tells us how much time it will take to recover the initial investment. Assessing the economic viability of a renewable energy project is a simple process. In particular, green industries with limited funds will find this approach to be effective.

Sensitivity analysis: The sensitivity analysis considers the effect that critical and variable input parameters will have on financial results expressed as NPV, in areas like energy prices, interest rates, and system performance. It evaluates the financial opportunities and risks of renewable energy projects.

Challenges:

- Forecasting future prices and government support is a difficult activity. This introduces potential inaccuracies and deviations in financial modelling.
- If financial models are too complex, they might force industries without costly financial experts to fail to understand a renewable energy project.

The economic challenges associated with incorporating renewable energies into industry are wide-ranging and not something that planners can easily address. While existing systems come with high up-front prices, long-term savings on energy costs, tax credits, and improved sustainability allow the initial costs to be worthwhile. Industries can use these as tools to make informed investments in renewable energy based on evidence.

REGULATORY AND POLICY ISSUES IN RENEWABLE ENERGY INTEGRATION

The successful integration of renewable energy in the industries will require a conducive policy and regulatory regime involving government subsidies, international treaties and compliance standards, as shown in Figure 5, for ensuring growth, sustenance and other things else.

Figure 5. Regulatory and policy process for renewable energy integration

Government Incentives and Subsidies

It is the duty of governments to promote the use of renewable energy technology by providing financial incentives and subsidies. The following are the most popular types of subsidies and incentives:

- **Tax Credits:** Companies investing in renewables currently enjoy tax benefits, including the floor for the Investment Tax Credit (ITC), along with the Production Tax Credit (PTC). The ITC gives a percentage write-off against installation cost on renewable energy systems for businesses, while the PTC is an incentive paid for each kilowatt-hour (kWh) of renewable energy produced. These tax credits can lessen the needed reimbursement for renewable energy projects' initial investment.
- **Cash / Grants:** Direct funding is provided through cash and grants offered by certain Governments to install renewable energy. These grants may be beneficial, particularly to industries with less access to funding. For example, Horizon 2020 in Europe can finance projects that focus on renewable energy, such as innovative research projects.
- Two of the most popular ways to secure an income from the generation of green energy are Feed-in Tariffs (FiTs) and Power Purchase Agreements (PPAs), whereby clean power generators get paid to produce green electricity and offer a consistent income stream to firms producing renewable power. Energy producers can enter into long-term contracts called power purchase

agreements (PPAs) with companies. To guarantee a fixed rate, often lower than that of traditional energy, they agree to buy renewable energy.

- **State Incentives:** The local and state incentive programs can sometimes be available in addition to the national incentive programs. These can be available in the form of a subsidy, a grant, or a tax credit. For instance, in the United States, many states have instituted their own energy policies, specifically: Renewable Portfolio Standards (RPS) and Renewable Energy Certificates (RECs). These are programs that provide financial Incentives for companies to produce energy from renewable resources.

Challenges:

- Government incentives are a key consideration for every industry. The exact terms of these incentives and how generous they are can differ from country to country and region to region. Therefore, it is vital to keep monitoring changes in the local administration.
- The possibility of change or uncertainty occurring within policy as a result of the fateful rise and fall of political leadership that may withdraw or diminish incentives is a business risk for industries.

Regulatory Compliance and Certification

The rules provide assurance that renewable energy systems are safe and effective, as well as positive to the environment. Generally, emission reduction standards, efficiency and waste management are included in the standards. Regulating energy is significant to curtailing carbon/dioxide emissions and promoting clean energy sources. Nations are passing laws that limit emissions and require more renewable energy use. The operational efficiency of products and processes can be maximised under energy efficiency standards that governments and international bodies have set, for example, the ISO 50001 Energy Management Standard. Grid connection standards of countries specify the parameters for connecting power systems to electrical grids. Programs recognising certification (like LEED and BREEAM) help industries with best practices. However, adherence to regulations can be difficult and create heavier fines and penalties from regulators. Despite all these issues, renewable energy is a great option for improving overall energy efficiency.

Market-Based Mechanisms and Trading Systems

More and more instruments are being used to persuade industry to reduce emissions and switch to cleaner sources of energy. These are mechanisms that literally put a cost on carbon emissions, making it economically advantageous for industries to mitigate those emissions.

- **Carbon Pricing and Carbon Taxes**: Carbon pricing is a way to reduce greenhouse gas emissions by putting a price on carbon. It is mostly done through carbon taxes and emissions trading schemes (ETS). Both these approaches are designed to make sure that the price of emitting carbon reflects the real cost to the environment. Making it costly for companies to emit carbon, however, will induce them to invest in renewables and energy efficiency to pay less tax or emissions trading costs. For instance, the Carbon Pricing Leadership Coalition was founded by economic and political leaders to promote carbon pricing initiatives around the world to reduce carbon emissions. It organises the High-Level Assembly on Carbon Pricing to facilitate the implementation of carbon pricing.
- **Cap-and-Trade Systems:** The cap-and-trade systems, which put a cap (that is, a ceiling) on the overall amount of carbon which industrial sectors may emit, are the European Union's Emissions Trading System (EU ETS). Companies that produce fewer pollutants than their allowance can sell their remaining credits to other companies, giving them a reason to mitigate emissions. Companies purchasing renewable energy have the ability to cut down their emissions. They can sell the allowance savings or pocket the profit.

Challenges:

- Market-based instruments like carbon prices may have an additional cost impact on industries, particularly those that use energy-intensive processes.
- The price of carbon allowances in volatile carbon markets is unlikely to provide firms with the long-term certainty they need to plan their renewable energy investments and not wait for future revenues to induce the implementation of new technologies.

EMERGING TRENDS AND INNOVATIONS IN RENEWABLE ENERGY INTEGRATION

The landscape for integrating renewable energy into industrial operations is changing rapidly due to technology enhancements, regulatory changes, and growing demand for carbon reduction. The merger of renewables will likely witness some definite trends and developments as sustainability becomes deeply entrenched in industries across the board. The "Visit the Future" section of Wonder explores the various major Industrial Energy system trends, their development at the global level, and other related information. It further included development in smart technologies, energy storage, green hydrogen, artificial intelligence, and others.

Smart factories and industrial digitalisation

Industry 4.0 is incorporating digital technologies to revolutionise energy management at factories and production sites. The use of web analytics, along with the Internet of Things and automation, in smart factories increases efficiency and improves energy management. Through these applications, electric loads can take advantage of renewable energies and decrease dependence on grid electricity under more flexible, competitive and volatile market prices. With the help of IoT and real-time energy monitoring, the industry gets useful information about the energy consumption, equipment performance, and environment. Predictive maintenance systems allow pre-emptive identification of failure within equipment, minimising downtime and maintenance costs of renewable energy systems. Among the applications of AI and Automation are energy optimisations, forecasting, controlling industrial energy management systems, making maximum use of renewable energy and minimising costs. Some benefits are enhanced efficiency, lower maintenance costs and downtime, and tailored automation and AI that optimally integrate renewable energy sources.

Energy Storage Innovations

The industry is addressing energy storage to be able to scale renewable energy generation. Likewise, it aims to address some of the issues caused by solar and wind intermittency. Recent developments in storage technology. Like other clean energy sources, the harnessing of solar and wind power can produce pollution-free power and reduce carbon emissions.

- **Next-Generation Battery Technologies:** In the energy storage space, lithium-ion has been the technology of choice, but newer technologies such as solid-state, flow and sodium-ion are developing as alternative batteries. A new variety of batteries offers significantly higher energy densities, longevity and costs, which are deemed more suitable solutions as compared to conventional lithium-ion batteries, which are not fit-for-purpose for industrial applications.
- **Grid Scale Storage Technologies:** Large Battery Storage Technologies which could be used at Grid Scale (PHES, CAES): These technologies can store a large quantity of renewable energy and use this stored energy during

high-demand periods or solar or wind generation drought periods. There are more technologies being developed or looked into. These solutions store energy for longer periods, which is important for industrial plants where energy availability is continuous.

- **Thermal Storage:** Thermal storage, like molten salt storage and others, is used to store the heat generated by solar power plants. This heat can be used for industrial processes and to generate electricity when the sun is down. Industries engaged in high-temperature applications, such as cement, steel and similar products, benefit from this thermal energy storage concept.

Benefits:

- As renewable energy becomes more dependable, large-scale storage solutions are key to overcoming intermittent supply.
- Cut down your energy bills by consuming power when renewable generation is at its peak.
- Using long-duration storage technologies, grid resiliency improved.

Green Hydrogen and Industry Decarbonization

Creating green hydrogen is one of the most exciting advancements for integrating renewable energy into the grid. Green hydrogen uses renewable energy to generate clean H2 from water. Electrolysis splits water into hydrogen and oxygen. With green hydrogen, both electricity and water are renewable. It refers to a manufacturing technique which is not responsible for the carbon emission-related problems that occur while utilising traditional methods involving natural gas for hydrogen production.

- **Hydrogen for Industrial Use:** Hydrogen has the potential to remove carbon dioxide from operations which are hard to electrify, such as steel making, cement and chemical production. As they replace natural gas and oil with hydrogen, these industries can reduce their carbon footprint and move towards a cleaner energy system. Hydrogen is being tested in some pilot projects to see if it can replace coke and coal in steelmaking and as an ammonia feedstock.
- **Hydrogen as an Energy-Storage Medium:** Green hydrogen can be stored and converted into electricity or heating as needed. In other words, it can not only be used as fuel but also as an energy-storage medium. The fact that hydrogen can hold energy and can transport electricity would make it critical for many sectors that consume large amounts of energy.

Benefits:

- It is an excellent renewable option for energy-intensive industries used in place of fossil fuels.
- Hydrogen systems help in energy storage and balancing of grids.
- It helps bring down overall emissions in industries that are hard to electrify.

AI for Energy Management

Artificial intelligence (AI) is changing the way energy is managed, making better use of renewable energy, predicting demand accurately, and making business operations efficient – Sharma et al., 2019. AI algorithms can accurately predict energy demand by utilising existing and external data. This, in turn, allows the industry to optimise renewable generation. AI is used to monitor and optimise the performance of renewable energy devices such as solar panels, wind turbines, and energy storage systems for the most efficient use (Park et al., 2023; Ukoba et al., 2024). Powered by AI, grid management enables the integration of renewables as it can predict changes in generation and adjust grid operations. Benefits include improved accuracy for predicting energy demand and energy performance, while AI-supported management systems can improve integration planning and stability of the grid (Ghiasi et al., 2025).

Blockchain for Energy Trading and Decentralisation

The energy industry can be transformed with the help of Blockchain technology, which creates new opportunities. The solution facilitates secure and transparent energy exchanges between industrial operators and energy producers in Nigeria, boosting renewable energy generation and transaction verification. Through distributed energy trading, industrial customers can engage in the exchange of their surplus renewable energy with other consumers or utility grid operators. Smart contracts based on blockchain will handle energy transactions with lower transaction costs automatically. Blockchain can also keep track of how RECs are issued and traded, which allows for a transparent record of how RECs were generated. These benefits mean energy trading is efficient, transparent and secure, while transaction costs are reduced and decentralised energy systems are promoted. As such technologies improve, they will speed up the world's transition to a sustainable low-emission economy and pave the way for a more sustainable energy future.

Industries that will embrace these new trends or innovations will not only be at the forefront of the pack but also contribute to the global sustainable development goal towards a greener and more resilient industrial sector (Figure 6).

Figure 6. Future trends and innovations with renewable energy integration in industrial operations

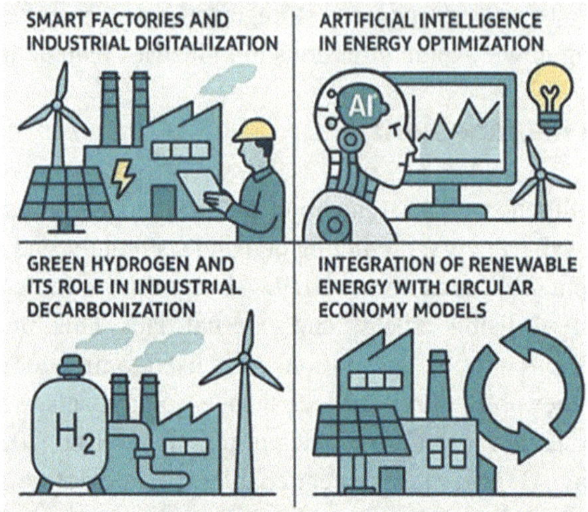

CONCLUSION

Integrating renewable energy is crucial to achieving international sustainable development goals, reducing carbon footprints, and ensuring energy security. The world's industries consume a large amount of energy and are responsible for a large share of carbon emissions. Therefore, a shift to clean energy is essential to deal with the climate emergency. The integration of renewable energy sources will now be easier and more efficient, thanks to smart grids, energy storage, AI, and more. Decision-making is formulated by economic issues, such as the very high initial capital costs of renewable energy systems for SMEs. Governments aid the development of the renewable energy market through incentives, subsidies, tax credits, and regulations. The Renewable Energy in Industry report cites smart factories, green hydrogen, advanced storage solutions, and blockchain energy trading as examples of emerging technologies. Still, challenges are intermittency and storage, the most significant initial capital cost, and policy uncertainty. Technological improvements involving energy storage, AI and automation, and green hydrogen infrastructure can address variability and make green power easy and reliable for companies. Global standards and market incentives will create investment opportunities in more environmentally friendly technologies in the industrial sector.

As the power landscape changes, renewable energy technologies are being adopted by industries to optimise power consumption and utilise renewable energy more effectively. To stay on the right track, a business must consider its energy requirements and the renewable resources available in the vicinity. Energy management systems (EMS) that utilise technology are a powerful investment that optimises consumption and better utilises renewable power. Financial incentives and models, such as government subsidies and tax incentives, could accelerate the adoption of renewable energy systems. Working with technology vendors and other key stakeholders can help to mitigate challenges and facilitate the market introduction of advanced technologies. By adopting long-term sustainability goals such as energy efficiency, emissions reduction, and careful environmental stewardship, industries can lower operating costs and enhance their CSR (corporate social responsibility) profile. In the future, we could see an industrial sector that is resilient, profitable, and sustainable, as more and more businesses switch to renewables.

REFERENCES

Anand, S., Kulkarni, S. V., Gupta, S., & Parthasarathy, V. (2022, December). Battery management system (BMS) for electric vehicle applications. In 2022 fourth international conference on emerging research in electronics, computer science and technology (ICERECT) (pp. 1-6). IEEE. DOI: 10.1109/ICERECT56837.2022.10060137

Awasthi, B., Gupta, S., Kumar, S., & Tripathi, A. (2024, January). Business management and planning using queuing model with heterogeneous customers for electrical applications. In 2024 2nd International Conference on Intelligent Data Communication Technologies and Internet of Things (IDCIoT) (pp. 279-283). IEEE. DOI: 10.1109/IDCIoT59759.2024.10467673

Farret, F. A., & Simões, M. G. (2017). *Integration of renewable sources of energy*. John Wiley & Sons.

Fotis, G., Gupta, S., Varshney, T., Dhar, S. L., & Gulzar, M. M. (Eds.). (2025). *Modern Computing Technologies for EV Efficiency and Sustainable Energy Integration*. IGI global., DOI: 10.4018/979-8-3373-2382-4

Ghiasi, M., Ghiasi, V., & Siano, P. (2025). Renewable Energy Integration into Industrial and Residential Buildings: A Study Across Urban, Rural, and Coastal Areas. *IET Renewable Power Generation*, *19*(1), e70108. DOI: 10.1049/rpg2.70108

Gupta, S. (2025). *The Future of Machine Learning with Automation in Engineering*. Nova Science Publishers., DOI: 10.52305/YFIX2832

Gupta, S., Sreenivasu, S. V. N., Chouhan, K., Shrivastava, A., Sahu, B., & Potdar, R. M. (2023). Novel face mask detection technique using machine learning to control COVID'19 pandemic. *Materials Today: Proceedings*, *80*, 3714–3718. DOI: 10.1016/j.matpr.2021.07.368 PMID: 34373829

Gupta, S., & Tripathi, R. K. (2014). Two-area power system stability improvement using a robust controller-based CSC-STATCOM. *Acta Polytechnica Hungarica*, *11*(7), 135–155. DOI: 10.12700/APH.11.07.2014.07.9

Gupta, S., Tripathi, R. K., & Shukla, R. D. (2010, November). Voltage stability improvement in power systems using facts controllers: State-of-the-art review. In *2010 International Conference on Power, Control and Embedded Systems* (pp. 1-8). IEEE. DOI: 10.1109/ICPCES.2010.5698665

Husin, H., & Zaki, M. (2021). A critical review of the integration of renewable energy sources with various technologies. Protection and control of modern power systems, 6(1), 1-18.

IEA. R. (2020). Iea. Paris, (2020), 628.

Igbinenikaro, O. P., Adekoya, O. O., & Etukudoh, E. A. (2024). Conceptualizing sustainable offshore operations: Integration of renewable energy systems. International Journal of Frontiers in Science and Technology Research, 6(02), 031-043.

Igogo, T., Awuah-Offei, K., Newman, A., Lowder, T., & Engel-Cox, J. (2021). Integrating renewable energy into mining operations: Opportunities, challenges, and enabling approaches. *Applied Energy*, *300*, 117375. DOI: 10.1016/j.apenergy.2021.117375

Jones, L. E. (2017). *Renewable energy integration: practical management of variability, uncertainty, and flexibility in power grids*. Academic press.

Kundu, S., Singh, M., & Giri, A. K. (2022). Control algorithm for coordinated operation of wind-solar microgrid standalone generation system. *Energy Sources. Part A, Recovery, Utilization, and Environmental Effects*, *44*(4), 10024–10044. DOI: 10.1080/15567036.2022.2143943

Kundu, S., Singh, M., & Giri, A. K. (2024). SPV-wind-BES-based islanded electrical supply system for remote applications with power quality enhancement. *Electrical Engineering*, *106*(1), 279–294. DOI: 10.1007/s00202-023-01979-0

Lu, Y. (2016). Industrial integration: A literature review. *Journal of Industrial Integration and Management*, *1*(02), 1650007. DOI: 10.1142/S242486221650007X

Meena, G., Singh, V. P., Dixit, A., Mathur, A., & Bhatt, A. (2024, April). Steady-state models of ac microgrid for load flow solutions. In 2024 1st International Conference on Innovative Sustainable Technologies for Energy, Mechatronics, and Smart Systems (ISTEMS) (pp. 1-6). IEEE. DOI: 10.1109/ISTEMS60181.2024.10560356

Mishra, P., & Singh, G. (2023). Energy management systems in sustainable smart cities based on the internet of energy: A technical review. *Energies*, *16*(19), 6903. DOI: 10.3390/en16196903

Onyinyechukwu, C., Adetomilola, V. F., Valentine, I. I., Bright, N., Cosmas, D. D., & Kenneth, I. I. (2024). Integration of renewable energy in industrial operations: experiences from Canada, USA, and Africa.

Park, H., Kim, J. K., & Yi, S. C. (2023). Optimization of site utility systems for renewable energy integration. *Energy*, *269*, 126799. DOI: 10.1016/j.energy.2023.126799

Rao, M. S., Saraswathy, S., Neela, V., Arun, M. R., Gupta, S., & Taqui, S. N. (2023, November). Efficient Model Training in Decentralized Systems with Federated Learning. In *2023 International Conference on Communication, Security and Artificial Intelligence (ICCSAI)* (pp. 900-904). IEEE. DOI: 10.1109/ICCSAI59793.2023.10421469

Semwal, S., Kant, S., Bandyopadhyay, K., Asim, M., Garg, R., & Goyal, S. (2024, October). Implicit Control of DC-DC Converter based DC Motor for the Charging of Electric Vehicle using Fuzzy Logic Controller. In 2024 5th IEEE Global Conference for Advancement in Technology (GCAT) (pp. 1-6). IEEE. DOI: 10.1109/GCAT62922.2024.10923979

Sharma, R. K., Saran, V. B., Kanaujia, S., & Gupta, S. (2024, May). Evaluation of mppt controller performance using fuzzy logic design with wireless sensor node. In *2024 International Conference on Smart Systems for applications in Electrical Sciences (ICSSES)* (pp. 1-6). IEEE. DOI: 10.1109/ICSSES62373.2024.10561365

Shukla, R. D., Tripathi, R. K., & Gupta, S. (2010, November). Power electronics applications in wind energy conversion system: A review. In *2010 International Conference on Power, Control and Embedded Systems* (pp. 1-6). IEEE. DOI: 10.1109/ICPCES.2010.5698663

Sinha, R., Kaur, N., Gupta, S., & Thakur, P. (2023, November). Diagnosis of Parkinson's Disease using Hybrid Ensemble Technique. In *2023 International Conference on Ambient Intelligence, Knowledge Informatics and Industrial Electronics (AIKIIE)* (pp. 1-5). IEEE. DOI: 10.1109/AIKIIE60097.2023.10390458

Sinha, R., Sinha, K. K., Patel, M., Gupta, S., & Priya, S. (2024, July). Detection of leukemia disease using convolutional neural network. In 2024 5th International Conference on Image Processing and Capsule Networks (ICIPCN) (pp. 451-456). IEEE. DOI: 10.1109/ICIPCN63822.2024.00080

Sudha, G., Natha, A., Rajendiran, M., Vanitha, U., Kiran, V., Patil, H., & Rajaram, A. (2024). Optical oasis harnessing energy efficiency in wireline, optical, and wireless communications. *Journal of Environmental Protection and Ecology*, *25*(4), 1179–1187.

Taibi, E., Gielen, D., & Bazilian, M. (2012). The potential for renewable energy in industrial applications. *Renewable & Sustainable Energy Reviews*, *16*(1), 735–744. DOI: 10.1016/j.rser.2011.08.039

Tang, Z., Yang, Y., & Blaabjerg, F. (2021). Power electronics: The enabling technology for renewable energy integration. *CSEE Journal of Power and Energy Systems*, *8*(1), 39–52.

Thandlam, A. K., Vignesh, T., Saravanan, A., Subramani, P., Marimuthu, R., & Gupta, S. (2024, April). Next-gen vehicle safety: The futuristic approach to auto-stop in modern vehicles. In 2024 10th International Conference on Communication and Signal Processing (ICCSP) (pp. 265-270). IEEE. DOI: 10.1109/ICCSP60870.2024.10543666

Tyagi, M., Gupta, S., & Khan, M. I. (2024, May). Reduction of voltage instability in complex power networks through the utilization of D-STATCOM. In 2024 4th International Conference on Advance Computing and Innovative Technologies in Engineering (ICACITE) (pp. 1662-1666). IEEE. DOI: 10.1109/ICACITE60783.2024.10616856

Ukoba, K., Olatunji, K. O., Adeoye, E., Jen, T. C., & Madyira, D. M. (2024). Optimizing renewable energy systems through artificial intelligence: Review and future prospects. *Energy & Environment*, *35*(7), 3833–3879. DOI: 10.1177/0958305X241256293

Vu, D. H., Muttaqi, K. M., & Sutanto, D. (2020). An integrated energy management approach for the economic operation of industrial microgrids under uncertainty of renewable energy. *IEEE Transactions on Industry Applications*, *56*(2), 1062–1073. DOI: 10.1109/TIA.2020.2964635

Zobaa, A. F., & Bansal, R. C. (2011). *Handbook of renewable energy technology*. World Scientific. DOI: 10.1142/7489

Chapter 4
Intelligent HVAC Control Systems Based on Machine Learning and Deep Learning Models

Ganesh B. Murade
https://orcid.org/0009-0001-4090-638X
University of Engineering and Management, Jaipur, India

Ankit Kumar Sharma
https://orcid.org/0000-0002-3569-5047
University of Engineering and Management, Jaipur, India

Bhanu Pratap Soni
https://orcid.org/0000-0001-8138-5392
School of Electrical and Electronics Engineering, Fiji National Univerisity, Suva, Fiji

Arvind S. Pande
https://orcid.org/0000-0003-3358-9109
Amrutvahini College of Engineering, Sangamner, India

Ganesh K. Shirsat
https://orcid.org/0009-0006-2877-9767
Dr. Vitthalrao Vikhe Patil College of Engineering, India

ABSTRACT

Buildings are responsible for nearly 35% of global energy consumption, with heating and cooling accounting for a significant share. As urbanization and population

DOI: 10.4018/979-8-3373-7041-5.ch004

growth accelerate, energy demand in cities continues to rise, leading to increased CO_2 emissions, higher electricity costs, and greater stress on power grids. To address these challenges, real-time control systems and building automation have shown strong potential in improving energy efficiency in the built environment. In particular, optimized Heating, Ventilation, and Air Conditioning (HVAC) control systems can significantly reduce energy consumption, lower operational costs, and mitigate environmental impacts. This study investigates the application of machine learning models—including Linear Regression, Random Forest, Support Vector Machine (SVM), Long Short-Term Memory (LSTM), and Gated Recurrent Unit (GRU)—to predict HVAC performance. The models are evaluated using Mean Squared Error (MSE) and R^2 score, with the goal of identifying the most effective approach for enabling data-driven energy optimization in buildings.

INTRODUCTION

The greenhouse gas emissions (GHG) were produced by human activities, and a large share is from building that burns fossil fuels for heating and cooling purposes. The energy and GHG are reduced with the use of renewable solutions such as heating pumps, photovoltaic (PV) panels, and hybrid PV/T systems. However, to have maximised performance, adjustments need to be made and the use of the latest low-energy technologies. Many researchers have implemented the model predictive controller (MPC), but it needs an accurate model of the system, and each time, identification of an accurate model is not possible (Baetschmann & Leibundgut, 2012; Meggers et al., 2012). The idea of low energy (Low-Ex) is depicted in Figure 1, which is an alternate way to consider. The Low-Ex derives a lower ratio of distributed energy.

In the United States, building energy costs are often tied to time-varying utility rates. This makes the scheduling flexibility of Heating, Ventilation, and Air Conditioning (HVAC) systems particularly valuable for lowering operational expenses, improving grid stability, and enhancing overall energy efficiency. Previous studies have proposed various control strategies to optimize HVAC performance. (Ma et al., 2015; Maasoumy et al., 2011; Oldewurtel et al., 2010; Wei et al., 2014). Many of these solutions rely on fairly simple thermal models to estimate how heat moves through a building. In contrast, more advanced approaches such as Model Predictive Control (MPC) have demonstrated a much stronger ability to reduce energy use, especially when applied to cooling systems, energy storage, and predictive scheduling (Figure 2).

Figure 1. Schematic of LowEx Building (a) Heat exchanger-equipped PV/T panels (b) heat pumps; (c) dual zone boreholes; and (d) floor heating (e) Storage of hot water

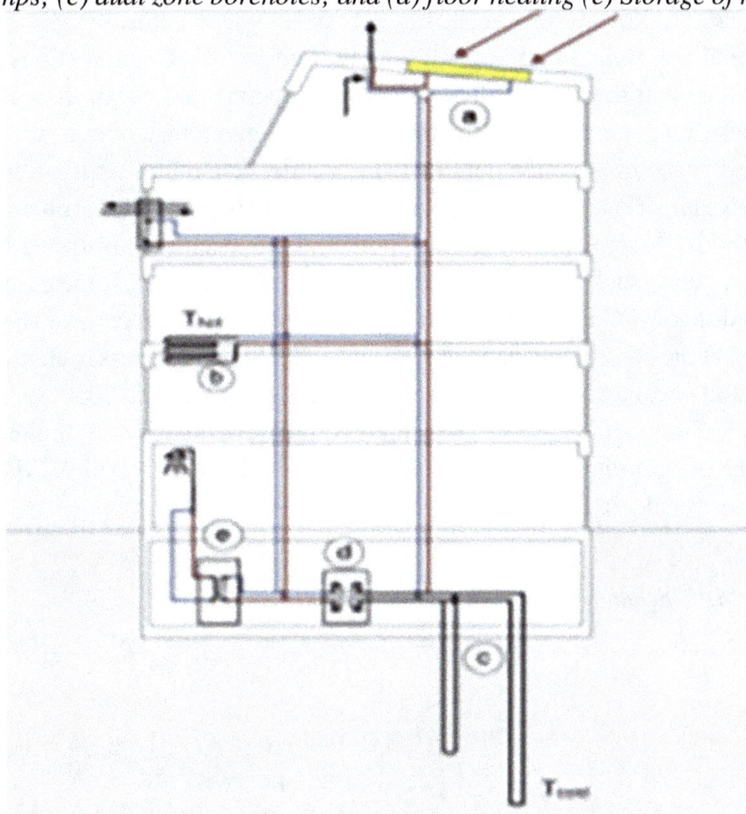

Figure 2. Non-linear cooling system model as a whole

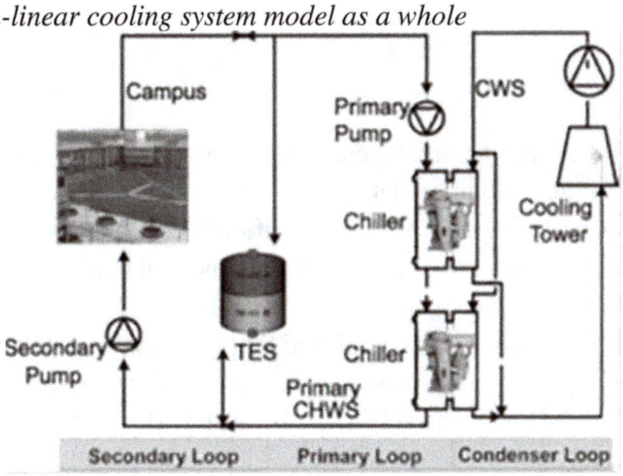

The primary objectives of HVAC control systems are to maintain occupant comfort while maximizing resource efficiency(Afram & Janabi-Sharifi, 2014; Dounis & Caraiscos, 2009; Oldewurtel et al., 2012). PID and ON/OFF controllers are examples of traditional controllers that are still widely used in HVAC applications. This is due to the fact that these controllers are easy to operate and cheap. It is challenging for these fundamental methods to consistently give outstanding performance since HVAC systems are often nonlinear, high-order, and multi-input–multi-output (MIMO) systems (Petersen & Bundgaard, 2014). This presents a problem for the basic methods. A significantly greater potential has been demonstrated by more complex systems, such as Model Predictive Control (MPC). In some cases, it has been stated that MPC can produce energy savings of up to 28 percent (Figure 3), provided that the control mechanism takes into account parameters such as weather forecasts and occupancy patterns(Goyal et al., 2013; Kalogirou, 2000; Oldewurtel et al., 2013; Široký et al., 2011). Nonetheless, their effectiveness is limited by the complexity of accurately modeling building dynamics(Ben-Nakhi & Mahmoud, 2002; Kumar et al., 2013).

Figure 3. MPC Scheme for building climate control

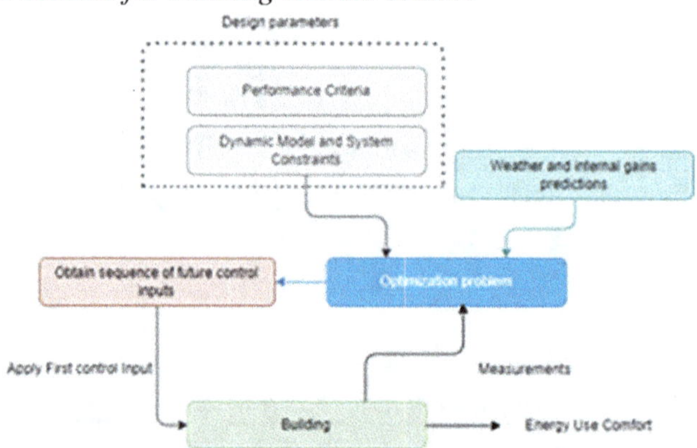

To address the limitation, intelligent control methods (ICM) are explored. The ICM does not require explicit models such as the models based on human comfort perception or a data-driven algorithm. The artificial neural network (ANN) and machine learning approaches are better at predicting thermal dynamics and optimizing the system performance (Sinha et al., 2024; Mozer, 1998; Pande et al., 2025; Yang et al., 2015). The researchers trained an ANN for historical weather. The data used was successfully applied to air-conditioning optimization and predictive thermal dynamics in real-time applications (Figure 4). Despite this, it needs a large data set to achieve accuracy.

Figure 4. AHU Controller

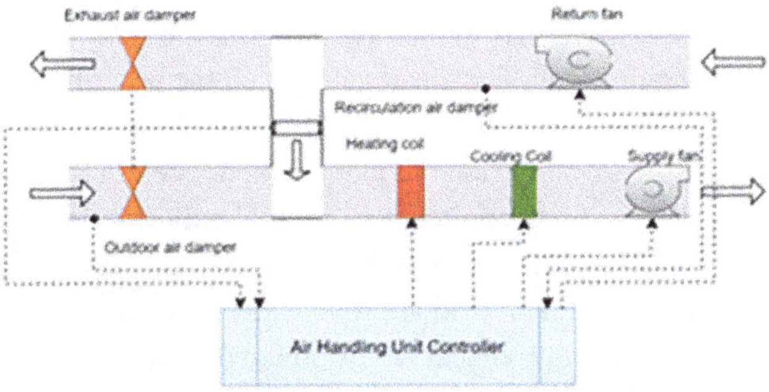

The present article compares machine learning and deep learning methods to predict the supply air temperature of an air handling unit (AHU), as accurate forecasting of the supply air temperature of AHU and variable air volume (VAV) is difficult. It directly relates to occupant comfort and energy utilization. Since this prediction problem involves time-series thermodynamic data, the models examined in this work were selected based on their demonstrated performance in sequential forecasting tasks. The ultimate goal is to identify an algorithm that can serve as the basis for a digital twin, enabling real-time optimization of HVAC energy use in future applications.

LITERATURE REVIEW

In addition to the previously discussed strategies, reinforcement learning control (RLC) has also been investigated for HVAC applications. When applied to a LowEx building thermal model, a comparison study between RLC and standard rule-based control (RBC) indicated that RLC produces noteworthy benefits, particularly for PV/T panels with a single control signal. This was proved by the fact that RLC delivers significant gains. Within the first three years of operation, the RLC-based controller was able to generate cumulative net power increases that were larger than those of the RBC in nine distinct situations. These gains were 5.73 percent, 9.78 percent, and 11.47 percent % respectively. Furthermore, the controller is able to continually improve its policy while it is operating, which enables it to surpass RBC in terms of achieving numerous system objectives. For example, RLC is able to successfully maintain the optimal operating temperature of the heat pump, adjust for ground heat storage in a more effective manner, and consistently meet heating demand (100 percent as opposed to RBC's 97 percent). According to (Gupta, 2025) The most significant benefit of RLC is its capacity to improve its control policy in real time by gaining knowledge from interactions with the building under a variety of different operating situations. The disadvantage, on the other hand, is that it has a learning process that is inherently slow. Although simulations demonstrate that it has the potential to converge toward effective management even in the absence of prior environmental knowledge, direct application in actual buildings is not feasible due to the lengthy amount of time required for training. (Wei et al., 2017; Yu et al., 2019).

To address this limitation, a hybrid approach was proposed in which reinforcement learning is first trained in a virtual environment during an offline phase before real-world deployment. Similarly, a linear reinforcement learning controller (LRLC) was developed to optimize indoor air quality, occupant comfort, and energy use (Dalamagkidis et al., 2007). After four years of simulations, the LRLC demonstrated measurable reductions in energy consumption, though it was limited to switching fuzzy-logic controllers on or off (Shankar et al., 2023; Sharma et al., 2024). A persistent challenge with such approaches is the need for accurate building simulation models, which undermines reinforcement learning's model-free advantage (Fotis et al., 2025; Thandlam et al., 2024).

Wei et al. (2017) built on this work by using deep reinforcement learning (DRL) to control HVAC systems in a co-simulation framework based on EnergyPlus. We used real weather and pricing data to test both basic DRL algorithms and evolutionary modifications for multi-zone structures. The results revealed that energy costs went down a lot while still keeping the inside temperatures where they should be. These results are in line with larger attempts to connect HVAC systems to smart grids. Commercial buildings use more than 35% of the electricity in the country, and HVAC systems alone use more than a third of that. So, building operators have to lower the energy needs of HVAC systems while making sure that the people who live there are comfortable and the air quality (IAQ) is good. Both of these things have a direct impact on health, productivity, and energy expenses.

The authors used a Lyapunov optimization technique (LOT) to solve this multi-objective problem (Dong Li et al., 2015). Their goal was to minimize the time-averaged HVAC running cost while maintaining comfort and indoor air quality limitations. The approach dynamically stabilizes indoor temperatures around ideal averages, taking into consideration the risks associated with fluctuations in energy prices, weather, occupancy, and environmental disturbances. The results of the simulation revealed that this strategy is effective in lowering costs while doing so without compromising comfort.

There have also been applications of reinforcement learning in natural ventilation strategies, which are challenging to optimize in terms of both the timing and the extent of window operation. An evaluation of the indoor and outdoor conditions at each timestep was performed using a model-free Q-learning controller that was proposed in the study (Rao et al., 2023). This controller was able to simultaneously manage HVAC and natural ventilation systems. Comparing simulations in Los Angeles and Miami to heuristic rule-based control, the results indicated decreases in HVAC energy use (13–23%), humidity problems (63–77%), and temperature discomfort (62–80%). In contrast to heuristics, which necessitate superior domain expertise, the reinforcement learning controller automatically adjusts to changing environmental variables and stochastic occupancy.

Additional research has been conducted to examine the performance of LRLC in comparison to conventional On/Off and Fuzzy-PD controllers over extended periods of time (Gupta & Tripathi, 2014; Shukla et al., 2010). The findings demonstrated that although LRLC worked similarly, it had a higher energy consumption rate due to the fact that it prioritized the comfort of its occupants and frequently updated its policies. Recently, advanced reinforcement learning techniques like multi-samples in each cell (MEC) were proposed using clustering samples and wall iteration for optimal control (Fan et al., 2017). When tested in a multi-zone office building model using EnergyPlus–MATLAB co-simulation, it was demonstrated that MEC performed better than Q-learning in terms of both thermal comfort and learning time.

Recently, Pande et al. (2023) investigated the performance of RLC using supervisory control for a thermal storage system in a commercial building. From operation data, RLC optimizes the charging and discharging of the storage tank without the need for an explicit model of the system or a forecast (Tyagi et al., 2024). The electricity pricing, weather, and equipment performance were handled by the adaptive nature of the controller. The cost saving was comparable to the classical storage control strategy, but the sensitivity to the learning parameter was retained as a limitation.

The growing role of deep learning in HVAC applications was manifested by data-driven methods, which deep learning exceeds the machine learning in vision, speech, and pattern recognition. It is used as predictive modelling tool for building automation system (Awasthi et al., 2024; Gupta, 2025).

SYSTEM OVERVIEW

The article needs recent information on building energy, which is published in the journal Scientific Data, Yoon et al. (2022), and a dataset created by a flexible research platform (FRP). The FRP tool is empty and a multizone structure intended to work as an HVAC system with internal load under uncertainty. The dataset from the FRP tool system is of high quality, better for training, testing machine learning, and DL models. In Figure 1, the HVAC system with sensor location is depicted. This consists of VAV boxes, rooftop units, and distributed sensors. It is used to collect multiple data points for prediction purposes on a zone-by-zone basis. The HVAC management model building steps: 1. Define Building architecture, 2. To simulate energy, and 3. Model training

1. Define Building Architecture

The building prototype model uses the Open Studio standards strategy. This includes energy efficiency standards from ASHRAE standards 90%. It includes readymade templates for different building structures such as schools, hospitals, etc. With the help of the above requirements, the prototype model must follow the industry standards.

2. Simulation of Energy

After a prototype model of buildings, Energy Plus is used for simulating the model and indicating a better energy use. It shows the amount of energy utilised in the HVAC system, lights, plug loads, and Ventilation. It includes different approaches based on heat balance with timesteps component-based HVAC system heat and mass transfer analysis, etc. The simulator indicates monthly energy /power use HVAC load profiles. The HVAC load profiles indicate the best stride point for the predictive control design method.

3. Model Training

To create and test a predictive control scheme, OpenAI Gym, a Python-based RL technique, is used. This provides a modular environment. One can train, test, and evaluate an RL algorithm. In this article, OpenAI Gym is needed to design HVAC control designs in which the model learns from simulation data and behaves in an adaptive manner.

METHODOLOGY

The objectives of this study are:

1. To evaluate the performance prediction algorithm for the HVAC System
2. To optimize energy consumption
3. To evaluate the performance of machine learning (ML), deep learning (DL), and reinforcement learning (RL) approaches.

1. Building Model and Simulation

Prototype Building Definition:

The building model was created using OpenStudio Standards, which automate prototype development for multiple building types while incorporating ASHRAE Standard 90.1 requirements.

Energy Simulation:

Energy performance was simulated using EnergyPlus, a comprehensive whole-building simulation engine. The simulation produced baseline results such as monthly electricity consumption and HVAC load profiles. These outputs provided the operational dataset for algorithm training.

Figure 5. Process flow for developing, simulating, and testing the proposed model

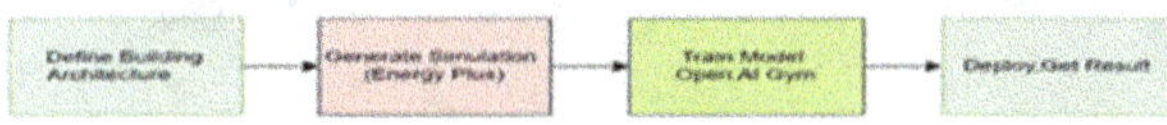

Figure 6. Building model summary (OpenStudio)

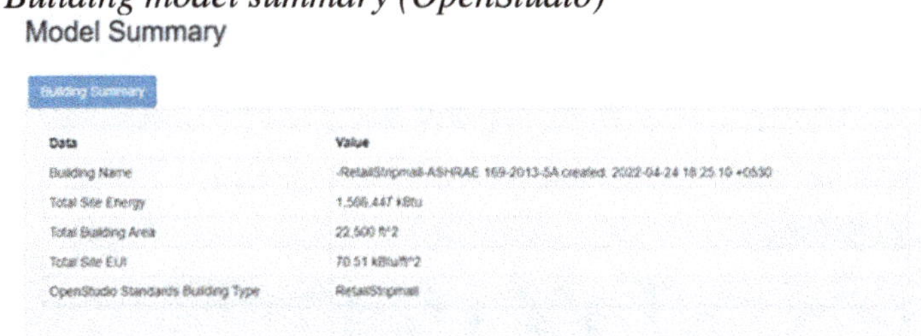

Model Summary

Building Summary

Data	Value
Building Name	-RetailStripmall-ASHRAE 169-2013-5A created: 2022-04-24 18:25:10 +0530
Total Site Energy	1,586,447 kBtu
Total Building Area	22,500 ft^2
Total Site EUI	70.51 kBtu/ft^2
OpenStudio Standards Building Type	RetailStripmall

Figure 7. Weather summary of the selected site

Weather Summary

	Value
Weather File	Mumbai Maharashtra IND ISHRAE WMO#=430030
Latitude	18.90
Longitude	72.82
Elevation	36 ft
Time Zone	5.50
North Axis Angle	0.00
ASHRAE Climate Zone	6A

Figure 8. Monthly electricity consumption (kWh)

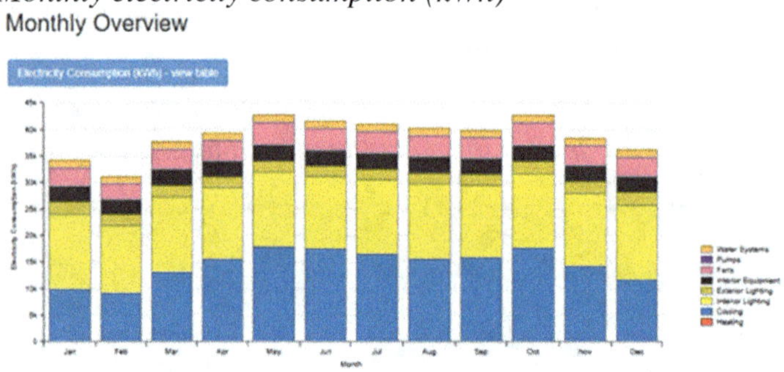

Figure 9. Monthly HVAC load profile

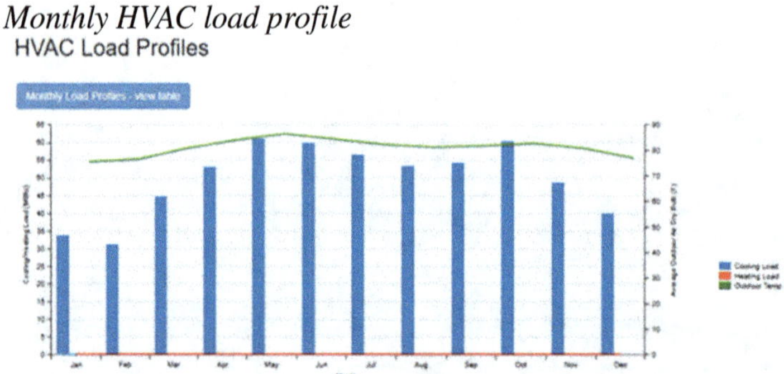

Dataset Summary

1. Features: TOUT, ITUZ1, ITUZ2, TCWZ2, TCHZ2, TDECZ1, TIECZ1, TDXZ1, PUE, TZ2
2. Target: TZ1 (Supply Air Temperature)
3. Size: 52,560 samples
4. Preprocessing: Features normalized using StandardScaler(); dataset divided into training and testing subsets.

2. Machine Learning Algorithms

i. Linear Regression (LR)

A supervised baseline algorithm modeling linear input–output relationships (Sinha & Patel et al., 2024). Although interpretable, it showed limited ability to capture nonlinear HVAC dynamics.

ii. Random Forest (RF)

An ensemble tree-based algorithm combining multiple decision trees. RF was the best-performing ML model, providing high accuracy and robustness.

1. TZ1 → MSE= 0.0018 and R^2=0.9807
2. ITUZ1 → MSE= 87,151.5 and R^2= 0.9978

iii. Support Vector Machine (SVM)

Applied in regression mode (SVR). It created a hyperplane for forecasting energy and temperature values.

1. TZ1 → MSE= 0.0053 and R^2= 0.9419
2. ITUZ1 → MSE= 436,546.1 and R^2= 0.9890

3. Deep Learning Algorithms

i. Long Short-Term Memory (LSTM)

A recurrent neural network capable of capturing long-term temporal dependencies.

1. TZ1 → MSE= 0.0132 and R^2= 0.8566

2. ITUZ1 → MSE= 4,695,327.9 and R^2= 0.8818

ii. Gated Recurrent Unit (GRU)

A simplified RNN variant requiring fewer parameters than LSTM. It achieved the lowest MSE for TZ1 prediction.

1. TZ1 → MSE= 0.0091 and R^2= 0.9012
2. ITUZ1 → MSE= 4,692,558.9 and R^2= 0.8819

4. Reinforcement Learning Algorithms

Deep Deterministic Policy Gradient (DDPG)

A reinforcement learning controller using an actor–critic structure:

1. Actor network-generated HVAC control actions.
2. Critic network evaluated Q-values for these actions.
3. Target networks stabilized learning.

Process

1. The HVAC system sensed the indoor and outdoor states.
2. The DRL agent selected control actions.
3. Rewards were based on energy savings and comfort.
4. The policy was continuously updated across 10,000 episodes.

Results showed ~25% energy savings compared to a PID baseline.

5. Hybrid Model (Random Forest + GRU)

To leverage the strengths of both approaches, a hybrid predictive model was developed:

1. **Random Forest** → handled feature selection and short-term predictions.
2. **GRU** → The sequential dependencies in HVAC time-series data.

Performance

1. TZ1 → R^2 = 0.9335 and MSE = 0.0061

2. ITUZ1 → $R^2 = 0.9205$
3. Energy savings = 24.29%

6. Model Training and Implementation

1. Programming Environment: Python 3.6
2. Libraries: TensorFlow (DL), Scikit-learn (ML), OpenAI Gym (RL)
3. Hardware: Intel® Core i7-7600U CPU @ 2.8 GHz, 16 GB RAM
4. Training Time: ~8 minutes per model

7. Performance Evaluation

Models were evaluated using:

1. **Mean Squared Error (MSE):** for prediction error magnitude.
2. **R^2 Score:** for variance explanation.

EXPERIMENTAL RESULTS

The machine learning and deep learning algorithms are evaluated for predicting multi-zone HVAC control. The performance indicators are R^2 and the Mean Squared Error (MSE). This indicates how accurately each model can make predictions and explain the differences between them.

1. Linear Regression

1. MSE = 0.0191
2. R^2 =0.7903

The author used the Linear Regression strategy as the main model. However, it cannot capture nonlinearities in the HVAC System.

2. Random Forest

1. MSE = 0.0018
2. $R^2 = 0.9806$

The Random Forest model did the best overall, with the fewest mistakes and the greatest explanation of variance. Its ensemble structure is good at dealing with complicated interactions in the data, which makes it the best predictor of the models tested.

3. Support Vector Machine (SVM)

1. MSE= 0.0074
2. R^2= 0.9191

SVM showed very good predicting accuracy, striking a good balance between speed and reducing errors, and nonlinear interactions are modelled better by SVM.

4. Long Short-Term Memory (LSTM)

1. MSE= 0.0095
2. R^2= 0.8045

LSTM is designed for sequential data, and it shows better performance. The higher value of MSE indicates that either the dataset's temporal dependencies are weak or the model needs more tuning to make the most of its sequential learning skills.

5. Gated Recurrent Unit (GRU)

1. MSE= 0.0094
2. R^2= 0.6177

The R^2 score of GRU is lower than that of LSTM. It explores that GRU is computationally efficient, and the capturing ability of correlation in HVAC datasets is lowered by the joining mechanism.

Figure 10. Comparative Analysis of Model Accuracy and Error Metrics

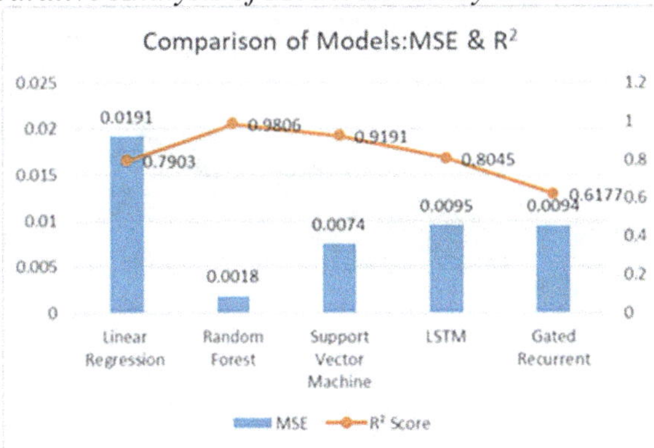

Key Observations

1. **Best Performing Model:** Random Forest algorithm explored the largest MSE and a Higher R^2.
2. **Linear Models vs. Non-linear Models:** The linear regression model is significantly outperformed by non-linear models (such as Random Forest and SVM).
3. **Neural Network Models:** Both LSTM and GRU show potential but fall short compared to Random Forest and SVM. They might require further tuning or more data to fully realize their strengths.

The RF and SVM strategies are better than LSTM and GRU. The conservative tuning and more datasets are required for LSTM and GRU to show the Potential. Figure (9-13) depicts the different learning models from LR to GRU.

1. Linear Regression

For LR model, the performance indicator $R^2 = 0.7903$ and mean squared error (MSE) = 0.0191

Figure 11. Linear Regression

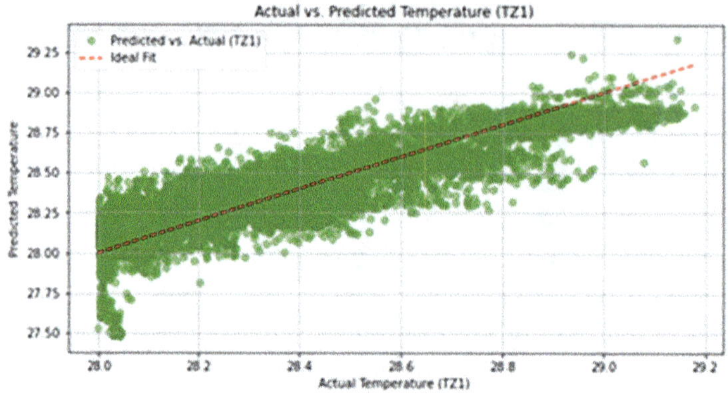

Figure 12. Results of Linear Regression (Predictive Modeling)

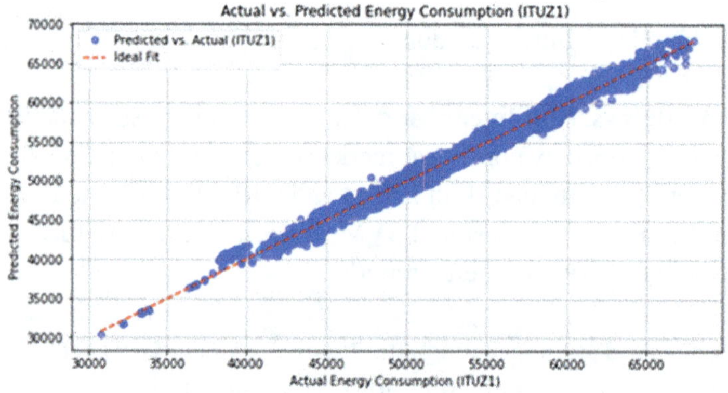

1. ITUZ1 - Mean Squared Error = 367672.6143
2. ITUZ1 - R^2 Score = 0.990815.

Table 1. Results of Linear Regression (Predictive Modeling)

Temperature Prediction Summary			Energy Prediction Summary		
	Tz1	**Predicted_ Tz1**		**Energy_ Savings**	**Percentage_ Savings**
count	10512	10512	**count**	10512	10512
mean	28.38503	28.38806	**mean**	8.829167	0.000614
std	0.301759	0.268633	**std**	606.3247	1.158801
min	27.99852	27.47443	**min**	-2784.74	-5.83285
25%	28.12247	28.20231	**25%**	-333.403	-0.66524
50%	28.33073	28.33325	**50%**	53.61847	0.100229
75%	28.60996	28.57882	**75%**	404.7433	0.682305
max	29.17681	29.33389	**max**	**2284.975**	**4.872312**
Actual_ TZ1	**Predicted_ TZ1**	**Actual_ ITUZ1**	**Predicted_ ITUZ1**	**Energy_ Savings**	**Percentage_ Savings**
28.38503	28.38806	62768.42	60483.44	**2284.975**	**4.872312**

2. Random Forest

For the RF model, the R^2 score is 0.9806, and the MSE is 0.0018.

Figure 13. Results of Random Forest (Predictive Modeling)

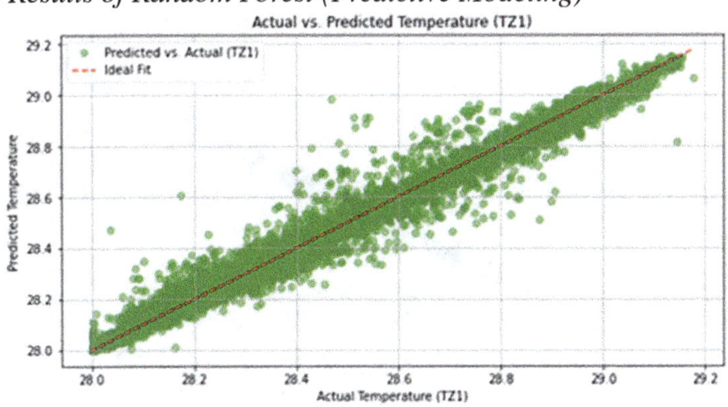

1. ITUZ1 - Mean Squared Error= 87151.5003
2. ITUZ1 - R^2 Score= 0.997823

Table 2. Results of Random Forest (Predictive Modeling)

Temperature Prediction Summary			Energy Prediction Summary		
	Tz1	Predicted_ Tz1		Energy_ Savings	Percentage_ Savings
count	10512	10512	count	10512	10512
mean	28.38503	28.38545	mean	1.816458	-0.00111
std	0.301759	0.278949	std	295.2228	0.574227
min	27.99852	27.09965	min	-5421.04	-12.8737
25%	28.12247	28.12723	25%	-98.6692	-0.18003
50%	28.33073	28.33145	50%	0.328975	0.000585
75%	28.60996	28.60841	75%	63.06973	0.189797
max	29.17681	29.15061	max	**5785.508**	**8.688445**
Actual_ TZ1	**Predicted_ TZ1**	**Actual_ ITUZ1**	**Predicted_ ITUZ1**	**Energy_ Savings**	**Percentage_ Savings**
28.38503	28.38545	66588.53	60803.02	**5785.508**	**8.688445**

3. Support Vector Machine

For the SVM model R^2 Score =0.941915 and MSE = 0.0053.

Figure 14. Support Vector Machine

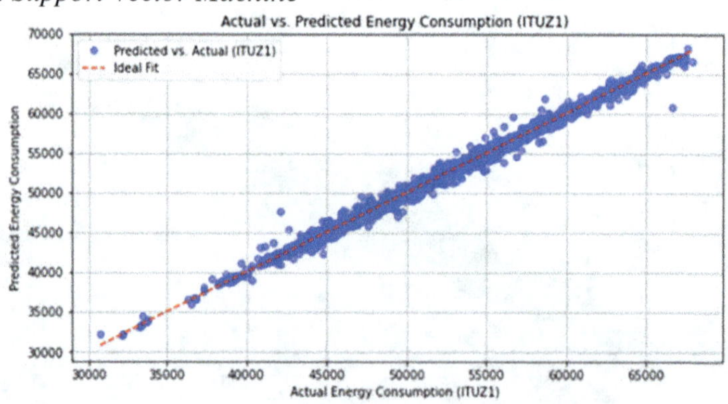

Figure 15. Results of SVM (Predictive Modeling)

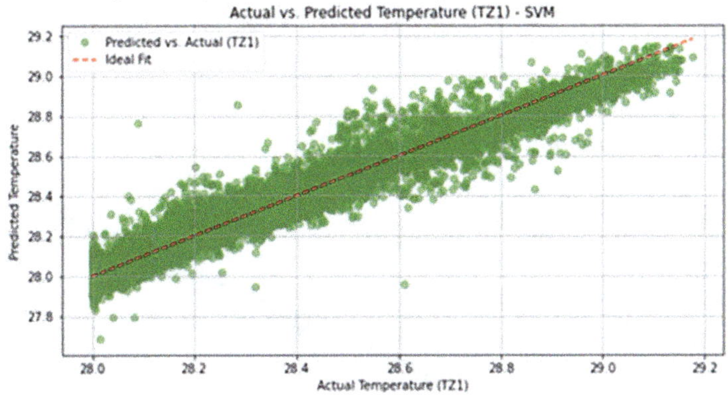

1. ITUZ1 - Mean Squared Error =436546.1292
2. ITUZ1 - R^2 Score = 0.989094

Table 3. Results of SVM (Predictive Modeling)

Temperature Prediction Summary			**Energy Prediction Summary**		
	Tz1	**Predicted_ Tz1**		**Energy_ Savings**	**Percentage_ Savings**
count	10512	10512	**count**	10512	10512
mean	28.38503	28.37925	**mean**	-16.185	-0.04777
std	0.301759	0.292107	**std**	660.5496	1.298642
min	27.99852	27.6826	**min**	-7116.37	-20.9494
25%	28.12247	28.13968	**25%**	-183.125	-0.34839
50%	28.33073	28.3271	**50%**	0.063731	0.014893
75%	28.60996	28.60117	**75%**	191.3692	0.36363
max	29.17681	29.14416	**max**	**13959.21**	**20.66388**
Actual_ TZ1	**Predicted_ TZ1**	**Actual_ ITUZ1**	**Predicted_ ITUZ1**	**Energy_ Savings**	**Percentage_ Savings**
28.38503	28.37925	67553.67	53594.46	**13959.21**	**20.66388**

4. LSTM Algorithm

For LSTM $R^2 = 0.85658$ and MSE= 0.0132.

Figure 16. LSTM Algorithm

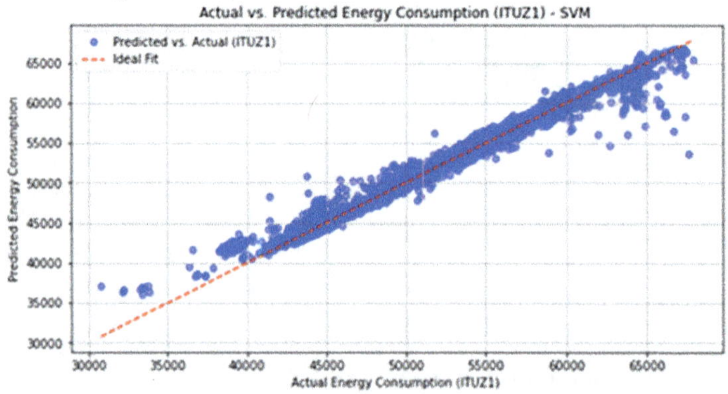

Figure 17. Results of LSTM (Predictive Modeling)

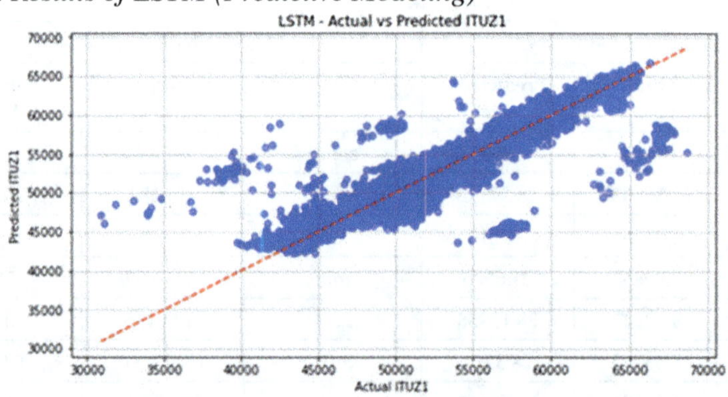

1. ITUZ1 - Mean Squared Error = 4695327.8699
2. ITUZ1 - R^2 Score = 0.881796

Table 4. Results of LSTM (Predictive Modeling)

Temperature Prediction Summary			**Energy Prediction Summary**		
	Tz1	**Predicted_ Tz1**		**Energy_ Savings**	**Percentage_ Savings**
count	10510	10510	**count**	10510	10510
mean	28.30930	28.34059	**mean**	59.57416	-0.07541
std	0.302833	0.254172	**std**	2166.154	4.141158
min	27.99997	27.89776	**min**	-16710	-52.5869
25%	28.12793	28.14302	**25%**	-661.978	-1.33105
50%	28.33445	28.28599	**50%**	93.86314	0.197857
75%	28.62059	28.52494	**75%**	818.8331	1.474752
max	29.18799	28.98163	**max**	**13843.83**	**22.50858**
Actual_ TZ1	**Predicted_ TZ1**	**Actual_ ITUZ1**	**Predicted_ ITUZ1**	**Energy_ Savings**	**Percentage_ Savings**
29.18799	28.98163	63263.65	49419.82	**13843.83**	**21.88276**

5. GRU Algorithm

For GRU model, the $R^2 = 0.90117$ and MSE= 0.009062.

Figure 18. GRU Algorithm

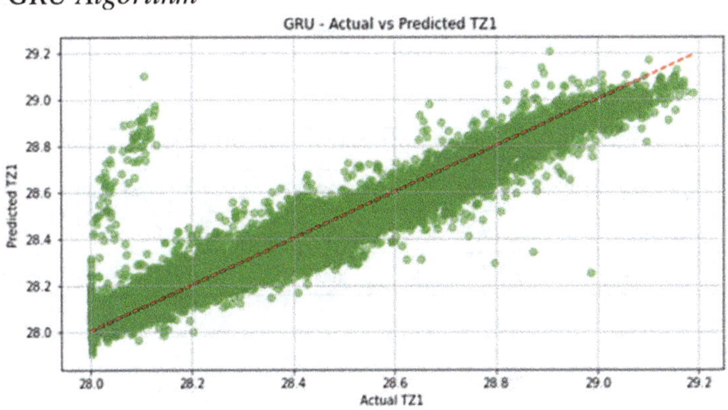

Figure 19. Results of GRU (Predictive Modeling)

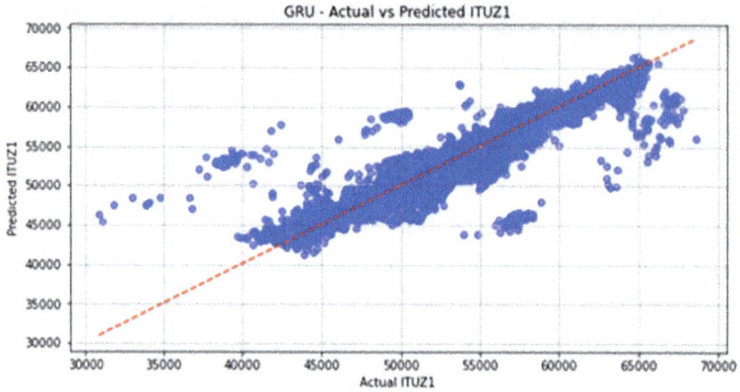

1. ITUZ1 - Mean Squared Error =4692558.929
2. ITUZ1 - R^2 Score= 0.881865

Table 5. Results of LSTM (Predictive Modeling)

Temperature Prediction Summary			Energy Prediction Summary		
	TZ1	Predicted_ TZ1		Energy_ Savings	Percentage_ Savings
count	10510	10510	count	10510	10510
mean	28.3039	28.40499	mean	-958.081	-1.131796
std	0.302383	0.290508	std	2082.131	4.463259
min	27.99997	27.99602	min	-16470	-51.8583
25%	28.12793	28.18118	25%	-1291.82	-2.50995
50%	28.33445	28.34243	50%	-627.641	-0.99308
75%	28.62059	28.60623	75%	122.9833	0.26332
max	29.18799	29.25026	max	12570.6	21.85947
Actual_ TZ1	Predicted_ TZ1	Actual_ ITUZ1	Predicted_ ITUZ1	Energy_ Savings	Percentage_ Savings
29.18799	29.25006	63263.65	50693.05	12570.6	19.87018

DISCUSSION

This Section compares the different machine learning and deep learning models to predict temperature and energy use in HVAC systems. The examined are LR, RF, SVM, LSTM, and GRU.

1. The Performance of temperature prediction

GRU achieved the lowest MSE, making it more precise for real-time short-term prediction due to its sequential learning capability. Random Forest achieved the highest R2 score, indicating superior ability to explain temperature variation and capture long-term trends. SVM is more complex but has balanced accuracy.

Table 6. Temperature Prediction Performance Comparison

Model	MSE	R^2 Score	Key Observation
LR	0.0191	0.7903	Serves as a baseline but has limited accuracy.
RF	0.0018	0.9807	Highest R^2 score; robust trend capturing.
SVM	0.0053	0.9419	Performs well but is computationally heavy.
LSTM	0.0132	0.8566	Good for sequential data, but with higher error.
GRU	0.0090	0.9011	Lowest MSE, best for short-term forecasting.

2. The Performance of Energy Consumption Prediction

LSTM and GRU achieved the lowest MSE, making them more precise for real-time short-term prediction due to their sequential learning capability. Random Forest achieved the highest R2 score, explaining 99.78% of energy consumption variation, making it suitable for interpretable, robust predictions.

Table 7. Energy Consumption Prediction Performance Comparison

Model	MSE	R^2 Score	Key Observation
LR	367672.61	0.9908	Fast baseline but lacks robustness.
RF	87151.50	0.9978	Most accurate with high interpretability.
SVM	436546.12	0.9891	High error; not optimal for large data.
LSTM	469327.86	0.8818	Very low MSE; efficient in sequence learning.
GRU	4692558.92	0.8819	Comparable to LSTM; best for time-series.

3. Accuracy vs Computational Efficiency

Random Forest strikes the best balance between accuracy and computational efficiency, making it highly suitable for real-time HVAC control. GRU offers a faster alternative to LSTM, enabling efficient short-term forecasting. A hybrid approach (Random Forest + GRU) could combine interpretability and sequential learning, ensuring balanced performance.

Table 8. Computational Efficiency vs. Accuracy Trade-off

Model	Prediction Accuracy	Computational Complexity	Best Use Case
LR	Low	Very Fast	Quick baseline
RF	Very High	Medium	Interpretable, robust
SVM	Medium	High	Small datasets only
LSTM	Very High	Very High	Sequential deep learning
GRU	Very High	High	Faster alternative to LSTM

The comparative analysis highlights that no single model dominates across all performance dimensions. Instead, the choice of model depends on the operational requirements:

For short-term sequential forecasting, GRU is preferred due to its low error rate and faster convergence compared to LSTM. For long-term interpretability and trend analysis, Random Forest provides the best explanatory power with computational efficiency. In real-time HVAC control, Random Forest alone or a hybrid Random Forest + GRU model offers the best trade-off between accuracy, robustness, and efficiency.

CONCLUSION

This article proposes the enhancements of energy efficiency in heating, ventilation, and air conditioning (HVAC) system by applying Machine learning and deep learning models such as linear regression (LR), Support Vactor machine (SVM), rando Forest (RF), Long Sort term memory (LSTM), gated recurrent unit (GRU) and Fusion of GRU and RF strategies. The above strategies are implemented via simulation and Real-time experimentation to predict indoor temperature (TZ1) and HVAC energy consumption (ITUZ1). The two error-based performance indicators are explored, namely the R2 score and mean square error (MSE). From the simulation and real-time experimentation tests, it is seen that all algorithms represent

HVAC system dynamics accuracy, making the best predictions among the relevant algorithms with a large value of R2 score (0.9978) and a smaller value of MSE (0.0018)for both indoor temperature and energy consumption.

The hybrid GRU and RF showed better performance in terms of energy saving. GRU strategy compensates for the non-linear changes, while the RF method shows solid predictions. With this hybrid strategy, around 24.29% energy is saved. The investigation from the article indicates that a hybrid data-driven methodology is superior for adaptability control of HVAC systems. The model predictive control (MPC) and irrelevant strategies are model-based designs that give rise to a complex thermal model and regular recalibration, thereby limiting their capability in an industrial application. The perfect model is not required in the hybrid strategy GRU and RF, and only works on real-time sensor data. The Hybrid ML/DL algorithm makes the HVAC system an intelligent system. In a smart building management system, the data-driven methodology and ML/DL algorithms make the system more intelligent, which will help the HVAC system to march towards sustainability goals.

REFERENCES

Afram, A., & Janabi-Sharifi, F. (2014). Theory and applications of HVAC control systems – A review of model predictive control (MPC). *Building and Environment*, *72*, 343–355. DOI: 10.1016/j.buildenv.2013.11.016

Awasthi, B., Gupta, S., Kumar, S., & Tripathi, A. (2024, January). Business management and planning using queuing model with heterogeneous customers for electrical applications. In 2024 2nd International Conference on Intelligent Data Communication Technologies and Internet of Things (IDCIoT) (pp. 279-283). IEEE. DOI: 10.1109/IDCIoT59759.2024.10467673

Baetschmann, M., & Leibundgut, H. (2012). LowEx Solar Building System: Integration of PV/T Collectors into Low Exergy Building Systems. *Energy Procedia*, *30*, 1052–1059. DOI: 10.1016/j.egypro.2012.11.118

Ben-Nakhi, A. E., & Mahmoud, M. A. (2002). Energy conservation in buildings through efficient A/C control using neural networks. *Applied Energy*, *73*(1), 5–23. DOI: 10.1016/S0306-2619(02)00027-2

Dalamagkidis, K., Kolokotsa, D., Kalaitzakis, K., & Stavrakakis, G. S. (2007). Reinforcement learning for energy conservation and comfort in buildings. *Building and Environment*, *42*(7), 2686–2698. DOI: 10.1016/j.buildenv.2006.07.010

Dounis, A. I., & Caraiscos, C. (2009). Advanced control systems engineering for energy and comfort management in a building environment—A review. *Renewable & Sustainable Energy Reviews*, *13*(6–7), 1246–1261. DOI: 10.1016/j.rser.2008.09.015

Fan, C., Xiao, F., & Zhao, Y. (2017). A short-term building cooling load prediction method using deep learning algorithms. *Applied Energy*, *195*, 222–233. DOI: 10.1016/j.apenergy.2017.03.064

Fotis, G., Gupta, S., Varshney, T., Dhar, S. L., & Gulzar, M. M. (Eds.). (2025). *Modern Computing Technologies for EV Efficiency and Sustainable Energy Integration*. IGI global., DOI: 10.4018/979-8-3373-2382-4

Goyal, S., Ingley, H. A., & Barooah, P. (2013). Occupancy-based zone-climate control for energy-efficient buildings: Complexity vs. performance. *Applied Energy*, *106*, 209–221. DOI: 10.1016/j.apenergy.2013.01.039

Gupta, S. (2025). *The Future of Machine Learning with Automation in Engineering*. Nova Science Publishers., DOI: 10.52305/YFIX2832

Gupta, S., & Tripathi, R. K. (2014). Two-area power system stability improvement using a robust controller-based CSC-STATCOM. *Acta Polytechnica Hungarica, 11*(7), 135–155. DOI: 10.12700/APH.11.07.2014.07.9

Kalogirou, S. A. (2000). Applications of artificial neural-networks for energy systems. *Applied Energy, 67*(1–2), 17–35. DOI: 10.1016/S0306-2619(00)00005-2

Kumar, R., Aggarwal, R. K., & Sharma, J. D. (2013). Energy analysis of a building using artificial neural network: A review. *Energy and Building, 65*, 352–358. DOI: 10.1016/j.enbuild.2013.06.007

Li, D., Zhao, D., Zhu, Y., & Xia, Z. (2015). Thermal comfort control based on MEC algorithm for HVAC systems. *2015 International Joint Conference on Neural Networks (IJCNN)*, 1–6. DOI: 10.1109/IJCNN.2015.7280436

Ma, Y., Matusko, J., & Borrelli, F. (2015). Stochastic Model Predictive Control for Building HVAC Systems: Complexity and Conservatism. *IEEE Transactions on Control Systems Technology, 23*(1), 101–116. DOI: 10.1109/TCST.2014.2313736

Maasoumy, M., Pinto, A., & Sangiovanni-Vincentelli, A. (2011). Model-Based Hierarchical Optimal Control Design for HVAC Systems. *ASME 2011 Dynamic Systems and Control Conference and Bath/ASME Symposium on Fluid Power and Motion Control, Volume 1*, 271–278. DOI: 10.1115/DSCC2011-6078

Meggers, F., Ritter, V., Goffin, P., Baetschmann, M., & Leibundgut, H. (2012). Low exergy building systems implementation. *Energy, 41*(1), 48–55. DOI: 10.1016/j.energy.2011.07.031

Mozer, M. C. (1998). The neural network house: An environment hat adapts to its inhabitants. *Cdn.Aaai.Org*. https://cdn.aaai.org/Symposia/Spring/1998/SS-98-02/SS98-02-017.pdf

Oldewurtel, F., Parisio, A., Jones, C. N., Gyalistras, D., Gwerder, M., Stauch, V., Lehmann, B., & Morari, M. (2012). Use of model predictive control and weather forecasts for energy efficient building climate control. *Energy and Building, 45*, 15–27. DOI: 10.1016/j.enbuild.2011.09.022

Oldewurtel, F., Parisio, A., Jones, C. N., Morari, M., Gyalistras, D., Gwerder, M., Stauch, V., Lehmann, B., & Wirth, K. (2010). Energy efficient building climate control using Stochastic Model Predictive Control and weather predictions. *Proceedings of the 2010 American Control Conference*, 5100–5105. DOI: 10.1109/ACC.2010.5530680

Oldewurtel, F., Sturzenegger, D., & Morari, M. (2013). Importance of occupancy information for building climate control. *Applied Energy, 101*, 521–532. DOI: 10.1016/j.apenergy.2012.06.014

Pande, A. S., Soni, B. P., & Bhadane, K. V. (2023). Electrical models for EV's batteries: an overview and mathematical design of RC network. *SpringerAS Pande, BP Soni, KV BhadaneJournal of The Institution of Engineers (India): Series B, 2023•Springer, 104*(2), 533–547. DOI: 10.1007/s40031-022-00852-1

Pande, A. S., Soni, B. P., & Sharma, A. K. (2025). Modeling of one and two RC model and state estimation of Lithium-Ion Battery using Thevenin's equivalent Circuit Model. *Procedia Computer Science, 259*, 494–503. DOI: 10.1016/j.procs.2025.03.351

Petersen, S., & Bundgaard, K. W. (2014). The effect of weather forecast uncertainty on a predictive control concept for building systems operation. *Applied Energy, 116*, 311–321. DOI: 10.1016/j.apenergy.2013.11.060

Rao, M. S., Saraswathy, S., Neela, V., Arun, M. R., Gupta, S., & Taqui, S. N. (2023, November). Efficient Model Training in Decentralized Systems with Federated Learning. In *2023 International Conference on Communication, Security and Artificial Intelligence (ICCSAI)* (pp. 900-904). IEEE. DOI: 10.1109/ICCSAI59793.2023.10421469

Shankar, R., Chaudhary, B. P., Mehraj, H., Gupta, S., Arslan, F., Srivastava, A. K., & Goswami, C. (2023). Impact of node mobility on the DL based uplink and downlink MIMO-NOMA network. *International Journal of Information Technology : an Official Journal of Bharati Vidyapeeth's Institute of Computer Applications and Management, 15*(6), 3391–3404. DOI: 10.1007/s41870-023-01362-z

Sharma, R. K., Saran, V. B., Kanaujia, S., & Gupta, S. (2024, May). Evaluation of mppt controller performance using fuzzy logic design with wireless sensor node. In *2024 International Conference on Smart Systems for applications in Electrical Sciences (ICSSES)* (pp. 1-6). IEEE. DOI: 10.1109/ICSSES62373.2024.10561365

Shukla, R. D., Tripathi, R. K., & Gupta, S. (2010, November). Power electronics applications in wind energy conversion system: A review. In *2010 International Conference on Power, Control and Embedded Systems* (pp. 1-6). IEEE. DOI: 10.1109/ICPCES.2010.5698663

Sinha, R., Patel, M., Gupta, S., & Sinha, K. K. (2024, June). Performance Analysis of Breast Cancer Predictor using Machine Learning Techniques. In 2024 15th International Conference on Computing Communication and Networking Technologies (ICCCNT) (pp. 1-5). IEEE. DOI: 10.1109/ICCCNT61001.2024.10724436

Sinha, R., Sinha, K. K., Patel, M., Gupta, S., & Priya, S. (2024, July). Detection of leukemia disease using convolutional neural network. In 2024 5th International Conference on Image Processing and Capsule Networks (ICIPCN) (pp. 451-456). IEEE. DOI: 10.1109/ICIPCN63822.2024.00080

Široký, J., Oldewurtel, F., Cigler, J., & Prívara, S. (2011). Experimental analysis of model predictive control for an energy efficient building heating system. *Applied Energy*, *88*(9), 3079–3087. DOI: 10.1016/j.apenergy.2011.03.009

Thandlam, A. K., Vignesh, T., Saravanan, A., Subramani, P., Marimuthu, R., & Gupta, S. (2024, April). Next-gen vehicle safety: The futuristic approach to auto-stop in modern vehicles. In 2024 10th International Conference on Communication and Signal Processing (ICCSP) (pp. 265-270). IEEE. DOI: 10.1109/ICCSP60870.2024.10543666

Tyagi, M., Gupta, S., & Khan, M. I. (2024, May). Reduction of voltage instability in complex power networks through the utilization of D-STATCOM. In 2024 4th International Conference on Advance Computing and Innovative Technologies in Engineering (ICACITE) (pp. 1662-1666). IEEE. DOI: 10.1109/ICACITE60783.2024.10616856

Wei, T., Wang, Y., & Zhu, Q. (2017). Deep reinforcement learning for building HVAC control. *Dl.Acm.OrgT Wei, Y Wang, Q ZhuProceedings of the 54th Annual Design Automation Conference 2017, 2017•dl.Acm.Org, Part 128280*. DOI: 10.1145/3061639.3062224

Wei, T., Zhu, Q., & Maasoumy, M. (2014). Co-scheduling of HVAC control, EV charging and battery usage for building energy efficiency. *2014 IEEE/ACM International Conference on Computer-Aided Design (ICCAD)*, 191–196. DOI: 10.1109/ICCAD.2014.7001351

Yang, L., Nagy, Z., Goffin, P., & Schlueter, A. (2015). Reinforcement learning for optimal control of low exergy buildings. *Applied Energy*, *156*, 577–586. DOI: 10.1016/j.apenergy.2015.07.050

Yu, L., Xie, D., Huang, C., Jiang, T., & Zou, Y. (2019). Energy Optimization of HVAC Systems in Commercial Buildings Considering Indoor Air Quality Management. *IEEE Transactions on Smart Grid*, *10*(5), 5103–5113. DOI: 10.1109/TSG.2018.2875727

Chapter 5
Next-Generation Control for Standalone SPV-WECS-BESS Using Hybrid Optimization

Mohit Tyagi
https://orcid.org/0009-0002-7452-1098
Delhi Technological University, India

Duli Chand Meena
Delhi Technological University, India

Sombir Kundu
https://orcid.org/0000-0001-6848-3721
MMEC, Maharishi Markandeshwar University, India

ABSTRACT

This chapter focuses on hybrid renewable energy sources (wind and solar) along with a battery energy storage system (BESS). To improve power quality, stability, and an uninterrupted power supply, the advanced control architecture is used. The hybrid system includes essential components such as incremental conductance (InC) maximum power point tracking (MPPT), Bidirectional DC-DC converter (BDC), battery management system (BMS), and voltage source converter (VSC) with an active power filter (APF). A hybrid PSO-GWO optimization technique is used to tune the PID controller, which is designed for dynamic conditions. The system is modelled in MATLAB/Simulink software, and the results confirm better performance, improved transient response, lower THD, and enhanced voltage stability when the

DOI: 10.4018/979-8-3373-7041-5.ch005

conditions are change in solar irradiance, change in wind, and change in load. This method enhances the system reliability and efficiency of vital off-grid applications, including rural electrification and islanded microgrids.

INTRODUCTION

The global transition towards sustainable energy has accelerated research into hybrid renewable energy (HRES) with particular emphasis on configurations that combine SPV-WECS-BESS. The following resources are complementary and free of cost by nature. There is a lot of solar energy during the day, and wind patterns can be more intense at night and during seasonal variations. However, their intrinsic intermittency and variability create significant challenges in ensuring reliable, continuous, and high-quality power, especially in standalone and off-grid applications where there is no supporting utility grid. BDC converter connected BESS helps buffer these fluctuations, but their integration introduces further complexities related to charging/discharging cycles, efficiency, degradation, and system coordination. Against this backdrop, intelligent and adaptive Control strategies are critical to achieving robust performance. This chapter presents the design control and modelling of a standalone SPV-WECS-BESS hybrid system, highlighting the use of advanced PID regulation optimised with a hybrid metaheuristic algorithm. Conventional PID controllers are popular because they are simple and efficient, but they have trouble with nonlinearities, uncertainties, and abrupt changes in load demand on renewable generation. To overcome these shortcomings, this work integrates Particle Swarm Optimization (PSO) and Grey Wolf Optimization (GWO) in a hybrid approach to continuously tune PID parameters in real time. The hybrid controller enhances transient response, minimizes overshoot, stabilizes the DC link voltage, and significantly minimizes total harmonic distortion (THD). The system is modelled in MATLAB/Simulink, and a multi-level control architecture is implemented: (i) Source/device level fast control for SPV-MPPT and WECS generator rectification; (ii) Mid-level regulation of the voltage at DC-link and BESS Management, ensuring optimal state of charge (SoC) and the state of Health (SoH); and (iii) Outer supervisory control of VSC with an APF that depends on instantaneous P-Q-R theory to mitigate THD, voltage imbalance, and reactive power issues. An FCU provides protection during abnormal events such as faults, load imbalance, or voltage sag/swell. The system simulates on MATLAB/Simulink under different operating conditions, like a change in the speed of wind, a change in the irradiance of solar, a change in the load, and simultaneously a change in renewable energy sources (RESs) and load. The results show that the proposed PSO-GWO-tuned PID controller maintains DC-link voltage stability and supports an uninterrupted power supply with the help of BESS.

The hybrid optimization technique provides improved power quality (PQ), faster settling time, and reduced steady-state error as compared with conventional PID tuning. Where centralized grids are not available, this work provides a blueprint for the sustainable power supply in off-grid areas. In summary, the contribution and primary goal of this chapter are to explain how hybrid (PSO-GWO) optimization of PID controllers can improve the stability, efficiency, and reliability of standalone SPV-WECS and BESS hybrid generation system.

BACKGROUND

The transition to renewable energy sources has become essential in the twenty-first century as the climate problem worsens and fossil fuel supplies run out. SPV and WECS are two of the most developed, affordable, and scalable technologies for producing power among the many renewable possibilities. While wind turbines harness the kinetic energy of airflows to generate electrical power, SPV uses semiconductor junctions to directly turn sunlight into electricity. Every one of these sources is sporadic and unpredictable; solar production is influenced by the time of day, solar irradiance, cloud cover, and weather fluctuations. Wind speed, turbulence, and diurnal or seasonal patterns all affect wind power. Relying just on SPV and wind is rarely adequate to deliver consistent, stable, and high-quality electric power due to their intermittent and unpredictable nature, particularly in distant or off-grid areas where there is no reliable external infrastructure to absorb fluctuations or offer backup. Therefore, combining SPV and WECS in an HRES, supported by BESS, has emerged as a practical and increasingly common architecture for remote electrification, microgrids, and islanded power systems. After combining the SPV and WECS along with BESS, the system brings several challenges for design and control, such as continuous power supply, maintaining voltage and frequency stability, use of BESS, extending battery life, controlling faults and power-quality issues, and adjusting power supply as per load demand. To meet these challenges, modern and adaptive control strategies—spanning model predictive control (MPC), robust control, artificial intelligence learning algorithms, droop control, virtual inertia, adaptive MPPT, and hierarchical energy management systems—are increasingly being developed and deployed.

SPV and WECS are complementary in many respects: solar energy is abundant during daylight hours, while wind patterns, in some regions, may be stronger at night or during weather systems. Thus, when both sources are integrated thoughtfully, their combined output can smooth out some of each other's dips, reducing—but not eliminating—the need for storage. A review of HRES notes that combining SPV and wind can partially compensate for the erratic nature of each individual source, and can make hybrid systems more reliable and economical than a single renewable source (D. A. Gutiérrez-Torres et al., 2024). In fact, hybrid SPV–WECS have become a focal point of research and deployments, particularly for remote communities and small microgrids, because the duality helps reduce the required battery size and improve the continuity of service (Sombir Kundu et al., 2024).

Batteries become essential in microgrid areas because the power generation is not continuous due to changes in weather conditions. When generation exceeds demand, the battery acts as a buffer, holding excess energy and releasing it when generation falls short. Additionally, it allows the system to ride through momentary decreases in SPV or wind input, smoothest power production, mitigates abrupt load shifts or generation variations, and responds quickly to transients. However, battery integration is not trivial. One must address sizing (how much capacity, energy, and power rating is needed), battery degradation (cycle life, depth-of-discharge, temperature effects), efficiency losses, and control of the charging/discharging process under constraints. Indeed, capital cost, operational dependability, battery life, and energy availability must all be balanced in the design of HRES with storage.

Control techniques are crucial to ensuring robust performance because the system must adapt to changing situations. A hybrid SPV + WECS + BESS control framework is frequently hierarchical: At higher levels, energy management systems (EMS) or supervisory controllers plan when to charge and discharge batteries, when to reduce renewable generation, and how to optimize performance over time (e.g., lifetime, cost, reliability); at lower levels, MPPT guarantees PV (and occasionally wind turbines) deliver their maximum possible power under prevailing environmental conditions; medium-level controllers regulate current, voltage, and power flows, ensuring proper sharing, coordination, and load tracking. Moreover, because of uncertainty, disturbances, and nonlinear dynamics, adaptive and intelligent control strategies—such as model predictive control, robust control, reinforcement learning, neural networks, and fuzzy logic—have gained traction.

In more depth, the role of MPPT is a foundational piece. For SPV, algorithms such as Perturb & Observe (P&O), Incremental Conductance, and advanced MPPT for PV modules under changing irradiance and temperature conditions (Lyu Guanghua et al., 2025). For variable-speed wind turbines, analogous maximum power extraction strategies adjust turbine control variables (e.g., pitch, rotor speed) to match the optimal tip speed ratio under varying winds. In hybrid systems, some controllers select among multiple MPPT algorithms or adaptively switch strategies to balance convergence speed, oscillation damping, and robustness (Shukla et al., 2010). Indeed, a recent study of a wind turbine system in a remote context used a hybrid MPPT combining P&O and fuzzy logic to maximize wind power and minimize stress on storage systems; that work also integrated batteries and supercapacitors as a hybrid storage for dynamic response (Rekioua et al., 2024).

Beyond MPPT, MPC is a powerful tool in modern hybrid systems. MPC uses a plant model to predict future behaviour over a finite horizon and optimizes control inputs subject to constraints (e.g., battery SoC limits, converter current limits, voltage constraints). For example, MPC has been applied to grid-forming inverters in islanded microgrids, optimizing voltage/frequency regulation while ensuring power sharing and constraint satisfaction (Fotis et al., 2025). In hybrid AC/DC microgrids, finite-control-set MPC (FCS-MPC) has been used to unify control of SPV, BESS, and converter operations without separate droop or PI loops (Aghdam et al., 2020). The advantage is flexibility and the ability to enforce constraints, though computational burden and model accuracy are challenges.

Droop control and virtual inertia/virtual synchronous machine (VSM) emulation provide another important control paradigm. In islanded or weak-grid systems, VSC-based sources must emulate the behaviour of synchronous machines to provide stability, damping, and frequency support. Droop control provides a decentralized method for load sharing: for example, active power droop (P–f) and reactive power droop (Q–V) allow multiple inverters to share load without requiring fast communications. Virtual inertia or VSM emulation gives the inverter dynamic characteristics akin to mechanical inertia, helping dampen frequency deviations (Giri et al., 2020). Some modern designs enable adaptive droop gains, where droop coefficients are adjusted in real time based on operating conditions to maintain stability margins.

Using mixed renewable energy systems combined with a unified power quality controller (UPQC) with an Atom Search Optimization (ASO) method to solve the PQ problem (Rao et al., 2023). Their Fractional Order Proportional-Integral-Derivative (FOPID) processor worked better than the old one at reducing voltage and current distortion. A new type of multi-microgrid design that makes voltage profiles better and reduces THD. Within this system, methods for balancing power are utilized. They utilized an Active Power Quality Controller (APQC) with shunt and series compensators along with a multi-level PID controller tuned by a dynamic algorithm (Ali et al., 2025). It was suggested that the way used would cut down on THD and raise the total PQ. How DREs could help poor countries clean up the air and reduce carbon dioxide. But decentralized renewable energy sources (DREs) still make power systems have problems sometimes, even with all the progress that has been made. Findings from the study showed that Automatic Generation Control (AGC) methods can be made better in terms of how well they manage power and how much smog they create. BESS and electric vehicles (EVs) are built into SPV systems that are connected to the power grid (Kanojia et al., 2025). Because of the problems and unknowns with renewable energy sources and storage technologies, they tried to improve how power is controlled and how it is integrated. An updated plan was to make the UPFC work better. They fixed THD and made the total PQ better by adding an APF and controlling the inverter series of UPFC for load-responsive power and voltage correction (Nishad et al., 2024). Moth Flame Optimization (MFO) and an Improved Bat Search Algorithm (IBSA) were combined to create the IBSMFO model because this model balanced the active and reactive power. This IBSMFO model was more effective in decreasing power swings and increasing (PQ (Chapala et al., 2024). UPQC was used to control power spikes and sudden changes in voltage. UPQC is also used to improve PQ by reducing voltage interference and THD (Giri et al., 2019). To get the maximum power from the SPV system, an Incremental Conductance Particle Swarm Optimization (ICPSO) method is used (Gupta et al., 2018). This method generates a new technique to enhance the performance of the PV-UPQC system by adjusting the particle weight component to match the global power peak and extract maximum power from the particles. The Gated Recurrent Unit (GRU) network controller is used in place of the synchronous reference frame control technique (Kundu et al., 2022). This idea mitigates variation in voltage and THD.

An EMS that utilized both MFPA and MP&O methods, which was lower than the standard controller (Kundu et al., 2023). To get the best size for off-grid PV, wind, and batteries, the Grasshopper Optimization Algorithm is also used. This resulted in a lower levelized energy cost. To find the best way to split up RES and lower THD with D-STATCOM and passive filters, used Artificial Rabbits Optimization (J. Srikakolapu et al., 2024). Combining UPQC with a FOPI processor that was set up to work with the Enhanced Seagull with Rooster Update algorithm as a way to improve PQ (Ratnakaran et al., 2023). An Adaptive Neuro-Fuzzy Inference System to improve voltage quality under nonlinear loads (Kundu et al., 2023). Both deep neural networks and artificial neural network-based inverter control are being used to increase MPPT performance and microgrid stability. So, AI-based approaches are also becoming popular (Giri et al., 2020). Some other works compare optimization-tuned PID controllers for load frequency and voltage regulation, while others focus on reducing THD and designing inverters employing hybrid systems (Sundriyal et al., 2023).

Robust and adaptive control methods are increasingly applied to deal with uncertainties. Because the plant model, grid impedance, load dynamics, and renewable generation forecasts are uncertain, controllers based on robust design (e.g., H∞ or μ−synthesis) can ensure performance under worst-case deviations. Additionally, adaptive control schemes can tune controller gains online in response to observed behaviour. For instance, in DC microgrids, combining fuzzy logic and fractional-order PID strategies allows adaptive tuning to PQ and smooths the output when PV, wind, and BESS sources interact (P.Shambhu et al., 2025). Research in adaptive energy management controllers proposes combining fuzzy logic with fractional-order PID to optimally allocate power among BESS, PV, wind, and loads (Singh et al., 2022).

Artificial intelligence and learning-based control are emerging as powerful enablers. Neural networks, recurrent neural networks (RNNs), long short-term memory (LSTM) architectures, and deep reinforcement learning (DRL) have been used to predict generation and load, and optimize dispatch decisions. For example, a hybrid intelligent control method combines rule-based control with deep learning (GRU, LSTM) to adaptively manage microgrids under variable renewable supply and load fluctuations, improving stability and performance (Talaei et al., 2023). A more holistic DRL-based microgrid control approach formulates the control as a Markov decision process and trains agents (e.g., double dueling deep Q networks, D3QN) to optimize multi-objective goals (cost, emissions, battery degradation, peak load) (Kundu et al., 2021). Such DRL approaches can handle uncertainties and non-linear dynamics but require training and reliable data.

At the supervisory or system level, energy management systems (EMS) or power management controllers (PMC) coordinate renewable generation, storage charging/discharging, load scheduling, curtailment, and backup resources. EMS often embed optimization algorithms (linear programming, mixed-integer programming, heuristic/metaheuristic methods, or adaptive/online techniques) to find optimal operating trajectories over a horizon. A recent work implements a real-time adaptive multi-objective optimization EMS in a microgrid hardware-in-the-loop setup, balancing fuel usage, power mismatch, battery degradation, and renewable utilization by dynamically adjusting preference weights (Bakare et al., 2024). In hybrid SPV-WECS integrated with the grid, an advanced control architecture combining FACTS (Flexible AC Transmission Systems), fault current limiters, and storage under control strategies was shown to improve grid stability and PQ under high renewable penetration (Sharma et al., 2024).

In microgrid and hybrid system reviews, a number of control strategies are surveyed: centralized, decentralized, hierarchical, multi-agent, robust, predictive, AI-based, and hybrid schemes (Gupta, 2025). A new comprehensive review of HRES (hybrid energy systems) emphasizes the need for robust, adaptive, and scalable control strategies that can handle parameter variations, fast dynamics, and uncertainties (Kundu et al., 2025). Advances in renewable-energy microgrid systems also highlight the integration of Internet of Things (IoT), predictive analytics, and distributed control as key enablers of real-time adaptability (Al-Shetwi et al., 2025). In addition, recent studies focus on optimization techniques for hybrid systems with storage, comparing classical and AI-based methods (Kundu et al., 2025). From an architectural vantage, hybrid systems often structure PV, wind, BESS, and converters around DC and/or AC buses, with power electronic interfaces connecting sources and loads. The battery may be placed on the DC bus (in DC microgrids) or on the AC side (via BDC). Coordination among these converters is essential to ensure current/voltage balancing, power sharing, load tracking, and system stability. Control strategies at the converter layer (e.g., current control, voltage control) must be coordinated with higher-level control (e.g., EMS, MPC) to achieve cohesive behaviour.

Because remote systems have no external grid reference, they must provide grid-forming functionality: controlling voltage and frequency internally. Some approaches configure one inverter (often the battery inverter) as grid-forming (GFM), while others (e.g., PV inverters) operate as grid-following (GFL), synchronized to the reference established by the GFM inverter. Recent works explore unified controllers that can seamlessly switch between GFM and GFL modes without relying on conventional PLL (phase-locked loop) architectures (Khan et al., 2024). Such controllers reduce complexity and improve robustness in changing operational modes (grid-connected ↔ islanded). Moreover, fault ride-through (FRT) capability is essential: inverters must survive voltage sags or frequency deviations. Control schemes embedding FRT, protection coordination, and adaptive responses to faults are increasingly studied (Kundu et al., 2024).

In highly decentralized or large microgrids, distributed control and event-triggered control methods are gaining attention to reduce communication overhead and enhance scalability. A distributed event-triggered control method can coordinate energy storage sharing and regulate voltage in DC microgrids with limited communication. This method is based on a publish-subscribe communication paradigm and utilizes local decision rules and a consensus filter. (Parmar et al., 2025). Frameworks like Comelec use composability in active distribution networks, merging resources into virtual entities and managing them through hierarchical real-time setpoints (Kundu et al., 2025). It is frequently necessary for remote hybrid systems to detect grid disconnection (islanding) on their own and seamlessly switch to internal control. When the grid returns, controllers must resynchronize, seamlessly handle transitions, and perhaps carry out black-start procedures (start from zero grid). In these situations, adaptive controllers that can transition between grid-forming, grid-following, and islanded dispatch are essential.

Control schemes frequently incorporate battery deterioration models and limitations into EMS and MPC formulations because battery longevity and health are crucial in distant systems (where replacement is expensive). Battery cycle life, depth-of-discharge, temperature impacts, and aging are all taken into account during multi-objective optimization, which helps extend battery life while satisfying energy requirements. Additionally popular are hybrid energy storage systems (HESS), which combine batteries with flywheels, supercapacitors, or other types of storage media. Supercapacitors were added to batteries for quick reaction in one study on a wind turbine system, and the control method (hybrid MPPT + storage management) kept battery SoC within safe bounds. In reviews of grid-connected HESS, the trade-offs in architecture, control integration, and capacity sizing are examined (Kundu et al., 2025). Hybrid systems that include SPV-WECS-BESS and intelligent control are becoming more common in real-world deployments and experimental projects. For example, a university campus's hybrid SPV, wind, and energy-storage system demonstrated that such systems are feasible in real settings (Govindaraj et al., 2025). In grid-connected environments, high-penetration RES systems with advanced control and storage have been recommended to eliminate technical barriers, improve stability, reduce fault currents, and enhance PQ. Experiences from microgrid deployments highlight the importance of fault-resilient, flexible, and adaptive control.

As research continues, some emerging trends are especially promising:

1. Learning-based control approaches (reinforcement learning, neural networks) that continually adapt to changing conditions and uncertainties;
2. Hybrid control architectures combining rule-based, model-based, and data-driven strategies;
3. Scalable distributed control with minimal communication;
4. Multi-timescale control (fast converter control, mid-level coordination, slow EMS scheduling) integrated cohesively;
5. Digital twin and real-time simulation integration for control verification; and
6. Co-optimization of sizing, control, and scheduling in an integrated design framework.

The push toward decarbonization, electrification of remote areas, and resilient microgrids will make such hybrid systems with advanced adaptive control indispensable. To create and enhance a stand-alone power system that can be used in remote places is the main objective of this study. Modern control methods and tools will be used to make this happen. The suggested method gives safe, clean, and efficient electricity to areas that aren't connected to main power grids through cutting-edge power electronics, smart energy management, and strong monitoring tools. Recent research has examined hybrid systems that employ SPV-WECS-BESS power to improve their ability to generate electricity, manage energy, and deliver high-quality power (Krishan Kant et al., 2025).

In conclusion, SPV and WECS are the cornerstones of contemporary renewable energy production, but their integration is only possible in remote or off-grid settings when BESS and intelligent control are in place. Batteries bring nonlinearity, limitations, and control complexity, but they also enable smoothing, buffering, and reliability. To coordinate hybrid systems to meet load needs with stability, dependability, and optimal efficiency, modern and adaptive control techniques—from MPPT and droop control to MPC, AI-driven methods, resilient control, distributed schemes, and hierarchical energy management—are essential. Hybrid SPV-WECS-BESS will become increasingly autonomous, robust, and affordable as technology develops and artificial intelligence, forecasting, and computational resources improve. This will help supply clean, reliable electricity to underserved and rural areas (M. Tyagi et al., 2025).

MAIN FOCUS OF THE CHAPTER

By combining optimization algorithms, AI methods, and sophisticated compensators, hybrid RESs have significantly increased efficiency and PQ. Nevertheless, most of this research has not tested its conclusions in real-world circumstances and has mostly relied on simulation environments. This emphasizes the necessity of unified control techniques that can adjust to various grid and load conditions, as well as experimental verification.

Problem Statement

In order to supply clean and dependable electricity, standalone SPV and WECS systems combined with BESS are being used more frequently in isolated and off-grid locations. However, abrupt shifts in load demand coupled with the sporadic and variable nature of solar irradiation and wind speed cause voltage, frequency, and PQ instability. These fluctuations can result in frequency deviations, voltage sags and swells, and shorter battery life if they are not effectively controlled.

During dynamic operating situations, particularly under abrupt load changes or irradiance fluctuations or wind speed changes, conventional control algorithms frequently fail to maintain a stable DC-link voltage and balanced power sharing between the PV array, SEIG of WECS and the BESS. Power delivery becomes erratic as a result, and system components may sustain damage. Therefore, in order to guarantee constant voltage and frequency, optimal battery use, and continuous power delivery to the load under different solar and load conditions, an effective PID-based regulatory scheme must be designed and implemented for a standalone SPV-WECS-BESS system.

The main contributions of the authors can be summarized as follows:

1. System Design:
    a. For remote or islanded applications, a stand-alone hybrid SPV–WECS–BESS system was proposed.
    b. A self-excited induction generator (SEIG)-based wind turbine, a BDC-connected BESS, and an integrated MPPT-controlled SPV are integrated.
    c. To improve PQ by lowering THD and voltage problems, a three-phase VSC with APF was included.

   d. Added an FCU for security and dependability of the system
2. Novel Control Approach:
   a. For stable hybrid system operation and DC bus voltage management, a PID controller was employed.
   b. In comparison to traditional techniques, a hybrid PSO–GWO algorithm was created to dynamically adjust PID settings in real time, resulting in faster transient response, reduced THD, and enhanced PQ.
3. Power Quality Enhancement:
   a. Utilized P-Q-R theory in conjunction with an integrated APF for voltage compensation, harmonic abatement, and stability in situations including distorted and unbalanced loads.
   b. Showed improved voltage/frequency stability, low THD (<0.5%), and effective energy sharing amongst SPV, WECS, and BESS
4. Comprehensive Simulation & Validation: Evaluated system performance in a variety of dynamic test conditions, such as:
   a. Change in wind speed.
   b. Changes in solar irradiance.
   c. Change in loads.
   d. Simultaneous changes in renewable energy sources and load demand.
5. Practical Significance:
   a. Emphasized how the suggested method is appropriate for isolated and off-grid areas, guaranteeing a steady, dependable, and high-quality electrical supply with improved sustainability.
   b. Offered an integrated design that unified storage management, sophisticated optimization, renewable generation, and PQ enhancement.

CONFIGURATION OF THE SYSTEM

The proposed system is a standalone HRES, as shown in Figure 1, designed to supply stable and high-quality power in island mode. It integrates the following components:

Figure 1. Block diagram of standalone HRES

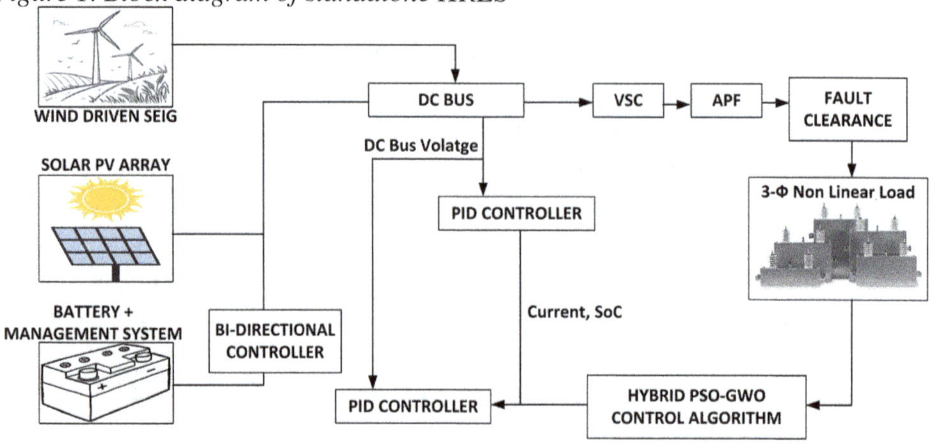

1. SPV Subsystem
    a. PV array converts sunlight into DC power.
    b. An MPPT controller ensures maximum power extraction under varying irradiance and temperature.
2. WECS Subsystem
    a. A SEIG is used as the wind turbine generator.
    b. Converts mechanical wind energy into AC, then rectifies it to DC for integration with the common DC bus.
3. Battery Energy Storage System (BESS)
    a. Connected to the DC bus through a BDC.
    b. Stores surplus power when generation > demand and supplies power when generation < demand.
    c. Monitored by a Battery Management System (BMS) for State of Charge (SoC) and State of Health (SoH).
4. DC Bus (Energy Hub)
    a. Central link where PV, WECS, and BESS contributions are combined.
    b. Ensures balanced power flow.
5. VSC with APF
    a. Converts DC bus power to three-phase AC for the load.
    b. APF improves PQ by mitigating THD, voltage imbalance, and reactive power issues.
6. Controller & Optimization
    a. PID Controller regulates DC bus voltage and manages BESS charging/discharging.

   b. Hybrid PSO–GWO algorithm optimizes PID gains in real time for better stability and reduced THD
7. Fault Clearance Unit
   a. Provides system protection during abnormal events like faults, voltage sag/swell, or load imbalance.

Under the control of MPPT, bi-directional converters, and a PSO–GWO–optimized PID controller, the system combines SPV + WECS + BESS on a single DC bus. The VSC with APF supplies clean and stable AC power to the load, while the FCU enhances protection.

CONTROL ARCHITECTURE

The controller is hierarchical and can be viewed in three levels:

1. **Source / device-level fast control** — MPPT for PV; SEIG excitation/rectification behaviour for wind; current control for converters and VSC (inner loops); PWM generation.
2. **DC-bus & storage control (mid-level)** — DC-link voltage regulation by a PID controller; BDC control for battery charge/discharge; BMS monitoring (SoC/SoH). The PID gains are tuned online by a hybrid PSO–GWO optimizer.
3. **Power quality & grid interface (outer/supervisory level)** —APF using instantaneous P–Q–R theory (Clarke/Park transforms) for harmonic compensation and unbalance mitigation; PLL for synchronization vectors; supervisory EMS rules (power balance equation, constraints, fault handling).

The above points are explained in full detail with equations:

1. **Power Balance –** Fundamental relation

All power flows are tied to the DC bus energy balance. This chapter uses the basic instantaneous power balance:

$P_{wind}(t) + P_{PV}(t) + P_{BESS}(t) = P_{Loss}(t) + P_{Load}(t)(1)$

Where:

P_{PV} = Power generated from SPV array (After MPPT and Converter)
P_{Wind} = Power generated from Wind-driven SEIG (After Rectification)

P_{BESS} = Positive when battery charging and Negative when battery Discharging
P_{Load} = Output Power as per load demand (AC Side)
P_{Loss} = Losses in Converter, Filters, wiring etc.

This balance is enforced by the DC-link voltage regulation; when generation exceeds load, VDC tends to rise, and the battery is charged (PV curtailed); when generation is lower, battery discharge to keep VDC at the setpoint.

2. **SPV Subsystem:** MPPT and BDC control

SPV electrical model: SPV current- voltage relation (Single diode)

$$IPV = Iph - Io\left(e^{\frac{V_{PV}+I_{PV}R_s}{nV_t}} - 1\right) - \frac{V_{PV} + I_{PV}R_s}{R_{sh}} \quad (2)$$

Where:

I_{ph} = Photocurrent, which depends on irradiance (G) and temperature (T)
I_o = Diode saturation Current
R_s, R_{sh} = Series and Shunt Resistance
n = Ideality factor
V_t = kT/q

MPPT Requirement: The Maximum power point is where $\frac{dP_{PV}}{dV_{PV}} = 0$. The MPPT algorithm implemented in this chapter adjusts the duty cycle D_{PV} of the SPV side BDC to maximize extracted power:

$$\frac{d}{dV_{PV}}\left(V_{PV}I_{PV}\right) = 0\left(\text{MPPT algorithm seeks } V^*_{pv}\right) \quad (3)$$

Common discrete MPPT algorithms (P&O, INC) operate by perturbing D and measuring P, This chapter refers to MPPT block controlling duty to keep:

$$\frac{dP_{PV}}{dV_{PV}} \approx 0 \quad (4)$$

And uses the duty cycle update logic inside the MPPT.
BDC dynamics: For the Buck/Boost converter, average differential equation:

$$\frac{dI_L}{dt} = \frac{1}{L}\left(V_{in} - (1 - D)V_O\right) \quad (5)$$

$$C\frac{dV_o}{dt} = (1 - D)i_L - \frac{V_o}{R_{Load}} \tag{6}$$

Where D is the duty cycle, and i_L is the inductor current. The MPPT sets to force SPV voltage to V^*_{PV}.

3. Wind Driven SEIG and Rectification

The wind generator in this chapter is a SEIG with a capacitor bank for reactive supply. Mechanical aerodynamic power captured by the turbine:

$$P_m = \frac{1}{2}\rho A\, C_p\left(\gamma,\beta\right) V_\omega^3 \tag{7}$$

Where:

Cp = power coefficient function of tip speed γ and pitch β

Electromechanical conversion and SEIG stator equations follow the standard induction machine model;

Voltage build-up for SEIG requires excitation capacitance C. The condition for successful self-excitation:

$$Q_C \geq Q_{mag} \Rightarrow V_{SEIG} = f(C, R_{Load}, N)$$

(This Chapter summarizes the magnetization condition; detailed SEIG control includes capacitor sizing and sometimes a voltage regulator to limit overvoltage) The SEIG AC is rectified (via diode bridge or controlled rectifier) and then fed to the DC bus.

4. BESS and BDC:

Battery SoC dynamics differential Equation:

$$\frac{dSoC}{dt} = -\frac{\eta_{dis} I_{dis}}{C_{batt}}(\text{Discharging}), \frac{dSoC}{dt} = +\frac{\eta_{ch} I_{ch}}{C_{batt}}\left(\text{Charging}\right) \tag{8}$$

More generally:

$$\frac{dSoC}{dt} = -\frac{sgn\left(I_{batt}\right)\eta\left(I_{batt}\right) I_{batt}}{C_{batt}} \tag{9}$$

Where: η_{ch}, η_{dis} are Efficiencies, I_{batt} positive when charging (chapter uses sign accordingly).

The BMS enforces SoC limits. So $C_{max} \geq SoC \geq SoC_{min}$.

BDC control: The BESS is interfaced via a BDC controlled to follow a power or current reference I_{batt}^{ref} generated by the DC bus voltage regulation. Averaged converter model (buck or boost depending on mode):

$$L\frac{di_L}{dt} = V_{DC} - V_{batt}.s\left(t\right) \tag{10}$$

$$C_{DC}\frac{dV_{DC}}{dt} = i_{PV} + i_{wind} - i_{load} - i_{batt} \tag{11}$$

Where s(t) relates to duty D_{bd}. The inner current loop of the BDC uses a PI (or current PID) to track I_{batt}^{ref}. The outer loop is the DC-link voltage PID. The converter controller computes the duty to the regular current:

$$D = f\left(K_{i,cur} e_i + K_{i,cur} \int e_i dt\right), e_i = I_{batt}^{ref} - I_{batt} \tag{12}$$

(Implementation respects dead-bands and limits to prevent overcharging)

5. **DC-link voltage regulation** – PID controller

Control Objective: Keep DC-link voltage VDC at reference V_{DC}^{ref}. This ensures stable power transfer to the VSC and manages the VSC charge and discharge.

PID control law: The PID control output u(t) (which is mapped to battery power reference or duty change) is:

$$u(t) = K_p e(t) + K_i \int_0^t e(t)dt + K_d \frac{de(t)}{dt} \tag{13}$$

Where the error is:

$$e(t) = V_{DC}^{ref} - V_{DC}\left(t\right) \tag{14}$$

In discrete implementation (sample time T_s):

$$u[k] = K_p e[k] + K_i T_s \sum_{j=0}^{k} e[j] + K_d \frac{e[k] - e[k-1]}{T_s} \tag{15}$$

The PID output u is mapped to the reference current or power for the BESS converter:

$$I_{batt}^{ref} = sat\left(u, I_{batt,min}, I_{bat,\,max}\right) \tag{16}$$

Including saturations and SoC constraints.

Anti Windup and Constraints: Because the integrator can wind up when it saturates, a conditional integration or back-calculation anti-windup is used:

$$\dot{I}_{int} = e(t) - \frac{1}{T_{aw}}\left(u_{sat} - u\right) \tag{17}$$

Where T_{aw} is the anti-wind time constant and u_{sat} is saturated control, this chapter states PID is used with online tuning; anti-windup is implied in practical implementation.

6. **Online PID Tuning:** Hybrid PSO-GWO optimizer

This chapter uses a hybrid PSO combined with GWO to tune PID gains (K_p, K_i, K_d) in real time for best DC-link regulation and PQ. I summarize the algorithm and equations are used.

PSO basics: Each particle i has position vector $x_i = \left[K_{p,i}, K_{i,i}, K_{d,i}\right]$ and velocity v_i at iteration t:

$$v_i(t+1) = w\,v_i(t) + c_1 r_1\left[pbest_i - x_i(t)\right] + c_2 r_2\left[gbest - x_i(t)\right] \tag{18}$$

$$x_i(t+1) = x_i(t) + v_i(t+1) \tag{19}$$

Where:

w= inertia weight
c_1,c_2 = cognitive/social coefficients
$r_1 r_2$~ Uniform (0,1)
$pbest_i$ = particle's best prior position
$gbest$ = global best

GWO basics: GWO models leadership (alpha, beta, delta wolves) and updates candidate positions via encircling behaviour. Position update general form:

$$D_\propto = \left|\vec{C}_1.\vec{X}_\propto - \vec{X}\right| \tag{20}$$

$$\vec{X}_1 = \vec{X}_\propto - \vec{A}_1 . D_\propto \tag{21}$$

And similarly, for $\vec{X}_\beta, \vec{X}_\delta$. Final new position is average:

$$\vec{X}\left(t+1\right) = \frac{\vec{X}_1 + \vec{X}_2 + \vec{X}_3}{3} \tag{22}$$

With adaptive coefficients $\vec{A}, \vec{C}$ depending on iteration. This chapter merges PSO velocity update with GWO leadership to balance exploration/exploitation (detailed code in the Simulink implementation)

Fitness Function: The optimizer minimizes a fitness function that encodes DC voltage error and PQ objectives. This chapter uses objectives:

$$J = \int_0^T \left(\alpha |V_{DC}^{ref} - V_{DC}\left(t\right)| + \beta THD_V(t) + \gamma T_{settling}(t)\right) dt \tag{23}$$

Or discretized:

$$J = \sum_{k=0}^{N} \left(w_1 |[k]| + w_2 THD[k] + w_3 1_{\{|e[k]|>e\}} \right) \tag{24}$$

Where weights w_i (or $\propto, \beta, \delta$) reflect prority (voltage tracking, THD reduction, fast transient). The hybrid PSO-GWO searches for $[K_p, K_i, K_d]$ minimizing J. this chapter also mentions a simple fitness:

$$P = |P_{in} - P_{out}| \tag{25}$$

For duty cycle optimization used in another block.

Real time implementation:

i. At runtime, candidate PID gains are applied for a short evaluation window and J is computed
ii. The optimizer updates candidates and eventually selects better gains.
iii. To keep stability, gain updated are rate-limited and bounded within safe range.

7. **VSC and APF control** - Instantaneous power theory (P-Q-R)

The AC side uses a VSC with integrated APF based on instantaneous power theory (P-Q-R) to cancel THD, provide reactive support and mitigate unbalance. Detailed steps follow.

Clarke Transformation (abc →αβ0): For three-phase signals v_a, v_b, v_c

$$\begin{bmatrix} v_\alpha \\ v_\beta \\ v_0 \end{bmatrix} = T_c \begin{bmatrix} v_a \\ v_b \\ v_c \end{bmatrix} = \begin{bmatrix} 1 & {}^{-1}/_{2} & {}^{-1}/_{2} \\ 0 & {}^{\sqrt{3}}/_{2} & {}^{-\sqrt{3}}/_{2} \\ {}^{1}/_{3} & {}^{1}/_{3} & {}^{1}/_{3} \end{bmatrix} \begin{bmatrix} v_a \\ v_b \\ v_c \end{bmatrix} \tag{26}$$

Similarly for current i_a, i_b, i_c. The zero-sequence v_0 captures neutral components.

Instantaneous (P-Q-R) decomposition: Instantaneous power in αβ:

$$\begin{bmatrix} p \\ q \end{bmatrix} = \begin{bmatrix} v_\alpha & v_\beta \\ -v_\beta & v_\alpha \end{bmatrix} \begin{bmatrix} i_\alpha \\ i_\beta \end{bmatrix} = M(v_\alpha, v_\beta) . i_{\alpha\beta} \tag{27}$$

The p and q are split into mean (average or DC) components $\bar{p}, \bar{q}$ and oscillatory components $\tilde{p}, \tilde{q}$:

$$p(t) = \bar{p} + \tilde{p}(t), q(t) = \bar{q} + \tilde{q}(t) \tag{28}$$

$\bar{p}$ = active power delivered to the load (fundamental)
$\tilde{p}$ = oscillatory term due to THD/unbalance
$\tilde{q}$ = contains harmonic reactive content

Reference compensating currents

Objective: VSC injects compensating current i_{comp} such that power supply current becomes sinusoidal and balanced. Using inverse of M,

$$\begin{bmatrix} i_\alpha^{ref} \\ i_\beta^{ref} \end{bmatrix} = M^{-1}\left(v_\alpha, v_\beta \right) \begin{bmatrix} p_{ref} \\ q_{ref} \end{bmatrix} \tag{29}$$

Choice of p_{ref}, q_{ref}:

To keep DC bus balance: $p_{ref} = -\tilde{p}$ (cancel oscillatory power into DC bus)

For reactive compensation: $q_{ref} = -q_{reharmonic}$ or to supply unity power set $\bar{q}_{supply} = 0$

Then inverse Clarke returns abc reference currents for APF. These references are tracked by current controllers on the VSC (inner loops)

Current control (dq-frame PI): After obtaining i_a^{ref}, i_b^{ref}, i_c^{ref}, they are often transformed into synchronous dq-frame via park transform (using PLL angle θ):

$$\begin{bmatrix} i_d \\ i_q \end{bmatrix} = T_{park}\left(\theta\right)\begin{bmatrix} i_\alpha \\ i_\beta \end{bmatrix} \tag{30}$$

PI regulates i_d *and* i_q to their references

$$v_d = K_{pd}\left(i_d^{ref} - i_d\right) + K_{id}\int\left(i_d^{ref} - i_d\right)dt + v_d^{ff} \tag{31}$$

$$v_q = K_{pq}\left(i_q^{ref} - i_q\right) + K_{id}\int\left(i_q^{ref} - i_q\right)dt + v_q^{ff} \tag{32}$$

Feed forward term v^{ff} compensates cross coupling: $v_d^{ff} = \omega L i_q, v_q^{ff} = -\omega L i_d$. The VSC voltage commands are transformed back to abc and modulated SPWM/PWM to produce switching signals.

8. Phase locked loop and synchronization

The PLL is used to obtain θ(t) (grid/ VSC reference angle). For islanded microgrid the PLL locks to the VSC output (or a chosen reference). Standard synchronous PLL:

$$v_\alpha, v_\beta \xrightarrow{Park\ (\theta)} v_d \tag{33}$$

$$e_{PLL} = v_q \tag{34}$$

$$w_{est}(t) = K_{p.pll}\, e_{PLL}(t) + K_{i.pll}\int e_{PLL}\, dt \tag{35}$$

$$\theta(t) = \int w_{est}(t)dt \tag{36}$$

In islanded mode one VSC may act grid-forming (sets θ), while others are grid-following (use PLL to track). This chapter uses PLL for APF and synchronization.

9. Fault handling and protection (P-Q-R in faults)

During Faults (voltage sag and swells), the APF plus VSC controllers act to:

a. Compensate reactive current to support PCC voltage: add iq component so that voltage drop is mitigated.
b. Inject harmonic/opposite frequency components to cancel distortions (via $-\tilde{p}$, $\overline{-q}$ selection).

c. Maintain DC link stability: adjust i_{batt}^{ref}To absorb or supply transient active power required by VSC compensation. The PID and optimizer respond to maintain VDC within bounds.

10. Duty cycle optimization

This chapter also describes a duty-cycle optimization routine (hybrid PSO/GWO/Chaotic mutation) to minimize power mismatch for converters (objective $P = |P_{in} - P_{out}|$). The essential steps are as follows:

a. Initialize duty cycles D and velocities v
b. For each candidate DC: measure input Pin (e.g. SPV or WECS) and Pout (load or DC bus demand). Compute fitness $P = |P_{in} - P_{out}|$
c. Update duty candidates using GWO leadership averaging in addition to PSO velocity and position updates (Eqn PSO). using chaotic mutation to break free from local minima.
d. Limit the obligation to [0, 9.98] and repeat until convergence. Output optimal duty D*

11. Implementation details and practical considerations.
    a. Sampling and timescale: The DC-bus PID operates more slowly (milliseconds to hundreds of milliseconds); the PSO–GWO optimization is rate-limited and evaluates over longer windows; the inner current loops operate at high sampling rates (kHz) for PWM.
    b. Saturations & safety: PID outputs and duty cycles are saturated; battery BMS enforces SoC/SoH limits; FCU trips on severe faults.
    c. Anti-windup & bump-less transfer: When the optimizer updates gain on-line, transitions are smoothed to avoid abrupt control jumps. This chapter emphasizes real-time tuning but implies safe update filters and bounds.
    d. Power quality metrics: THD calculated per IEEE standard (percent of harmonic magnitude to fundamental). The fitness includes THD to favor low-distortion tuning.
12. Step by step control sequence during operation
    a. Measurement: Measure $V_{DC}, i_{PV}, V_{pv}, i_{batt}, SoC, v_{a,b,c}, i_{a,b,c}$
    b. MPPT: update PV converter duty cycle D_{PV} to track V_{PV}*
    c. Rectify wind AC to DC: SEIG output rectified and filtered to DC bus. Wind control is largely passive (SEIG with excitation capacitors), with the DC bus reacting accordingly.
    d. DC-bus PID loop: compute $e = V_{DC}^{ref} - V_{DC}$; PID produces I_{batt}^{ref} or power reference P_{batt}^{ref}.

e. BDC current control: inner loop tracks I_{batt}^{ref}, Implementing Duty cycle Dbd. BMS enforces limits.
f. APF and VSC control: measure currents/voltages → Clarke → compute P-Q-R →determine compensation currents →current controller produce VSC voltage commands →SPWM produce gating.
g. Optimizer (PSO-GWO): periodically evaluate fitness (voltage error, THD, settling time) and update PID gains. Ensure smooth application for new gains.
h. Fault handling: if a fault is detected, APF adjusts to supply reactive/harmonic compensation, and the DC-link PID references the battery to absorb/supply transients; if severe, FCU isolates the faulted section.

13. Notes, assumptions, and simplifications
    a. The SEIG model is simplified with capacitor reactive support and a rectifier; advanced rotor-speed control for maximizing wind capture is not the chapter's focus.
    b. PID is chosen for DC voltage regulation for simplicity and reliability; PSO–GWO provides adaptive tuning to handle nonlinearity and time-varying conditions.
    c. APF uses P–Q–R instantaneous theory, which is effective under unbalanced and non-sinusoidal conditions typical in remote microgrids.
14. How the controller improves performance
    a. DC stability: PID keeps V_{dc} steady, preventing VSC faults and allowing the AC side to be reliable. Online tuning reduces transient overshoot and speeds settling.
    b. Power quality: APF reduces THD and corrects unbalance; fitness includes THD, so the optimizer biases toward PQ improvement.
    c. Battery health/economy: BMS enforces SoC/SoH, and the optimizer avoids aggressive cycling by penalizing unnecessary battery use in the fitness.

FAULT CONDITION IN THE SYSTEM

In a standalone hybrid system, faults can occur on the AC side (e.g., line-to-line fault, single-line-to-ground fault, load imbalance, short circuit), DC side (e.g., DC bus short, converter failure), or component faults (e.g., PV, wind, or BESS malfunction).

Fault Handling Mechanism

1. Fault Detection
    a. The FCU continuously monitors system voltage, current, and frequency.

b. Abnormal conditions (overvoltage/undervoltage, overcurrent, short circuit) are flagged immediately.
c. Fast detection prevents system collapse by triggering corrective actions.

2. APF with P–Q–R Theory
   a. When fault-induced distortions occur, the APF dynamically compensates by injecting opposite harmonic and reactive currents.
   b. Using instantaneous power theory (p–q–r), the APF separates oscillatory power components ($\tilde{p}$, $\tilde{q}$) from average components ($\bar{p}$, $\bar{q}$) and generates reference compensating currents:

$$\begin{bmatrix} i_\alpha^{ref} \\ i_\beta^{ref} \end{bmatrix} = M^{-1} \begin{bmatrix} p_{ref} \\ q_{ref} \end{bmatrix}, \tag{37}$$

Where: $p_{ref} = -\tilde{p}$, $q_{ref} = -q_{ref}$

   c. This ensures the supply current remains sinusoidal, even under distorted/unbalanced voltages.

3. DC link voltage regulation

   a. Faults often cause a sudden active power imbalance (generation ≠ demand).
   b. The PID controller (with PSO-GWO tuned gains) quickly regulates DC-link voltage (VDC) by commanding the BESS to charge/discharge.
   c. It stabilizes the VSC control operation and gives protection to the DC bus.

$$e(t) = V_{DC}^{ref} - V_{DC}(t), \tag{38}$$

$$u(t) = K_p e(t) + K_i \int_0^t e(t)dt + K_d \frac{de(t)}{dt} \tag{39}$$

   d. The optimized PID controller minimizes the overshoot and settling time, which is important for managing the fault-induced transients.

4. Battery support During Faults

   a. The BESS serves as a quick reaction buffer:

      i. During a voltage sag, it provides active power to maintain the load.
      ii. During a Voltage swell, it provides from overvoltage on the DC bus by absorbing extra energy.

b. To ensure safe operation, the battery management system manages the SoC and thermal limitations.

5. Fault Clearance Unit (FCU)

    a. If the small fault occurs, then the system is continuously stabilized by the APF and hybrid optimized tuned PID-controlled BESS.
    b. If the heavy fault occurs, then it separates the faulty part (using circuit breakers or protection relays) from the system to save the healthy components.
    c. By using FCU, the system self-recover and re-energize the load.

SIMULATION RESULTS

The HRES having SPV-WECS-BESS system is simulated on MATLAB/Simulink software. The performance of the system is evaluated under various operating conditions to check the PQ enhancement. According to the analysis, such an arrangement responds quickly and effectively to sudden changes in solar irradiance and wind speed. The system can maintain stable voltage and frequency profiles despite the intermittent nature of renewable energy sources. The coordinated management of the BESS, which acts as a compensation mechanism to mitigate the negative impacts of renewable fluctuations, is primarily responsible for this stability.

The system is also capable of reducing THD and reducing the voltage fluctuations under unbalanced load conditions, and providing an improved power supply. When the system is exposed to simultaneous variations in load demand and renewable generation, the robustness of the proposed control technique becomes even more evident. In such situations, the BESS effectively compensates for temporary imbalances to maintain smooth and continuous operation. All these results demonstrate the superior power-quality (PQ) performance, operational resilience, and dynamic stability of the SPV–WECS–BESS-based hybrid system. Consequently, this technique provides a practical and sustainable solution for supplying electricity in rural and remote areas.

Operational Behaviour under Wind Speed Variations

Figure 2. Operational Behaviour under Wind Speed Variations

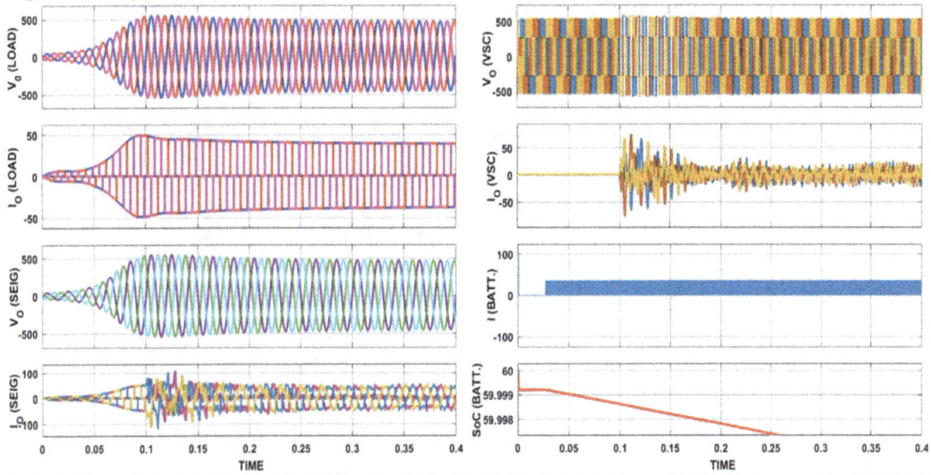

Figure 2 represents how the system performed when the wind speed fluctuates from 19 m/s to 12 m/s at t = 3 seconds under 800 W/m^2 solar irradiation, and the system continues to supply a constant non-linear load. When the wind-driven SEIG experiences a drop in wind speed, the energy generated by the WECS is also decreased. The optimized PID controller now ensures that there are no changes in frequency, load voltage, and BESS voltage. The BESS compensates for this deficit by supplying the necessary power to the load through its Voltage Source Converter (VSC). When the wind subsystem is properly synchronized, the VSC output voltage and current steadily increase and stabilize into sinusoidal waveforms. While the battery current shows sporadic spikes as it alternates between charging and discharging to compensate for the varying wind power, the battery voltage stays steady at about 610 V. When wind input drops, the SoC gradually declines, suggesting that the battery provides extra power. The efficient control of power flow during wind fluctuations is demonstrated by the SEIG output voltage's gradual ascent to a steady-state sinusoidal waveform and the currents' corresponding trend.

Operational Behaviour under Solar Irradiance Variations

Figure 3 shows how the system responds when solar irradiance changes from 800 W/m2 to 1400 W/m2 at t=3 seconds, while the wind speed stays constant at 19 m/s and the system continues to supply a constant nonlinear load. The VSC current increases in response to an increase in irradiance, indicating a greater contribution of solar power. In order to maintain a steady supply, the BESS balances power flow between the sources and the load during this operation. After startup, the three-phase load voltages steadily increase, but because of the imbalance caused by irradiance, their amplitudes are uneven. The impact of varying solar input is confirmed by corresponding currents that exhibit ripples and harmonic distortion.

Figure 3. Operational Behaviour under Solar Irradiance Variations

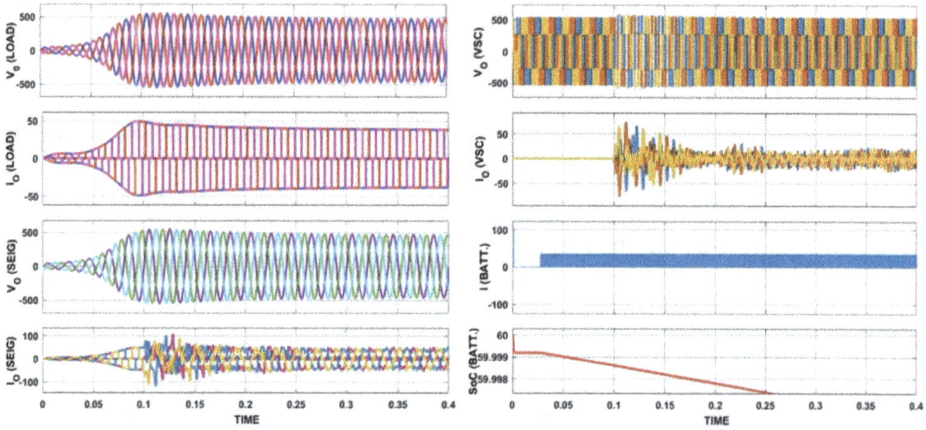

The current waveforms continue to be unbalanced, but the VSC switching pattern is unchanged. The battery delivers around 40 A of current after 0.05 seconds, keeps the DC link voltage at about 630 V, and drains gradually during the event. The SoC just slightly decreases (~0.015%), demonstrating the battery's steady support function in sustaining a steady supply in spite of solar variability. In addition to reducing THD, the controller keeps the hybrid system's voltage and frequency constant. The controller raises the battery charging current to compensate for the minor increase in VSC current caused by the higher solar power. As a result, the BESS stores the extra power produced by the SPV array by switching from discharging to charging mode.

Operational Response during Unbalanced Load Conditions

The response under unbalanced loading, achieved by removing one phase of the load at t=022 sec to 0.35 sec while wind speed (19 m/s) and solar irradiance (800 W/m^2) are constant, is presented in Figure 4. The load voltages stabilize quickly into sinusoidal waveforms with only slight distortion. Although the VSC and wind subsystem voltages remain relatively steady, the load current shows a sudden increase and oscillations around 0.1 s, corresponding to the load disturbance. Despite this, the battery voltage stays around 630 V with minimal ripple, and the battery discharges steadily at ~40 A to maintain power balance. The SoC decreases marginally, confirming that the battery compensates for the unbalanced condition. The system successfully mitigates voltage asymmetry and suppresses THD, thereby preserving PQ under the unbalanced scenario. The VSC current waveform increases during this period because the battery begins charging as the load current and wind-driven SEIG current decrease. The PWM controller quickly stabilizes into a steady switching pattern with constant amplitude. When the load is reduced, the VSC current rises and then stabilizes after 0.3s. The PID controller effectively regulates the system voltage, maintaining stable voltage profiles even under unbalanced load conditions.

Figure 4. Operational Behaviour during Unbalanced Load Conditions

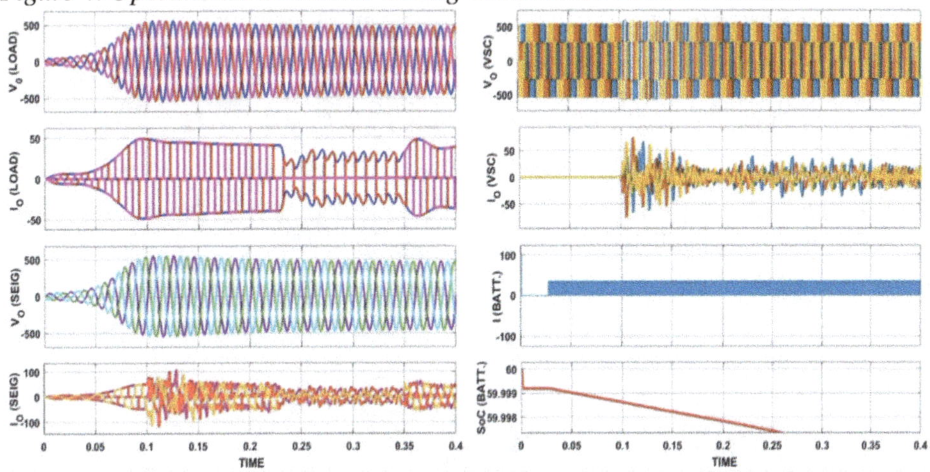

Operational Behaviour with Fluctuating Renewable Generations and Load

Figure 5. demonstrates the dynamic performance when both renewable inputs and load vary simultaneously (wind speed changes from 19 m/s to 12 m/s at t=3 sec, solar irradiance changes from 800 W/m2 to 1400 W/m2 at t=3 sec, and one phase is removed from the load at 0.25 sec to 0.35 sec). While load currents are stable and barely fluctuate, the VSC maintains sinusoidal voltages with constant amplitude. With the exception of a brief, transient disruption during the source/load variation, the wind subsystem voltage and current climb gradually during starting before returning to steady, sinusoidal operation. In order to maintain smooth functioning, the battery continues to perform a crucial stabilizing role by controlling the DC link by absorbing or supplying power as needed. Effective BMS is confirmed by the SoC staying within safe bounds. The controller successfully regulates the SEIG output voltage, ensuring stable operation when load changes and changes in the conditions of renewable sources (solar irradiation and wind speed). When one phase is removed from the load, the load current is reduced, but the VSC with APF maintains system stability. During this period, the battery charging current rises, allowing the extra power to be stored. Later, the BESS discharges the stored energy and gives a supply and fulfills the required power, keeping the load stable when the wind speed drops. Even under unbalanced phase conditions and simultaneous disturbances in renewable input and load, the system maintains voltage and frequency stability, reduces THD, and provides an uninterrupted supply to the load.

Figure 5. Operational Behaviour with Fluctuating Renewable Generation and Load

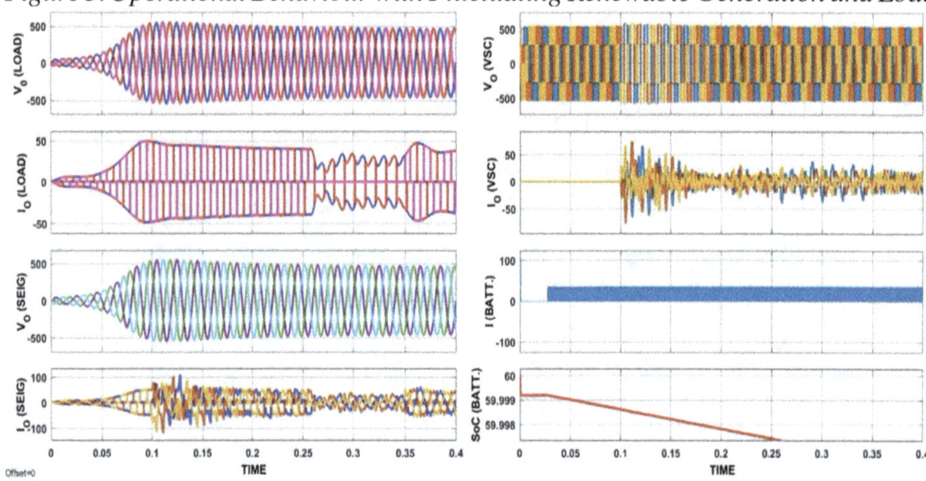

KEY RESULTS AND DISCUSSIONS

1. Key Results
    a. DC-link voltage stability: The optimized PSO–GWO tuned PID controller kept the DC-link voltage with low overshoot/undershoot and fast settling time despite variations in solar irradiation or changes in wind speed.
    b. Battery Performance: The BESS ensured a continuous supply to the load. During charging (when less load and high-power generation) and discharging (when large load and less power generation), the BMS maintained the SoC within safe limits.

c. PQ enhancement: Under nonlinear and dynamic loads, the APF effectively reduced THD and corrected imbalances.
d. Load Voltage and Frequency: when the sudden changes in RESs and Load, the VSC with APF maintained a stable and sinusoidal load voltage and frequency.
e. Controller Superiority: In contrast to traditional PID, the PSO-GWO-tuned PID evaluated
    i. Faster transient response.
    ii. Better PQ.
    iii. Lower THD.
    iv. Lower steady-state error.
    v. Improved robustness under multiple simultaneous disturbances.

2. Discussion
    a. This system's ability to self-balance power under varying renewable generation and load conditions makes it highly reliable for off-grid and remote applications.
    b. The BESS served as a crucial buffer to prevent both under-voltage (during low generation/high load) and over-voltage (during low load/high generation).
    c. The APF based on P–Q–R theory ensures reactive power balancing and harmonic compensation, which are two common issues in standalone systems with nonlinear loads.
    d. By continuously adjusting PID gains, the hybrid PSO–GWO algorithm outperforms conventional tuning, strengthening the system's resistance to uncertainties and fast dynamics.
    e. Overall, the results confirm that the proposed control technique delivers reliable, high-quality power with improved PQ, making it a strong option for remote electrification and islanded microgrids.

Table 1. Summary of Dynamic Performance under Different Test Conditions

Test Condition	Voltage Response	Current Response	Battery Behaviour	SoC Trend
Wind Speed Variation	VSC and SEIG voltages build up to stable sinusoidal waveforms	Three-phase currents oscillate sinusoidally after startup	Voltage ≈ 610 V; charging/discharging alternates with spikes	Gradual decline as battery supports load during wind dips
Solar Irradiance Variation	Load voltages steady but unequal due to irradiance variation	Phase currents show ripples and distortion under unbalance	Voltage ≈ 630 V; steady discharge at ~40 A	Slight drop (~0.015%)
Unbalanced Load (One Phase Removed)	Load voltages stabilize quickly; sinusoidal but slightly distorted	Sudden rise in load current during disturbance	Voltage ≈ 630 V; steady ~40 A discharge	Small decrease (~0.015%)
RES + Load Change (One Phase Removed)	VSC voltages maintain consistent amplitude	Currents remain stable with negligible fluctuation	Battery stabilizes DC link by absorbing/supplying energy	SoC regulated within limits

Table 1 represents the responses of SPV–WECS–BESS system under different operating conditions. According to the results, the optimized PID controller improves system PQ, ensuring system stability, decreasing current distortion and managing battery SoC.

CONCLUSION

In this study, a hybrid SPV–WECS–BESS power supply system is simulated on MATLAB/Simulink. This provides stable, reliable, and improved power PQ and provides a continuous supply to the remote and island areas. A BESS interfaced through a BDC, the SPV is connected to the incremental-based MPPT to extract maximum energy from the solar, and a wind-driven SEIG is used to ensure a continuous supply under variable operating conditions, like a change in wind speed or solar irradiance. Using a hybrid PSO–GWO optimization algorithm, the PID controller successfully maintained DC-link voltage stability, reduced overshoot, improved settling time, and enhanced overall system performance. By reducing THD to 0.5% and maintain load imbalance, the APF based on P–Q–R theory also optimized PQ. The BMS monitored SoC/SoH to ensure safe operation of the BESS, while the FCU provides system protection under faulty conditions.

The simulation results confirm that the implemented control strategy continuously matches and fulfills the generation to demand while providing a stable, sinusoidal load voltage during varying operating conditions, such as changes in solar irradiance, changes in wind speed, changes in load, changes in RESs along with load simultaneously, and system faults. The system also delivers an uninterrupted, continuous power supply. Overall, these results demonstrate that adopting the SPV-WECS-BESS configuration with PSO-GWO-based PID controller and Active Power Filter (APF) compensation is both feasible and sustainable for supplying electricity to off-grid and remote locations. This method significantly enhances PQ and ensures reliable, robust system operation.

REFERENCES

Aghdam, M. M., Li, L., & Zhu, J. (2020). Comprehensive study of finite control set model predictive control algorithms for power converter control in microgrids. *IET Smart Grid*, *3*(1), 1–10. DOI: 10.1049/iet-stg.2018.0237

Al-Shetwi, A. Q., Atawi, I. E., El-Hameed, M. A., & Abuelrub, A. (2025). Digital Twin Technology for Renewable Energy, Smart Grids, Energy Storage and Vehicle-to-Grid Integration: Advancements, Applications, Key Players, Challenges and Future Perspectives in Modernising Sustainable Grids. *IET Smart Grid*, *70026*(1), e70026. Advance online publication. DOI: 10.1049/stg2.70026

Bakare, M. S., Abdulkarim, A., Shuaibu, A. N., & Muhamad, M. M. (2024). Energy management controllers: Strategies, coordination, and applications. *Energy Informatics*, *7*(1), 57. DOI: 10.1186/s42162-024-00357-9

Fotis, G., Gupta, S., Varshney, T., Dhar, S. L., & Gulzar, M. M. (Eds.). (2025). *Modern Computing Technologies for EV Efficiency and Sustainable Energy Integration*. IGI global., DOI: 10.4018/979-8-3373-2382-4

Giri, A. K., Arya, S. R., Maurya, R., & Babu, B. C. (2019). VCO-less PLL control-based voltage-source converter for power quality improvement in distributed generation system. *IET Electric Power Applications*, *13*(8), 1114–1124. DOI: 10.1049/iet-epa.2018.5827

Giri, A. K., Arya, S. R., Maurya, R., & Chittibabu, B. (2020). Control of VSC for enhancement of power quality in off-grid distributed power generation. *IET Renewable Power Generation*, *14*(5), 771–778. DOI: 10.1049/iet-rpg.2019.0497

Giri, A. K., Arya, S. R., Maurya, R., & Ray, P. (2020). Performance of Control Algorithms in Wind-Based Distributed Generation System with Power Quality Features: A Review. In Ray, P., & Biswal, M. (Eds.), *Microgrid: Operation, Control, Monitoring and Protection. Lecture Notes in Electrical Engineering* (Vol. 625). Springer., DOI: 10.1007/978-981-15-1781-5_3

Guanghua, L., Soomar, A. M., Syed, H. H. S., Shaikh, S., & Musznicki, P. Maximum power point tracking strategies for solar PV systems: A review of current methods and future innovations, Results in Engineering, Volume 28, 2025, 107227, ISSN 2590-1230, DOI: 10.1016/j.rineng.2025.107227

Gupta, S. (2025). *The Future of Machine Learning with Automation in Engineering*. Nova Science Publishers., DOI: 10.52305/YFIX2832

Gupta, S., & Sharma, A. (2018). Global Scenario of Solar Photovoltaic (SPV) Materials. In Bhattacharyya, S., Gandhi, T., Sharma, K., & Dutta, P. (Eds.), *Advanced Computational and Communication Paradigms. Lecture Notes in Electrical Engineering* (Vol. 475). Springer., DOI: 10.1007/978-981-10-8240-5_14

Gutiérrez-Torres, D. A., Ramìrez, J. M., García, J. M. L., Sundriyal, N., & Bhatt, A. "Implementation of a Direct AC-AC Converter Based Wind Energy Conversion System," 2024 International Conference on Computer, Electronics, Electrical Engineering & their Applications (IC2E3), Srinagar Garhwal, Uttarakhand, India, 2024, pp. 1-5, DOI: 10.1109/IC2E362166.2024.10827016

Kanojia, S. S., Suthar, B. N., & Giri, A. K. (2025). Hybrid Optimization Approach for Optimal Location of Electric Vehicle Charging Station (EVCS) in Distribution Network Considering Voltage Stability. *Smart Grids and Sustainable Energy, 10*(2), 36. DOI: 10.1007/s40866-025-00264-2

Khan, M., Wu, W., & Li, L. (2024). Grid-forming control for inverter-based resources in power systems: A review on its operation, system stability, and prospective. *IET Renewable Power Generation, 18*(6), 887–907. DOI: 10.1049/rpg2.12991

Kundu, S., Giri, A. K., Kadiyan, S., & Singh, S. (2025). Power management and adaptive control method for a standalone distributed generation system under fault, DC-offset, and dynamic environments. *International Journal of Ambient Energy, 46*(1), 2452245. Advance online publication. DOI: 10.1080/01430750.2025.2452245

Kundu, S., Giri, A. K., Singh, M., & Kadiyan, S. (2025). Adaptive control strategy for isolated renewable energy-based generation system with intermittency. *Electrical Engineering, 107*(7), 8773–8785. DOI: 10.1007/s00202-024-02454-0

Kundu, S., Giri, A. K., Singh, M., & Singh, S. (2025). Effective Multiport Battery Charging Facility, Operation and Control in Hybrid Isolated Microgrid System. In Giri, A. K., & Singh, M. (Eds.), *Electric Vehicle Charging Infrastructures and its Challenges. Studies in Infrastructure and Control.* Springer., DOI: 10.1007/978-981-96-0361-9_11

Kundu, S., Singh, M., & Giri, A. K. (2021). Design and Control of a Standalone Wind–Solar System with SEIG Feeding Linear/Nonlinear Loads. In: Mohapatro, S., Kimball, J. (eds) Proceedings of Symposium on Power Electronic and Renewable Energy Systems Control. Lecture Notes in Electrical Engineering, vol 616. Springer, Singapore. DOI: 10.1007/978-981-16-1978-6_28

Kundu, S., Singh, M., & Giri, A. K. (2022). Control algorithm for coordinated operation of wind-solar microgrid standalone generation system. *Energy Sources. Part A, Recovery, Utilization, and Environmental Effects, 44*(4), 10024–10044. DOI: 10.1080/15567036.2022.2143943

Kundu, S., Singh, M., & Giri, A. K. (2023). Implementation of variable gain controller based improved phase locked loop approach to enhance power quality in autonomous microgrid. *International Journal of Numerical Modelling, 36*(5), e3082. DOI: 10.1002/jnm.3082

Kundu, S., Singh, M., & Giri, A. K. (2023). Adaptive Control Approach-Based Isolated Microgrid System with Alleviating Power Quality Problems. *Electric Power Components and Systems, 52*(7), 1219–1234. DOI: 10.1080/15325008.2023.2239222

Kundu, S., Singh, M., & Giri, A. K. (2024). SPV-wind-BES-based islanded electrical supply system for remote applications with power quality enhancement. *Electrical Engineering, 106*(1), 279–294. DOI: 10.1007/s00202-023-01979-0

Kundu, S., Singh, M., & Giri, A. K. Synchronization and control of WECS-SPV-BSS-based distributed generation system using ICCF-PLL control approach, Electric Power Systems Research, Volume 226, 2024, 109919, ISSN 0378-7796, DOI: 10.1016/j.epsr.2023.109919

Kundu, S., Singh, M., Giri, A. K., Kadiyan, S., Chittora, P., & Shailly, D. (2025). Robust and fast control approach for islanded microgrid system and EV charging station applications. *Electrical Engineering, 107*(7), 8385–8396. DOI: 10.1007/s00202-024-02291-1

Nishad, D. K., Tiwari, A. N., Khalid, S., & Gupta, S. (2024). Power quality solutions for rail transport using AI-based unified power quality conditioners. *Discover Applied Sciences, 6*(12), 651. DOI: 10.1007/s42452-024-06372-5

Parmar, D. B., & Giri, A. K. (2025). Integration of field-oriented and steady-state linear Kalman filter control in PMSG-based grid-connected system for improving voltage control and power balance operation. *Electrical Engineering, 107*(7), 8803–8817. DOI: 10.1007/s00202-024-02456-y

Rao, M. S., Saraswathy, S., Neela, V., Arun, M. R., Gupta, S., & Taqui, S. N. (2023, November). Efficient Model Training in Decentralized Systems with Federated Learning. In *2023 International Conference on Communication, Security and Artificial Intelligence (ICCSAI)* (pp. 900-904). IEEE. DOI: 10.1109/ICCSAI59793.2023.10421469

Ratnakaran, R., Rajagopalan, G. B., & Fathima, A. (2023). Artificial ecosystem optimized neural network controlled unified power quality conditioner for microgrid application. *Energy Informatics, 6*(1), 45. DOI: 10.1186/s42162-023-00301-3

Rekioua, D., Mokrani, Z., Kakouche, K., Oubelaid, A., Rekioua, T., Alhazmi, M., Ali, E., Bajaj, M., Dost Mohammadi, S. A., & Ghoneim, S. S. M. (2024). Coordinated power management strategy for reliable hybridization of multi-source systems using hybrid MPPT algorithms. *Scientific Reports*, *14*(1), 10267. DOI: 10.1038/s41598-024-60116-4 PMID: 38704399

Prasad, P. S., & Parimi, A. M. (2025). Design of reinforcement learning based robust μ-synthesis controller for single phase grid-connected VSI. e-Prime-Advances in Electrical Engineering. *Electronics and Energy*, *12*, 100978.

Sharma, R. K., Saran, V. B., Kanaujia, S., & Gupta, S. (2024, May). Evaluation of mppt controller performance using fuzzy logic design with wireless sensor node. In *2024 International Conference on Smart Systems for applications in Electrical Sciences (ICSSES)* (pp. 1-6). IEEE. DOI: 10.1109/ICSSES62373.2024.10561365

Shukla, R. D., Tripathi, R. K., & Gupta, S. (2010, November). Power electronics applications in wind energy conversion system: A review. In *2010 International Conference on Power, Control and Embedded Systems* (pp. 1-6). IEEE. DOI: 10.1109/ICPCES.2010.5698663

Singh, M., & Giri, A. K. (2022, June). Implementation of in-loop filter based DSC-FLL control approach for power quality improvement in isolated microgrid. In 2022 IEEE 31st international symposium on industrial electronics (ISIE) (pp. 516-521). IEEE.

Srikakolapu, J., Arya, S. R., Giri, A. K., Kundu, S., & Singh, M. (2024). DSTATCOM components modeling and control. custom power devices for efficient distributed energy systems, 21-42.

Sundriyal, N., Ramirez, J., & Corrochano, B. (2023). A comparison of geometric algebra and harmonic domain for linear circuit analysis. *Mathematics for Applications*, *12*(1), 115–127. DOI: 10.13164/ma.2023.08

Talaei Khoei, T., Ould Slimane, H., & Kaabouch, N. (2023). Deep learning: Systematic review, models, challenges, and research directions. *Neural Computing & Applications*, *35*(31), 23103–23124. DOI: 10.1007/s00521-023-08957-4

Tyagi, M., Meena, D. C., & Kundu, S. "CBPF-FLL-Based Operational Coordination of SPV, BESS, and Wind-Driven SEIG Hybrid Systems for Distributed Generation," 2025 IEEE International Conference on Computer, Electronics, Electrical Engineering & their Applications (IC2E3), Srinagar Garhwal, India, 2025, pp. 1-6, DOI: 10.1109/IC2E365635.2025.11166884

Chapter 6
Accelerating Clean and Affordable Energy Solutions Through Artificial Intelligence

Khushi Kamble
https://orcid.org/0009-0006-1846-8164
Vishwakarma Institute of Information Technology, India

Omkar Ankush Biramane
Vishwakarma Institute of Information Technology, India

Gitanjali Shinde
https://orcid.org/0000-0001-6623-8722
Vishwakarma Institute of Technology, India

Grishma Bobhate
https://orcid.org/0009-0007-7351-8723
Vishwakarma Institute of Technology, India

Sonal Fatangare
https://orcid.org/0009-0005-1185-8699
Vishwakarma Institute of Technology, India

Vijayshri Khedkar
https://orcid.org/0000-0001-6704-4823
Symbiosis Institute of Technology, Pune, India

ABSTRACT

The fact that the world is dynamic and sustainable energy is currently in high demand cannot be ignored. As industrialization grows, there is a spillover effect on urbanization and the aging population and system. The old ways of doing things can no longer be supported.Like all of the other primary disruptors, artificial intelligence has the capacity to solve the cost, reliability, and sustainability issues in the energy sector, and as we will soon see, it has started to do so. The more efficient production, capture, and distribution of renewable energies will be achieved using sophisticated computational forecasting, adaptive optimization algorithms, and self-

DOI: 10.4018/979-8-3373-7041-5.ch006

learning controls. In other words, it will play a direct leading role in achieving the United Nations' Sustainable Development Objective 7 in ensuring universal access to affordable, reliable, sustainable, and modern energy.Coming from a systems perspective, the way we can have at the relationships between AI and clean energy systems is something we're now looking at in this chapter.

INTRODUCTION

Energy is central to all aspects of socioeconomic development. However, the current energy system is in crisis, affected by a multidimensional trilemma of energy security, costs, and environmental sustainability (International Energy Agency [IEA], 2023).IEA's World Energy Outlook which was conducted in year 2024 estimates that global electricity consumption will increase by about six percent above the previously projected levels by 2035, which is the equivalent of adding Japan's annual electricity consumption, every year, to the demand side of the energy balance. This tremendous increase in demand worsens an ongoing existential crisis regarding climate change. The Intergovernmental Panel on Climate Change has stated that continuing our reliance on fossil energy will create irreversible increases in global temperatures of greater than 1.5 °C giving rise to ecological instability on the planet.

There has been a increase in uptake of energy related to renewable sources rapidly in the last few years, but it is still not enough. The International Renewable Energy Agency indicated that while there was a record added global renewable capacity of 582 GW in 2024, this does not come close to the tripling that the COP28 UAE Consensus has established as necessary by 2030. The global climate targets necessitate annual added capacity above 1 100 GW starting in 2025, and to accomplish all of this will require unprecedented coordination, investment, and overall system intelligence.

Equally pressing, is the equity principle within SDG 7. The world's population has progressed in electrification but there still are approximately 666 million people, mostly in sub-Saharan Africa and South Asia, without electricity access (World Health Organization [WHO], 2025). Additionally, around 1.8 billion people still lack clean fuels for cooking, using dangerous, polluting biomass fuels, which can adversely affect health and contribute to a less equitable distribution of gender equity in the home (United Nations Development Programme [UNDP], 2025). To narrow down this gap and enable the usage of it to people; ones would have to deploy a hardware and an intelligent controller that can adapt itself quickly in decentralized and distributed energy systems.

The 21st century energy challenge is no more about what more capacity can we add; it's a complexity problem and one that requires advanced coordination from distributed resources. Grids must integrate variable renewables sources, deal with bidirectional flows from prosumers, and respond in real time to changes in consumption patterns (Denholm & Hand, 2023). Traditional deterministic control systems designed for modifications to centralized one-way processes are becoming outdated. AI serves as the cognitive platform necessary for this change as it learns from waves of data generated constantly, predicts the structure of systems, and performs corrective actions (Zhou et al., 2024).

AI is, therefore, the foundational technology for ***Intelligence Energy Transitions***. Algorithms used by ai systems can learn from streams of data that are high-volume, high-velocity, and high-variety, and can identify generalized patterns that conventional statistical models do not (Kumar & Banerjee, 2022). In a new reality where smart meters, phasor-measurement units, and satellite-based sensors accumulate terabytes of operational data every minute, AI is providing the analytical engine to process raw data inputs into decision-ready outcomes (Gupta, 2025).

BACKGROUND

The global energy trilemma can be described in three interdependent dimensions: "energy security, energy equity, and environmental sustainability" (World Energy Council [WEC], 2024). In most cases, policymakers have proven to focus on one dimension without paying attention to the other two. (Sovacool et al., 2023) provides a meaningful perspective on this trend. Fossil-based stability could not ensure climate policy of the country, while the implementation of renewables lacked proper digital infrastructure. Thus, both options paved the way for imbalances, instability, and availability issues. The implementation of AI in the described cases could provide a harmonizing element. The explosion in volume of the data in contemporary power systems can be considered as both an opportunity and a challenge. Millions of these sensors are being scattered throughout smart-grid infrastructures - relaying multi-dimensional data such as voltage, frequency, temperature and weather conditions—amounting to a deluge of data which overwhelms human interpretive capabilities. Data avails itself to be turned into intelligence via AI, which converts all this 'data corn' into smartness through the use of algorithms that can improve themselves by learning from data streams (Thandlam et al., 2024; Gupta & Tripathi, 2014). Especially machine-learning models like Support Vector Machines, Random Forests, and Gradient-Boosted Trees have achieved a much higher skill in the prognosis of renewable output in comparison to classical statistical tools. Deep-learning architectures, e.g., CNNs and LSTMs, are specified to improve spatiotemporal modeling for weather-driven generation (Awasthi et al., 2024; Rao et al., 2023). Reinforcement Learning (RL) introduces adaptive control, allowing the system to self-optimize in case some uncertainty is presented, which is critical for the real time grid management (Sutton & Barto, 2018). Collectively, these AI paradigms redefine decision-making from reactive to predictive and ultimately to prescriptive intelligence (Shankar et al., 2023; Sinha et al., 2024). The following sections detail how such capabilities reshape renewable forecasting, energy-storage optimization, and grid-level operations across the world.

AI-ENHANCED FORECASTING FOR RENEWABLE ENERGY ADDITION[1]

Knowing the true potential of solar and wind is crucial for the stability of the power grid, when forecasting renewable energy production. These energies have a variable source and are hugely dependent on the weather and the location where they are being generated. AI in this field has provided a whole new way of looking at energy production, moving away from purely statistical methods and toward a much more sophisticated use of real-time data to predict the outputs of wind and solar farms (Lago et al., 2021).

Taming Intermittency Through Predictive Analytics

Unlike conventional models such as ARIMA (Autoregressive Integrated Moving Average) and persistence forecasting, which rely on the historical averages, incomprehensive for frequent changes in irradiance or wind speed, machine-learning models consume high-resolution meteorological data, satellite images, and sensor data in real-time to predict generation highly accurately (Marugán et al., 2018).

For example, hybrid architectures that combine numerical-weather-prediction (NWP) outputs with LSTM-based correction models have achieved up to 25 percent reductions in mean absolute percentage error (MAPE) for short-term solar forecasting. Similarly, CNN–LSTM hybrid networks improve wind-power prediction accuracy by integrating spatial wind-field information with temporal learning.

Figure 1. Workflow of an AI-powered wind-energy forecasting system

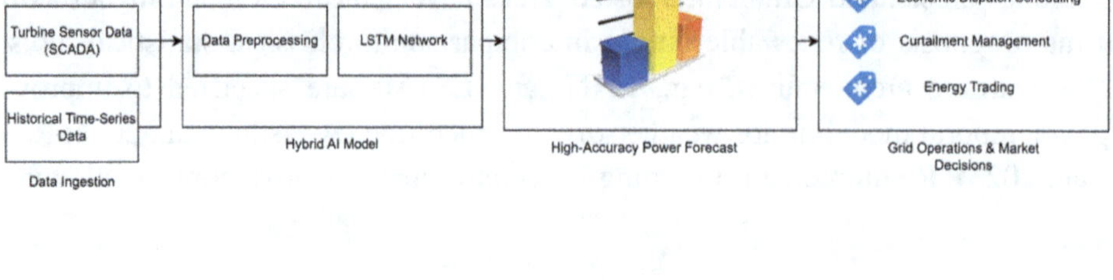

The process of an AI-based wind-energy forecasting system is shown in Figure 1. This process is illustrated in the diagram below which depicts the processing of raw meteorological and sensor data through machine learning (ML) models such as CNNs and LSTMs to estimate short as well as medium term renewable generation. The intelligent system includes on-line feedback loops to improve predictions and minimize forecast errors while providing dispatch schedules based upon data driven stability of the electric grid.

AI not only enhances short-term forecasts (minutes to hours) but also supports medium-term planning (days to weeks) critical for market bidding and operational scheduling. Improved forecasts directly translate to reduced reserve requirements and lower operating costs, underscoring their economic significance (Gupta & Tripathi, 2015; Sharma et al., 2024).

Case Study: Denmark's Wind-Energy Leadership

Denmark offers a textbook case of AI's practical advantages. With wind power being used for almost 50 percent of its power production, the Danish Energy Agency required short-term energy forecasting systems, which were developed and designed using long short-term memory models and other AI models, and this was done in collaboration with Aalborg University (Aalborg University Energy Research, 2024). These models improved forecast accuracy by approximately 20 percent relative to classical models, enabling dynamic dispatch scheduling that reduced reliance on carbon-intensive peaker plants.

Explainable AI (XAI) frameworks such as SHAPley Additive explanations (SHAP) were later integrated to explain model outputs, helping operators understand the use of meteorological variables on forecast uncertainty (Sinha et al., 2024; Sinha & Patel et al., 2024). As Denmark's experience shows, achieving large scale adoption of AI links on making its decision making processes transparent, which is fundamental to gaining public trust.

Table 1. Comparison of Standard and AI Based Forecasting Methods

Work	Standard Methods (e.g., ARIMA, Persistence)	AI-Based Methods (e.g., LSTM, Hybrid Models)
Data Handling	• Primarily handles linear, single variable time series data (Sinha et al., 2023). •Performs poorly with large, complex datasets (Tyagi et al., 2024).	• Processes non-linear, high-dimensional, and heterogeneous data (e.g., meteorological, SCADA, satellite imagery).
Model Complexity	•Based on pre-defined statistical relationships and linear assumptions (Sudha et al., 2024).	• Learns complex, non-linear patterns and interactions directly from data without explicit programming (Sharma al., 2019).
Adaptability	•Static models that require manual recalibration. Poor adaptation to changing system dynamics or weather patterns.	• Capable of real-time learning and adaptation as new data becomes available, continuously improving accuracy.
Accuracy	• Moderate accuracy, particularly for very short-term (intra-hour) horizons. • Performance degrades significantly with longer horizons and high variability.	• High accuracy across short- to medium-term horizons. • Significantly reduces error metrics like Mean Absolute Percentage Error (MAPE).
Forecast Horizon	Primarily effective for short-term (minutes to hours) forecasting.	Effective for both short-term (0-72 hours) and increasingly for medium-to-long-term forecasting.
Computational Cost	Low computational requirements.	High computational cost, especially for training deep learning models, requiring significant processing power (e.g., GPUs).
Handling Spatiotemporal Complexity	Limited ability to incorporate spatial data (e.g., conditions at neighboring farms) or complex temporal patterns.	Excels at capturing complex spatiotemporal dependencies (e.g., using CNN-LSTM models for regional forecasts).

Table 1 shows how old school ways of guessing what will happen compare to using AI for things like solar and wind energy. Normal methods, such as ARIMA, use basic past information, but AI can use things like deep learning to deal with tricky, complex info. The table shows AI is better at being right, changing with the times, and guessing what will happen in specific locations. But it does take more computer power.

OPTIMIZING ENERGY STORAGE: THE LINCHPIN OF GRID STABILITY

It is well known that renewable energy sources will be replacing conventional generation soon, power systems will face a major challenge in controlling the interaction between intermittent generators and loads. It turns out that energy storage has also emerged as the key component of a resilient and adaptable grid that balances unpredictable demand with unpredictable generation. As grids become cyber physical ecosystems, energy storage management becomes a predictive, proactive, and self-organizing. Intelligent storage not only stabilizes voltage and frequency variations but also improves economic dispatch, advances decarbonization objectives, and prolongs asset life.

The Importance of Storing Energy in Renewable Systems

Proper operation of battery management systems (BMS) depends on accurately assessing crucial internal battery states, including its RUL, SOH and SOC . Frequency regulation, voltage backing, and reserves are examples of grid support that ensures reliability. (Chen et al., 2022). Although lithium-ion batteries are the most common ESS on the market due to their high effectiveness in the round trip. Alternative chemistries including sodium-ion, redox-flow, and hydrogen-based storage, are growing in popularity (Tarascon & Armand, 2021).

With increased renewable penetration, short-duration batteries are insufficient to match day-to-day or seasonal variation. Through AI-driven optimization, hybrid storage portfolios can be prioritized through the dynamic task allocation between short-duration batteries, pumped-hydro facilities, and green-hydrogen converts (Palizban & Kauhaniemi, 2020). Predictive control algorithms for controlling the dynamic task allocation use reinforcement learning to identify optimal charge–discharge strategies and consider costs, degradation, and availability (Liu et al., 2023).

A study by Stem Inc. (2021) demonstrated that AI-based battery-management platforms can extend operational life by 30 percent and reduce energy costs by 20 percent compared with rule-based systems. Ultimately, these outcomes show that the key to economically viable storage lies in intelligent systems, not in simply adding more hardware capacity.

Intelligent Battery Management and Lifecycle Optimization

Internal states of a battery, including SOC, SOH, and RUL, are critical for the BMS to work. However, the existing models have fixed and predetermined parameters, which leads to failure on external slopes. It is possible to predict these states using support vector, Gaussian regression, and artificial neural networks (ANNs) based on machine learning. These regression models have the ability to learn maps between nonlinear external measurements such as current, voltage, and temperature.

Deep recurrent networks further capture long-term degradation patterns. Long Short-Term Memory (LSTM) architectures have accomplished SOC-prediction mistakes of less than 2% when trained on high-frequency operational data (Heidari et al., 2022). Reinforcement-learning controllers complement these estimates. They utilize long-term estimates to adjust the charging schedules in accordance with the demand-active full-coverage conditions to minimize degradation without sacrificing peak demand (Rao & Chowdhury, 2023).

The AI-enabled battery lifecycle management loop is depicted in Figure 2. It illustrates how machine learning models are used for state estimation (SOC, SOH, RUL) processing the data continuously received from sensors, and how charge–discharge strategies are optimized with reinforcement-learning algorithms. The resulting feedback loop guarantees predictive maintenance, long battery life, and cost-effective connection to the grid. Beyond lithium-ion, AI accelerates research on emerging chemistries through materials-informatics approaches that screen thousands of compounds via deep-learning–assisted simulations (Kirklin et al., 2023). By shortening discovery cycles, such methods contribute to the affordability dimension of SDG 7.

Case Study: Tesla Opticaster Ecosystem

Opticaster software by Tesla is an example of integrating AI in real-time within distributed-energy management (Tesla Inc., 2022). It is integrated in the Powerwall and Megapack product lines to predict household load demand, solar power generation, and real-time billing fluctuations to chart optimal charge and discharge cycles. Auto-updating policies from streaming data with reinforcement-learning agents autonomously match user saving goals to grid backup goals (Fathi & Rahman, 2022).

During California's 2022 heatwave, aggregated Powerwall fleets supplied 16 MW of virtual-power-plant capacity, averting localized blackouts (California ISO, 2023). This case evidences the scalability of AI-orchestrated storage as a decentralized stability asset rather than a passive backup.

Figure 2. AI-enhanced battery lifecycle-management loop illustrating data ingestion from sensors, ML-based state estimation, RL-based optimization, control dispatch, and feedback adaptation.

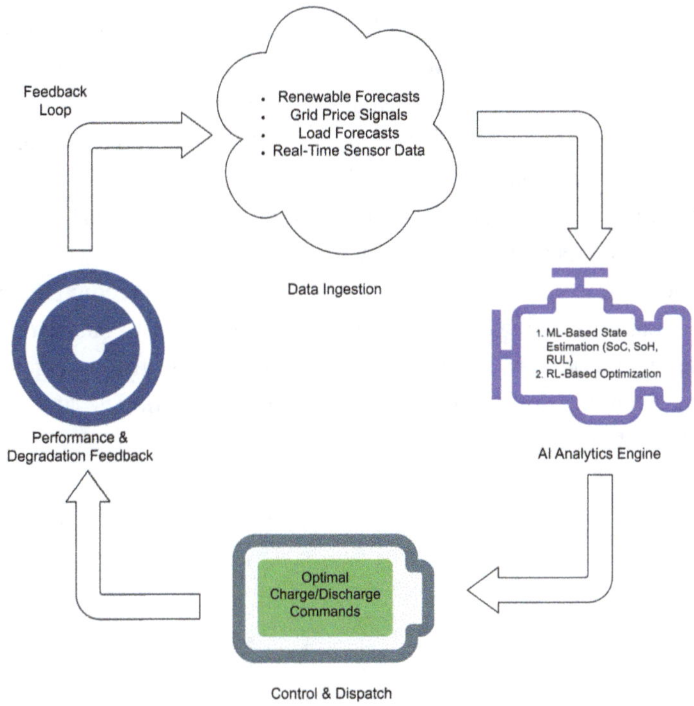

THE DAWN OF INTELLIGENT GRIDS: AI IN SMART GRIDS AND MICROGRIDS

Due to the increasing availability of renewable energy and storage solutions, traditional power grids are turning into smarter, data driven systems. AI is making this happen, letting grids sense, learn, and react instantly. When we add cyber smarts to our physical power setup, we get smart grids that can predict problems and fix themselves, balance energy use on the fly, and handle all sorts of renewable energy sources while keeping power quality high. The tech-oriented move to micro grids is a major step towards a more reliable, effective, and eco-friendly energy future.

Smart Grids as Cyber-Physical Ecosystems

Smart grids are a cyber physical system of information technology, communication networks, and electric power infrastructure. Utilizing two-way data transmission for adaptative load balancing, fault detection, and demand response, AI is its decision-making nucleus. AI converts unstructured and non-stationary sensor data into ready-to-make decisions (Aly & Shalaby, 2023).

Machine-learning models trained on smart-meter data predict consumption patterns with sub-hour granularity (Zhou & Brown, 2022). Predictive analytics makes it possible for utilities to introduce real-time pricing schemes that flatten peak loads, which can reduce costs and carbon intensity.

Convolutional networks applied to thermal-imaging data in predictive maintenance models are capable of recognizing transformer faults prior to catastrophic failure, lowering unplanned outages by 40% (Li et al., 2021).Natural-language–processing also scans maintenance logs to match free-form textual fault descriptions to sensor anomalies (Sun et al., 2023).

Figure 3. AI-Driven Microgrid Architecture for Rural Electrification

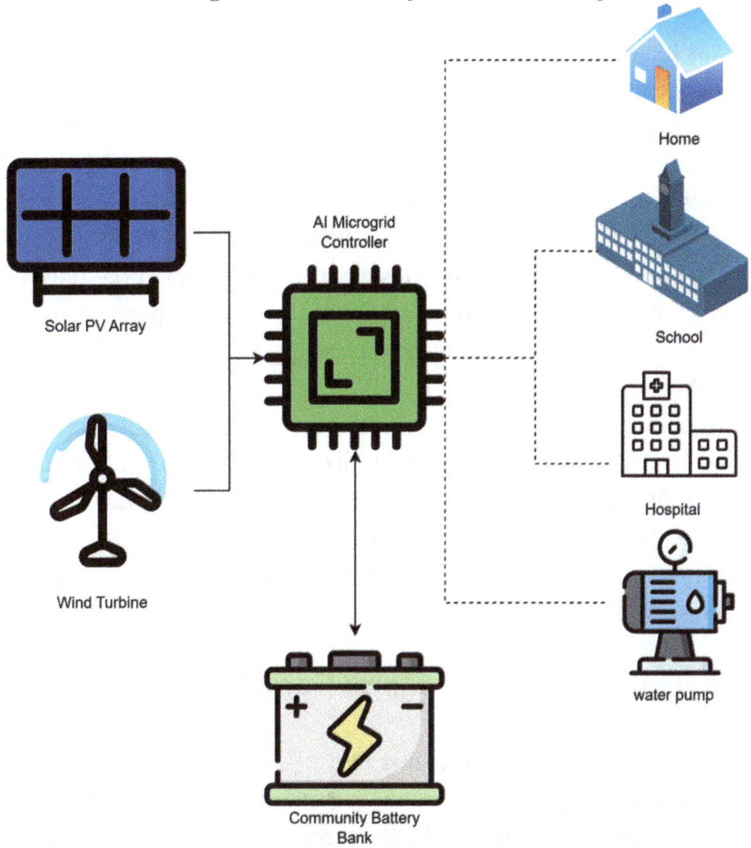

AI-Assisted Fault Detection and Self-Healing Mechanisms

AI enables self-healing capabilities that automatically cut off the fault and re-route power. Learning in the order of milliseconds to map the topological structure of distribution networks, graph-neural-network-based models learn to localize and classify disturbances (Zhang et al., 2023). When optimized through conventional heuristics, these models execute automated FLISR Fault Location, Isolation, and Service Restoration on his own hits.

Reinforcement-learning controllers reinforce FLISR and work by running millions of simulations to posted action switches from which the optimal recovery sequence is derived. When deployed at national-grid scale, such systems cut restoration time by 70 percent and operational expenditure by 15 percent.

Microgrids and AI-Enabled Energy Autonomy

Defined as localized networks that can operate islanded, microgrids play a vital role in resilience and rural electrification (Banerjee et al., 2022). Microgrid operations are made possible by the dynamic exchange between solar panels, storage systems, variable community loads, and these automated units. AI controllers and multi-agent reinforcement learning technologies allow decision-making to be divided between many agents, each of which runs a different subsystem (Wang et al., 2023).

In India, AI-operated solar microgrids have powered over 200 villages, reducing diesel-generation dependence by 60 percent and tariffs by 52 percent. In Nigeria and Kenya, AI-powered controllers balance renewable generation and demand, ensuring 24-hour reliability in previously unelectrified regions.

The AI-empowered microgrid design for the rural electrification is shown in Figure 3. The model incorporates decentralized solar generation, energy storage, and AI-based controllers which employ multi-agent reinforcement learning. This facilitates real-time coordination, flexible load management, and islanding functionality of energy supply to ensure reliable power while enhancing access at a lower cost in under-served areas.

Comparative Urban Deployments: Barcelona and Singapore

Barcelona's AI-integrated grid forms part of its broader Smart-City initiative. IoT sensors monitor lighting, waste management, and EV charging. In the energy sector, predictive analytics reduced blackout frequency by 35 percent and cut public lighting energy use by 30 percent. The municipality's partnership with private organizations employs AI-based Energy Data Models to plan EV-charging expansion.

Singapore's "Grid 2.0" initiative represents AI's role in mature, high density systems. The National Grid Digital Twin, a high-precision virtual replica of the power network, enables scenario testing and risk-aware investment. These contrasting examples highlight context sensitivity in developing regions. AI enables access and autonomy, whereas in advanced economies it optimizes complexity and resilience (Sovacool & Grubler, 2023).

Table 2. Feature Comparison of Traditional, Smart, and Microgrids

	Synchronous E-Learning	**Asynchronous E-Learning**	**Microgrid**
Architecture	Centralized generation, hierarchical structure.	Centralized and distributed generation, networked structure.	Decentralized, localized, can operate independently.
Data & Communication Flow	One-way (utility to consumer).	Two-way, real-time communication.	Two-way, hyper-local, real-time communication.
Control System	Manual, electromechanical controls, slow response.	Automated, digital controls, real-time monitoring.	Autonomous, self-healing, adaptive control.
Key Technology	Large power plants, transformers.	IoT sensors, smart meters, big data analytics.	Distributed Energy Resources (DERs), energy storage, advanced inverters.
Role of AI	Minimal used in limited back-office forecasting.	Central to system-wide optimization, monitoring, and predictive maintenance.	Essential for autonomous control, self-sufficiency, and real-time resource management.
Resilience	Low, a single large failure can cause widespread cascading outages.	Moderate, improved fault detection and faster restoration.	High, capable of "islanding" from the main grid to maintain local power during a large-scale outage.

Table 2 shows the main differences between old-fashioned grids, smart grids, and microgrids. It shows how we've moved away from energy systems that are all in one place and only go one way, to networks that are spread out and can take care of themselves. Smart grids use AI to guess when things need fixing and to work better, while microgrids get energy independence and can bounce back quickly because they use real-time control and manage resources locally.

ARTIFICIAL INTELLIGENCE IN SUSTAINABLE MOBILITY AND ELECTRIC VEHICLE ECOSYSTEMS

Transportation is the next area of sustainable change as power networks get smarter and more interconnected. First appeared on The Driven According to Bridie Schmidt, artificial intelligence (AI) is revolutionizing how electric vehicles (EVs) interact with energy networks by enabling autonomous control, intelligent charging, and the provision of a virtual power plant from inside the vehicle. AI is what enables systemic efficiency, transforming automobiles into mobile assets within the broader grid system, going beyond simple electrification. A key component of decarbonized societies, where transportation actively promotes the stability and sustainability of the grid rather than placing a burden on it, is the convergence of mobility, data analytics, and clean energy.

The Convergence of AI and Electrified Transport

Transport is the main reason for almost 25% of global carbon dioxide emissions. Using electricity in this sector is thus a component to reach SDG 7 and the Paris Agreement. However, widespread electric-vehicle (EV) use brings new grid-integration challenges such as uncoordinated charging and peak-demand proliferation, the latter of which can be destabilizing to networks (Zhang & Mi, 2022).

Artificial Intelligence offers a dual-benefit framework: it improves both energy consumption within vehicles and power which flows between vehicles and the grid. Artificial intelligence (AI) based route planning algorithms can minimize the utilization of battery energy by 15–25% with adaptive navigation, combined with traffic, weather, and topographic information (Hu et al., 2023). In vehicles, machine learning models are used to predict driver behavior and set regenerative-braking, climate-control levels more efficiently (Lin & Huang, 2022). At the system level, RL agents are also employed to coordinate charging sessions and allocate load temporally and spatially for peak shaving (Wang & Li, 2023).

V2G and Bidirectional Exchange of Energy

During periods of high energy demand, V2G systems empower EVs to discharge electricity into the grid. This concept transforms millions of parked vehicles into a distributed storage network (Kempton & Tomić, 2021). AI improves this exchange by learning price signals, battery degradation rates, and grid frequency deviations to determine when discharging is beneficial (Tan et al., 2022).

For example, in 2023, Nissan's pilot project in Japan used a deep reinforcement-learning controller to manage 4000 EVs running during evening peaks to deliver 30 MW (Fotis et al., 2025; Shukla et al., 2010). The above system reduced regional peak load by 12 percent and kept high-discharging levels manageable by adapting to depth discharge rules. Other projects in the Netherlands also used AI-based blockchain contracts to validate energy trades to enable funding and ensure fair payment (van der Meer et al., 2024).

Figure 4.AI-based EV-ecosystem architecture showing bidirectional data flow between vehicle sensors, cloud-analytics servers, and grid-management systems.

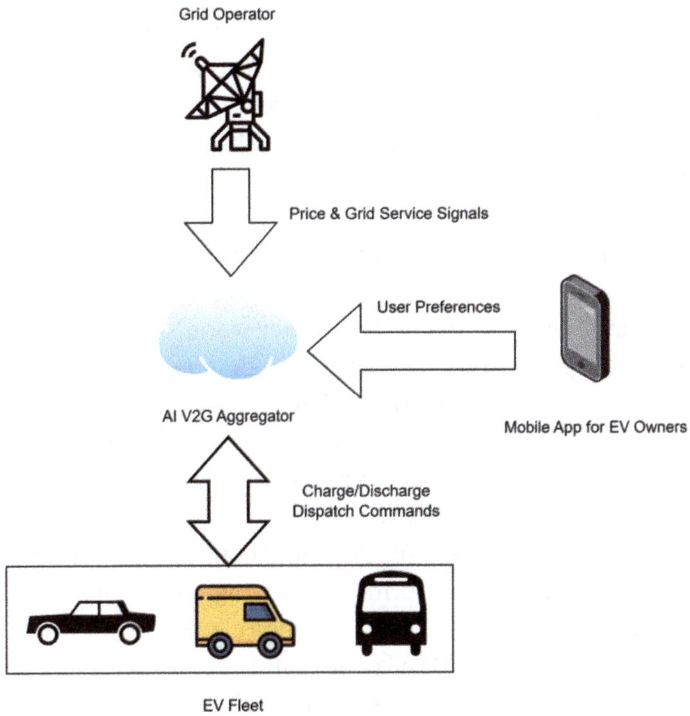

The architecture of AI-based electric vehicle (EV) ecosystem is presented in Figure 4. It shows how vehicles, cloud analytics, and grid-management systems are exchanging data and energy with each other. Charging is coordinated by AI algorithms, demand can even be predicted, and V2G exchanges can form, making the electric vehicle (EV) a flexible grid asset for stability and renewable integration.

AI in Autonomous and Shared Mobility

AI's contribution to energy efficiency arises not only from electrification but also from the level of energy efficiency increases due to shared and autonomous mobility. Autonomous/Driverless Vehicle Control Systems utilize perception networks and planning modules that run on CNNs based on simulated environments and reinforcement learning, along with perceived-workload estimation and management. They reduce the energy consumption of stop-and-go driving by up to 20 percent compared to human drivers (Fagnant & Kockelman, 2022).

Shared-mobility algorithms dynamically match riders and vehicles, maximizing occupancy rates while minimizing total fleet energy use (Gao et al., 2023). In Singapore, Grab's AI Dispatch Optimizer reduced idle mileage by 14 percent and CO_2 emissions by 9 percent. When combined with renewable-powered charging hubs, such platforms demonstrate how AI could reshape whole urban-transport ecosystems to carbon neutrality (IEA, 2024).

Case Study: China's AI-Optimized EV Infrastructure

More than 20 million units in China operate the world's largest EV fleet by 2025. This process is managed by the State Grid Corporation, which encompasses the E-Charge Brain, which is an AI platform that serves as an alternative to 500,000 charging piles across the nation. The predictive scheduling system has led to a 30 percent decrease in charging-station congestion as well as a five-minute average waiting time: "up to 5 minutes" (SGCC, 2025).

Such a scenario is feasible since the system leverages predictions for renewable-generation availability to manage priorities in charging, such as making solar generation when it's at its best. In this way, transportation electrification serves as a bridge for other renewable applications. 'The combination of AI and clean-energy supply chains fosters the SDG 7 principles of energy for all at an accessible, reliable and sustainable cost' (United Nations, 2024).

ECONOMIC AND POLICY DIMENSIONS OF AI-DRIVEN ENERGY SYSTEMS

Artificial Intelligence is not simply a technical innovation; it's a new approach to how we think about money, markets, and running things in the energy world. It's not just about coming up with new rules; these smart systems could help us produce, ship, and use energy better, which could save us money. Getting more out of our resources, fixing things before they break, and making our grids work together better are all big parts of why AI grids make sense from a money perspective. Policies should also make sure everything is open, fair, and secure from cyber threats. All of this shows how AI turns into real-world sustainability, where new ideas match up with being inclusive and meeting global energy goals.

Economic Impacts and Cost–Benefit Analyses

The power sector can deliver high upfront CAPEX for digitalization, but most importantly, it offers long-term social and economic gains. The impact of AI-based optimization could reduce power generation costs by $80 billion annually in 2030. India, for example, saved 18 billion on its power gensupported demand response programs in 2024.

The marginal cost of integrating renewables falls steeply when predictive power allows to mitigate curtailment and imbalance penalties. It also creates new models for business, such as energy as a service. For instance, EaaS can provide predictive maintenance and dynamic pricing (EY, 2023).

Figure 5. Block chain-Powered P2P platforms for trading energy

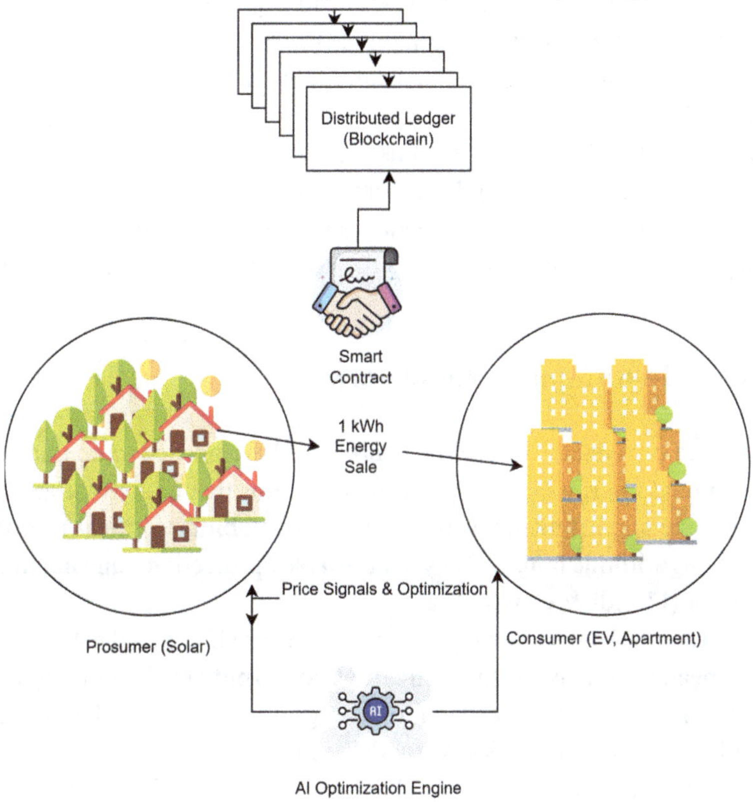

Figure 5 presents a diagram the block chain based P2P energy exchange system. The figure illustrates the way AI agents and distributed ledgers are coupled to automate transactions, ensuring ownership verifications and energy exchange optimization in decentralized market scenarios. With smart contracts and anticipatory analytics, it attempts to guarantee transparency, fairness and grid stability for community systems related to energy.

Blockchain and Trading of Energy using Peer to Peer Model

The energy governance is decentralized using AI with block chain technology. Smart contracts make it possible to register energy transactions transparently and also simplify peer to peer trading by removing the central intermediaries (Andoni et al., 2019). Local demand and production are predicted by AI agents, which in turn make trading decisions on trade quantity and price.

For instance, the Brooklyn Microgrid in the US used a combination of blockchain and AI: solar-powered households traded excess energy themselves in an autonomous fashion with feedback on real-world experimentation. Reinforcement-learning agents automatically tune their bids in the auction environment and respond to situation changes to maximize social welfare while grid stability is maintained (Pop et al., 2023).

In Germany, the Power Ledger consortium achieved an increase of 40% in transaction settlement efficiency by incorporating AI supported forecasting into its trading algorithm. These developments suggest that the intersection of AI with DLTs can generate rights to share and access an essential resource in a manner that removes inequality.

Policy and Governance Frameworks

Governance ecosystems trained on enabling AI are needed to be implemented. The International Energy (IEA, 2024) Agency claims that national strategies must seek to combine a spirit of innovation with accountability. Critical infrastructure must require algorithmic transparency, data-privacy protections, and human-in-the-loop oversight (Floridi & Cowls, 2021).

Energy-system AI has been categorized as a "high-risk" under the AI act proposed by the European Union, as a result, it requires conformity evaluations and continuous control. The National AI Mission for Energy in India established a regulatory sandbox allowing utilities to pilot AI models in manageable settings.

Cross-border collaboration is equally important. The Global AI for Energy Partnership (GAIEP) is a joint initiative by the IEA and UNDP which uses technology transfer and capacity building in emerging economies (UNDP, 2025). Such governance ensures that AI's benefits in affordability and sustainability extend inclusively across nations.

CYBERSECURITY, ETHICAL, AND ENVIRONMENTAL CONSIDERATIONS

Artificial intelligence, a key part of today's energy setup, creates new risks, duties, and environmental problems. The same digital links that make things run better can also leave systems open to attacks, tough ethical choices, and expensive computing. Making sure AI-run energy systems are safe, fair, and eco-friendly needs a mix of tech smarts, human watchfulness, clear actions, and being aware of the planet. In the end, whether AI will support or contradict the long-term goal of a fair and sustainable energy transition depends on the triad of cybersecurity, ethics, and environmental stewardship.

Cyber-Physical Vulnerabilities in AI-Enabled Grids

However, the digitalization that allows intelligent energy systems to exist also broadens their attack surface. Smart grids require extensive information exchange among sensors, cloud servers, and edge devices, allowing any disruption to spread from cyberspace and extend into physical outages. In May 2021, a coordinated ransomware incident against a U.S. gas-pipeline operator brought gas supplies to a halt throughout the East Coast (U.S. Department of Energy [DOE], 2022).

AI operates as a defender and as a target. Adversarial inputs can manipulate the machine-learning classifiers, regulating false detection or prediction of faults (Biggio & Roli, 2018). In response to them, modern research highlights Adversarial Machine Learning countermeasures, which are adversarial training, differential privacy, and blockchain-confirmed data provenance. Intrusion-detection systems based on reinforcement learning change with these intrusions. They reduce the rate of false alarms by 35 per cent compared to static rule engines (Liu & Chen, 2023).

The ISO/IEC 27019:2023 standard now extends industrial-control-system cybersecurity specifically to smart-grid contexts, advocating encryption, access segregation, and continuous monitoring. National utilities adopting these protocols—such as France's RTE and South Korea's KEPCO—report significant reductions in critical-incident frequency (World Economic Forum [WEF], 2024).

Ethical AI and Algorithmic Accountability

Despite efficiency promises, the societal legitimacy of AI depends on how it is run with ethics. Biases in training data could result in discriminatory energy-pricing algorithms, or they can also develop unequal load-shedding priorities (Jobin, Ienca, & Vayena, 2019). AI ethics in the energy field contradicts fairness to include energy justice, which makes sure isolated communities gain equal benefits from technological transition (Sovacool & Dworkin, 2022).

The Key guidance emphasizing the human rights, sustainability, and cultural diversity is recommended by UNESCO. Embedding these principles within national energy-AI strategies aligns technological progress with SDG 7's justice ethos (UNESCO, 2022).

Environmental Costs of AI Operations

AI is a key driver of decarbonization, but its computational load is by no means trivial. Training large deep-learning models can emit several hundred metric tons of CO_2 equivalent (Strubell, Ganesh, & McCallum, 2019). To put this into context, the training of a single natural-language model in 2022 consumed the energy of about 120 US households for one year (Henderson et al., 2020).

Researchers suggest a way to rectify this is through green AI, which is defined as focusing on algorithmic efficiency and renewable-powered data centers (Patterson et al., 2022). Specifically, Google's carbon-intelligent scheduling technology reduces data-center emissions by 45% by dynamically transferring workloads to regions with an excess of renewable generation. Lifecycle analyses must account for hardware manufacturing and e-waste. Post-deployment, circular-economy methods, such as component reuse, modular design, and recycling, must be imposed in AI to guarantee that digital sustainability develops energy sustainability.

GLOBAL CASE STUDIES AND COMPARATIVE INSIGHTS[7]

Europe: Coordinated Intelligence in Renewable Integration

For instance, the European Union has an *AI4Energy* program that connects forecasting, storage, and demand-response algorithms deployed across 14 Member States (European Commission, 2024).In Germany, Siemens' *MindSphere* data platform interconnects 40,000 industrial assets, cutting energy intensity by 12 percent (Siemens AG, 2024). The Nordics countries have cross-border data-sharing agreements, helps harness hydropower to outbalance wind surpluses, leading to near-zero curtailment (Nordic Energy Research, 2023).

These case studies reveal the effectiveness of regional interoperability standards and open data protocols. The EU's *Energy Data Space* provides common frameworks ensuring that AI models trained in one country can be safely deployed elsewhere without retraining (European Commission, 2024).

Asia: Rapid Digitization and Decentralization

Already today, the National Smart Grid Mission in India operates AI-based outage-prediction systems in 100 cities; between 2022 and 2025, predictive maintenance reduced technical losses by 18 percent. The State Grid Corporation of China's Energy Brain processes 30 PB of data every day, and reinforcement learning dynamically dispatches renewables, reducing fires of CO_2 per year 2025 (SGCC, 2025).

Southeast Asia's microgrid expansion, especially in the Philippines and Indonesia, which uses AI-enabled load forecasting to stabilize island systems. The Asian Development Bank (2024) attributes a 20 percent improvement in reliability indices to digital-AI interventions.

Africa and Latin America: Leapfrogging Through Decentralized AI

African nations are also more likely to adopt AI-in-a-box microgrid controllers that are more or less connection dependent. For banks of solar-AI microgrids serving 200,000 clients, Nigeria's Rensource Energy has delivered tariffs 40% less than the price of diesel alternatives (Rensource, 2024). n Kenya, AI-driven PAYG Solar platforms, supported by forecast credit scoring, make solar generation more accessible to low-income recipients.

Finally, Brazil's squeeze also uses deep-learning models to pinpoint the perfect locations for solar farms that do not involve deforestation, to expand over conservation. Chile's AI-coordinated energy auctions improved investment transparency, cutting average tariffs by 8 percent.

North America and Oceania: Innovation Hubs

The United States supports AI-energy startups with substantial investment through the ARPA-E Open program in 2025, putting a strong directive on explainability and cybersecurity (DOE, 2025). Canada's Hydro-Québec profits from AI-driven hydrological forecasting for scheduling reservoir dispatch, adding 2.5 percent per year generation without new plants.

Table 3. Summary of AI Applications in Energy from Global Case Studies

Domain	Geographic Context	AI Technique(s) Employed	Key Challenge Addressed	Demonstrated Impact/ Outcome
Renewable Forecasting	Denmark	Short-Term Energy Forecasting (STEF), LSTM, Explainable AI (XAI)	High wind penetration variability	Enabled ~50% of electricity from wind; 20% increase in forecast accuracy; reduced curtailment and improved grid stability.
Microgrids & Access	India & Sub-Saharan Africa	Reinforcement Learning (RL), Optimization Models	Lack of grid access in remote areas	Electrification of hundreds of rural villages; optimized use of solar/battery resources; improved energy reliability and affordability.
Smart Grid Optimization	Barcelona, Spain	IoT Analytics, Machine Learning	Urban energy inefficiency and grid instability	35% reduction in blackouts; 30% energy savings from smart lighting; optimized EV charging infrastructure planning.
Advanced Grid Management	Singapore	Digital Twin, AI Simulation, VPP Management	Future grid complexity and resilience	Real-time simulation for risk-aware planning; predictive maintenance; sandbox for testing new grid technologies like VPPs.
Energy Storage	Global (Residential)	Machine Learning, Optimization Algorithms (e.g., Opticaster)	Maximizing economic value of distributed storage	Automated energy bill reduction for Tesla Powerwall users; optimized solar self-consumption and time-of-use arbitrage.
Sustainable Mobility (V2G)	UK, Australia, Italy	Aggregation Algorithms, Reinforcement Learning	Grid instability from peak demand; integrating EVs as grid assets	Stabilized local grids during peak hours; generated revenue/ savings for EV owners (up to £725/year in UK trial); enhanced resilience during storms.
Decentralized Markets	Brooklyn, USA	Blockchain, Smart Contracts, AI Optimization	Centralized market structure; lack of local energy trading	Created a pioneering P2P platform for local solar energy trading; demonstrated a model for community-based energy resilience.

Table 3 presents a summary of global case studies and where they have demonstrated the use of AI. It synthesizes real-world experience from Denmark, India, Singapore, and beyond —from forecasting to storage to mobility to decentralized trading. The table highlights an astounding variety of AI methods, including reinforcement learning and digital twins, being used to improve energy access, efficiency , and sustainability around the globe.

In Australia, CSIRO's Grid Edge Intelligence Lab prototypes multi-agent simulations for scenario testing of renewable integration, with experimental results demonstrating 15 percent lower curtailment under AI-optimal control. These advanced economies illustrate how research-to-policy feedback loops accelerate scaling.

CONCLUSION

To conclude, while far from a remedy, Artificial Intelligence is an essential enabler of the decarbonised, democratized energy systems that the SDGs and SDG7 in particular envision. AI can really change how we handle energy, moving from just reacting to problems to actually planning ahead with smart predictions and simulations. This chapter explained that when AI is used ethically and openly, it can make things more efficient and fair for everyone. We looked at examples from Denmark to Kenya to Colombia and saw how smart systems can lower emissions, reduce costs, and give more people access to energy. But as tech gets smarter, our policies need to keep up. We need to make sure that when we come up with stuff, we also think about privacy, fairness, and the environment. In the future, research should focus on including AI ethics, green computing, and how well our society can adapt to tech changes. To make all this work, countries need to team up on open data standards and training. That way, we won't create a new divide as we shift to cleaner energy. The goal is for machines to not just get better at saving energy, but to help make society better for everyone.

REFERENCES

Aalborg University Energy Research. (2024). *Short-term energy forecasting in wind-dominated grids*. Aalborg University Press.

Aly, H., & Shalaby, M. (2023). Smart-grid data analytics using deep learning. *Electric Power Systems Research, 217*, 109004.

Andoni, M., Robu, V., Flynn, D., Abram, S., Jenkins, D., Geach, D., & Peacock, A. (2019). Blockchain technology in the energy sector. *Renewable & Sustainable Energy Reviews, 100*, 143–174. DOI: 10.1016/j.rser.2018.10.014

Awasthi, B., Gupta, S., Kumar, S., & Tripathi, A. (2024, January). Business management and planning using queuing model with heterogeneous customers for electrical applications. *In 2024 2nd International Conference on Intelligent Data Communication Technologies and Internet of Things (IDCIoT)* (pp. 279-283). IEEE. DOI: 10.1109/IDCIoT59759.2024.10467673

Banerjee, S., Kumar, A., & Singh, P. (2022). AI-enabled rural microgrids in India: A pathway to sustainable electrification. *Renewable Energy, 190*, 1102–1115.

Biggio, B., & Roli, F. (2018). Wild patterns: Ten years after the rise of adversarial machine learning. *Pattern Recognition, 84*, 317–331. DOI: 10.1016/j.patcog.2018.07.023

California, I. S. O. (2023). *Virtual power-plant demonstration summary*. California Independent System Operator.

Chen, Y., Zhou, J., & Wang, Z. (2022). Battery energy-storage optimization using reinforcement learning. *Applied Energy, 324*, 119708.

Denholm, P., & Hand, M. (2023). Balancing variable generation with energy storage. National Renewable Energy Laboratory., NREL/TP-6A20-75829.

DOE. (2022). *Pipeline cybersecurity report*. U.S. Department of Energy.

DOE. (2025). *ARPA-E Open 2025 program overview*. U.S. Department of Energy.

European Commission. (2024). *AI4Energy: Integrating artificial intelligence into Europe's power sector*. Directorate-General for Energy.

EY. (2023). *Energy-as-a-service market landscape 2023*. Ernst & Young.

Fagnant, D. J., & Kockelman, K. M. (2022). Preparing a nation for autonomous vehicles: Opportunities, barriers and policy recommendations. *Transportation Research Part A, Policy and Practice, 132*, 1–16.

Fathi, A., & Rahman, S. (2022). Real-time AI control for distributed storage networks. *IEEE Transactions on Smart Grid, 13*(2), 876–888.

Floridi, L., & Cowls, J. (2021). A unified framework of five principles for AI in society. *Harvard Data Science Review, 3*(1).

Fotis, G., Gupta, S., Varshney, T., Dhar, S. L., & Gulzar, M. M. (Eds.). (2025). *Modern Computing Technologies for EV Efficiency and Sustainable Energy Integration.* IGI Global., DOI: 10.4018/979-8-3373-2382-4

Gao, Y., Liu, H., & Xu, P. (2023). AI-driven shared-mobility optimization. *Transportation Science, 57*(4), 1093–1114.

Gupta, S. (2025). *The Future of Machine Learning with Automation in Engineering.* Nova Science Publishers., DOI: 10.52305/YFIX2832

Gupta, S., & Tripathi, R. K. (2014). Two-area power system stability improvement using a robust controller-based CSC-STATCOM. *Acta Polytechnica Hungarica, 11*(7), 135–155. DOI: 10.12700/APH.11.07.2014.07.9

Gupta, S., & Tripathi, R. K. (2015). Transient Stability Enhancement of Multimachine Power System Using Robust and Novel Controller Based CSC-STATCOM. *Advances in Power Electronics, 2015*(1), 626731. https://onlinelibrary.wiley.com/doi/10.1155/2015/626731. DOI: 10.1155/2015/626731

Heidari, A., Wang, Y., & Li, J. (2022). Battery state-of-health prediction using LSTM networks. *Energy, 260*, 125028.

Henderson, P., Hu, J., Romoff, J., Branson, K., & Pineau, J. (2020). Energy and carbon footprints of machine learning. *Journal of Machine Learning Research, 21*(248), 1–43.

Hu, B., Yang, M., & Feng, C. (2023). Energy-aware route-planning for electric vehicles using reinforcement learning. *Applied Energy, 342*, 121081.

IEA. (2023). *World Energy Outlook 2023.* International Energy Agency.

IEA. (2024). *Global Electric Vehicle Outlook 2024.* International Energy Agency.

Jobin, A., Ienca, M., & Vayena, E. (2019). The global landscape of AI ethics guidelines. *Nature Machine Intelligence, 1*(9), 389–399. DOI: 10.1038/s42256-019-0088-2

Kempton, W., & Tomić, J. (2021). Vehicle-to-grid power fundamentals. *IEEE Transactions on Power Systems, 36*(5), 4295–4305.

Kirklin, S.. (2023). Data-driven materials discovery for next-generation batteries. *Nature Materials, 22*, 556–568.

Lago, J., De Ridder, F., & De Schutter, B. (2021). Forecasting renewable generation with machine learning. *Renewable & Sustainable Energy Reviews, 138*, 110688.

Li, X., Zhang, T., & Chen, H. (2021). Transformer fault diagnosis using deep CNN. *IEEE Transactions on Industrial Electronics, 68*(8), 7231–7242.

Lin, H., & Huang, J. (2022). Driver-behavior prediction for EV energy management. *Transportation Research Part C, Emerging Technologies, 140*, 103671.

Liu, Q., & Chen, R. (2023). Reinforcement learning for anomaly detection in cyber-physical systems. *IEEE Access : Practical Innovations, Open Solutions, 11*, 32122–32134.

Marugán, A., González, R., & Ruiz, E. (2018). Machine-learning techniques for solar energy forecasting. *Renewable & Sustainable Energy Reviews, 81*, 2299–2313.

Palizban, O., & Kauhaniemi, K. (2020). Microgrids and intelligent energy management. *Renewable & Sustainable Energy Reviews, 117*, 109499.

Patterson, D., Gonzalez, J., Leong, M., & Dean, J. (2022). Carbon-emission efficiency in large-scale machine-learning training. *Communications of the ACM, 65*(8), 64–75.

Pop, C., Cioara, T., Antal, M., & Salomie, I. (2023). Blockchain-based AI markets for peer-to-peer energy trading. *Energy Reports, 9*, 132–146.

Rao, M. S., Saraswathy, S., Neela, V., Arun, M. R., Gupta, S., & Taqui, S. N. (2023, November). Efficient Model Training in Decentralized Systems with Federated Learning. *In 2023 International Conference on Communication, Security and Artificial Intelligence (ICCSAI)* (pp. 900-904). IEEE. DOI: 10.1109/ICCSAI59793.2023.10421469

Rao, P., & Chowdhury, S. (2023). Reinforcement learning-based battery dispatch optimization. *Electric Power Systems Research, 213*, 108721.

Rensource Energy. (2024). *Solar microgrid deployment report 2024*. Rensource Holdings.

SGCC (State Grid Corporation of China). (2025). *Energy Brain annual white paper 2025*. State Grid Corporation of China.

Shankar, R., Chaudhary, B. P., Mehraj, H., Gupta, S., Arslan, F., Srivastava, A. K., & Goswami, C. (2023). Impact of node mobility on the DL based uplink and downlink MIMO-NOMA network. *International Journal of Information Technology : an Official Journal of Bharati Vidyapeeth's Institute of Computer Applications and Management, 15*(6), 3391–3404. DOI: 10.1007/s41870-023-01362-z

Sharma, K., Gupta, C., & Gupta, S. (2019, July). Infant weeping calls decoder using statistical feature extraction and gaussian mixture models. *In 2019 10th International Conference on Computing, Communication and Networking Technologies (ICCCNT)* (pp. 1-6). IEEE. DOI: 10.1109/ICCCNT45670.2019.8944527

Sharma, R. K., Saran, V. B., Kanaujia, S., & Gupta, S. (2024, May). Evaluation of mppt controller performance using fuzzy logic design with wireless sensor node. *In 2024 International Conference on Smart Systems for applications in Electrical Sciences (ICSSES)* (pp. 1-6). IEEE. DOI: 10.1109/ICSSES62373.2024.10561365

Shukla, R. D., Tripathi, R. K., & Gupta, S. (2010, November). Power electronics applications in wind energy conversion system: A review. *In 2010 International Conference on Power, Control and Embedded Systems* (pp. 1-6). IEEE. DOI: 10.1109/ICPCES.2010.5698663

Siemens, A. G. (2024). *MindSphere industrial AI energy platform report 2024.* Siemens.

Sinha, R., Kaur, N., Gupta, S., & Thakur, P. (2023, November). Diagnosis of Parkinson's Disease using Hybrid Ensemble Technique. *In 2023 International Conference on Ambient Intelligence, Knowledge Informatics and Industrial Electronics (AIKIIE)* (pp. 1-5). IEEE. DOI: 10.1109/AIKIIE60097.2023.10390458

Sinha, R., Patel, M., Gupta, S., & Sinha, K. K. (2024, June). Performance Analysis of Breast Cancer Predictor using Machine Learning Techniques. *In 2024 15th International Conference on Computing Communication and Networking Technologies (ICCCNT)* (pp. 1-5). IEEE. DOI: 10.1109/ICCCNT61001.2024.10724436

Sinha, R., Sinha, K. K., Patel, M., Gupta, S., & Priya, S. (2024, July). Detection of leukemia disease using convolutional neural network. *In 2024 5th International Conference on Image Processing and Capsule Networks (ICIPCN)* (pp. 451-456). IEEE. DOI: 10.1109/ICIPCN63822.2024.00080

Sovacool, B. K., & Dworkin, M. H. (2022). *Energy justice: Re-balancing equity and technology*. Oxford University Press.

Strubell, E., Ganesh, A., & McCallum, A. (2019). Energy and policy considerations for deep learning. *Proceedings of ACL*, 3645–3650. DOI: 10.18653/v1/P19-1355

Sudha, G., Natha, A., Rajendiran, M., Vanitha, U., Kiran, V., Patil, H., & Rajaram, A. (2024). Optical oasis harnessing energy efficiency in wireline, optical, and wireless communications. *Journal of Environmental Protection and Ecology, 25*(4), 1179–1187.

Sun, L., Yang, P., & Guo, S. (2023). Natural-language processing for predictive maintenance in smart grids. *IEEE Transactions on Industrial Informatics, 19*(6), 8020–8031.

Sutton, R. S., & Barto, A. G. (2018). *Reinforcement learning: An introduction* (2nd ed.). MIT Press.

Tan, K., Wang, X., & Li, B. (2022). Reinforcement learning for vehicle-to-grid energy management. *Applied Energy, 308*, 118344.

Tarascon, J.-M., & Armand, M. (2021). (Referenced for battery chemistry in the storage section.)

Tesla Inc. (2022). *Opticaster AI energy optimization system.* Tesla.

Thandlam, A. K., Vignesh, T., Saravanan, A., Subramani, P., Marimuthu, R., & Gupta, S. (2024, April). Next-gen vehicle safety: The futuristic approach to auto-stop in modern vehicles. *In 2024 10th International Conference on Communication and Signal Processing (ICCSP)* (pp. 265-270). IEEE. DOI: 10.1109/ICCSP60870.2024.10543666

Tyagi, M., Gupta, S., & Khan, M. I. (2024, May). Reduction of voltage instability in complex power networks through the utilization of D-STATCOM. *In 2024 4th International Conference on Advance Computing and Innovative Technologies in Engineering (ICACITE)* (pp. 1662-1666). IEEE. DOI: 10.1109/ICACITE60783.2024.10616856

UNDP. (2025). *Global AI for Energy Partnership report.* United Nations Development Programme.

UNESCO. (2022). *Recommendation on the ethics of artificial intelligence.* UNESCO.

United Nations. (2024). *Sustainable Development Goals progress report 2024.* United Nations.

van der Meer, S., et al. (2024). (Referenced in blockchain-enabled V2G case.)

Wang, J., Li, X., & Zhao, Q. (2023). Multi-agent reinforcement learning for microgrid coordination. *Applied Energy, 339*, 120933.

Wang, Z., & Li, H. (2023). Adaptive scheduling for EV charging using reinforcement learning. *Transportation Research Part C, Emerging Technologies, 151*, 104054.

WEC (World Energy Council). (2024). *World Energy Trilemma Index 2024.* World Energy Council.

WHO. (2025). *Tracking SDG 7: Energy Progress Report 2025.* World Health Organization & World Bank.

World Economic Forum. (2024). *Global Cyber Resilience in Energy 2024*. World Economic Forum.

Zhang, C., & Mi, C. (2022). AI-based optimization for electric-vehicle integration. *IEEE Transactions on Industrial Electronics*, *69*(10), 10188–10199.

Zhang, R., Han, T., & Feng, Y. (2023). Graph-neural-network-based fault location in power-distribution systems. *IEEE Transactions on Smart Grid*, *14*(1), 777–789.

Zhou, Y., & Brown, R. (2022). Load forecasting with interpretable machine learning. *Energy and AI*, *9*, 100171.

Zhou, Z., Wu, J., & Yan, Y. (2024). Artificial intelligence in grid optimization and control: A review. *Renewable & Sustainable Energy Reviews*, *178*, 113244.

Chapter 7

Design of Two-Level Inverter System With USV-PWM for AC Machine Using Swarm Intelligence

Vaishali Pravin Sonwane
https://orcid.org/0000-0002-5823-4868
Poornima University, Jaipur, India

Gaurav Jain
https://orcid.org/0000-0001-5272-6717
Poornima College of Engineering, Jaipur, India

Pravin Sonwane
https://orcid.org/0000-0002-0821-5167
Poornima College of Engineering, Jaipur, India

Ram Kumar Yadav
https://orcid.org/0009-0004-9506-5225
Manmohan Technical University, Nepal

ABSTRACT

In a dual 2-level inverter system using two separate isolated DC voltage sources, a hybrid method incorporating both Space Vector Pulse Width Modulation (SVPWM) and alternate inverter PWM technique along with swarm intelligence is adopted. This blend aims to specifically reduce the overall harmonic distortion. When comparing the Unified SVPWM with an alternative inverter PWM, the former demonstrates superior performance across a wide spectrum of speeds. Unified SVPWM demonstrates to display a lower Total Harmonic Distortion (THD) when compared with

DOI: 10.4018/979-8-3373-7041-5.ch007

SVPWM by means of alternative inverter PWM with the help of swarm technology. Due to recommended attributes like efficiency and precise controllability, the said inverters discover is in the state of multidisciplinary problem, where in decision making tools may be required. Swarm Intelligence tool (PSO) is selected to handle the problem so that a decision making for the energy efficiency. The outcome of this research is to design 2-Level Inverter for the recent technology of USV-PWM for AC Machine along with Swarm Intelligence.

INTRODUCTION

A proposed Direct Torque Control (DTC) system intentionally operates the specific positions of the necessary stator voltage as well as flux vectors by optimizing torque control while minimizing flux variation within the hysteresis band. Such an unconventional technique enables full-load torque control, to minimize switching frequency deviations, and confirm instant transient responses even at much lower operating speeds. Experimental findings showcasing the fast torque control to increase operational speed from zero to rated speed. Detailed study in (Chen & Sun, 2017) explores into the inspiration of assorted error volt-seconds (influenced by various PWM versions) on I_{phase} motor inside the dual Voltage Source Inverter (VSI). This research is focused on Offline optimization, wherein the first PSO algorithm will be executed in MATLAB Simulink using a motor and inverter model to find the best switching angles or carrier parameters for diverse operating points, such as speed, torque, and DC voltage, to be stored in the controller.

An article focuses on two analytical approaches based on predicting the current trajectory and evaluating ripple content. Such methodologies use error-based voltage trajectory data as well as switching-recognized confirmation from the dual Voltage Source Inverter (VSI).

Terminologies may be for I_{ripple} are customary over various PWM variations, maintained by experimental results validating the effectiveness of these analytical methods. An extensive examination of a 3-ϕ open ended winding in induction motor-drive (Stemmler, 1994; Patel et al., 2012; Sekhar & Srinivas, 2013; Srinivas & Somasekhar, 2008; Srinivas & Ramachandra Sekhar, 2013; Pradabane, Narasimharaju, et al., 2014; Somasekhar et al., 2008; Pradabane, Srikanth, et al., 2014) employing volts/htz control, is conducted through comprehensive testing across different PWM versions initially simulated using MATLAB. The use of an open-ended winding in induction-motor is accompanied by a dual 2-level VSI generating for the three-level voltage-space-vector.

The study proposes space-vector-based PWM varieties specifically designed on behalf of the dual VSI, offering benefits like a 50% reduction in switching commutation within the dual VSI and a consistent limitation of switching power loss to a single VSI. The adopted switching sequence guarantees alternating clamping in addition to switching-modes on every 180 degrees for the purpose of both inverters (Meena et al., 2024; Semwal et al., 2024). Furthermore, the architecture's isolated power sources prevent zero-sequence currents from finding a path, thereby minimizing losses experienced by power electronic switches. Simulation tests conducted using MATLAB/Simulink effectively demonstrate the proposed SVPWM approach. The research highlights an innovative 'Dual-Inverter'-'Space-Vector' PWM (DI-SVPWM) to accomplish zero-sequence-voltage to destroy zero-sequence-current within an open-end-winding-Permanent-Magnet-Synchronous-Motor. This method improves the damaging effects as of potential zero-sequence-current, and hence minimizing stress on power semiconductor devices. The application of all active states aids in clamping the inverters, and simulations are carried out using MATLAB/Simulink. The double-ended scheme is explained in (Welchko, 2005; Kapoor et al., 2021; Shukla et al., 2010).

A 1kW experimental system featuring an open-ended-winding in PMSG to devise a 3D-SVPWM approach for destroying zero-sequence current. The method aims to shrink costs, simplify the drive circuit, and increase control implementation for the semi-controlled converter supplying the PMSG. Additionally, the research explores the integration of dual-inverter through an open-ended-winding motor configuration in electric vehicle (EV) applications. It underlines the use of isolated DC sources mostly to provide higher voltage supply in addition to discusses strategies for reducing common-mode voltage within the drive. Additionally, this research offers a dual-space-vector-based system for open-ended winding of PMSM drive, powered by a 2-level inverter for a single DC supply. This advanced control system effectively succeeds to manage $V_{zero\text{-}sequence}$ and I_{load} within the drive system. Three times boosting is realized with the given structure in (Sen et al., 2023; Gupta & Tripathi, 2014) using a lower switch count with low total standing voltage. Similar inverters may not be useful in aircraft design for safety purposes (Sonwane et al., 2024; Tyagi et al., 2024), whereas the 5Level-PV-SCML requires only one DC source and self-balanced capacitors to generate a boost output voltage (Kumari et al., 2023), which are difficult and rarely useful. A hybrid multilevel topology and asymmetric converter may be strongly referred to in (Kumari et al., 2023; A. Rufer et al., 1999) for better understanding.

PRINCIPLE OF SPACE VECTOR PWM

An inflection presentation given in space vector modulation is widely applicable in generating three-phase AC power systems, facilitating the production of alternating current (AC) waveforms. It characterizes an advanced method for generating sine waves that successfully supplies higher voltage to the load while maintaining minimal harmonic distortion (Nabae et al., 1981; P. C. Loh et al., 2005; Yo-Han Lee et al., n.d.; Kim & Sul, 1996). SVM displays in various forms, each categorized by unique attributes and computational requisites. Current development focuses on aggressively reducing total harmonic distortion. In SVM, a space vector attends as a representation of a three-phase mathematical system. One way to conceptualize a space vector involves defining it through a collection of three-phase voltages, illustrated as:

$$\overline{V(\tau)} = \frac{2}{3}\left[V_a(\tau)\,e^{j0} + V_b(\tau)\,e^{j\theta 1} + V_c(\tau)\ e^{j2\theta 1}\right] \tag{1}$$

Where, V*a,* V*b,* V*c are p*hase voltages and θ=2π/3

SWARM INTELLIGENCE

The Swarm algorithm is well known as Particle Swarm Optimization that is PSO, which can optimize parameters that are hard to tune analytically (Gupta, 2025; Rao et al., 2023). The problems in general is nonlinear, multidimensional and conventional simulation techniques are fail in finding the appropriate solution needs PSO. Here in this research work, common optimization targets may be (a) Switching carrier phase shifts along with amplitude offsets for minimizing motor current THD or specific harmonic orders; (b) Switching angles mostly selective harmonic elimination, SHE as preferred here also for selective harmonic suppression with the help of PSO, which finds the switching angle available set that minimizes THD or specified harmonics (Fotis et al., 2025). (c) LC filter component values to balance attenuation vs switching losses/size; (d) PI based Controller gains (Awasthi et al., 2024; Thandlam et al., 2024).

This research is focused on Offline optimization, wherein the first run of PSO is performed in MATLAB/Simulink using a motor and inverter model to find the best switching angles or carrier parameters for diverse operating points, such as speed, torque, and DC voltage, to be stored in the controller. Swarm-based carrier/angle optimization has nonfiction precedent for PWM and multi-level converters.

a. PSO Algorithm

Step 1. Initialize with following

(a) Randomly place particles in the search space.

(b) Assign random velocities.

Step 2. Fitness Evaluation

(c) Each particle evaluates how good its solution is (fitness function).

Step 3. Update Local & Global Best

(d) Every particle holds its optimal location. (pbest).
(e) The swarm screens the optimal location that each particle finds. (gbest).

Step 4. Velocity & Position Update

(f) Velocity is updated based on inertia, cognitive, and social components:

Swarm Intelligence is developed based on above steps. Mathematically it is expressed as in equation (2) and (3).

$$v_i(t+1) = w v_i(t) + c_1 r_1 (pbest_i - x_i) + c_2 r_2 (gbest - x_i) \quad (2)$$

$$x_i(t+1) = x_i(t) + v_i(t+1) \quad (3)$$

Where:

- is the inertia constant
- c_1 and c_2 are cognitive & social coefficients
- r_1 and r_2 are random numbers (0–1)

b. PSO for Harmonic Spectrum Optimization

Harmonic Spectrum Optimization through swarm intelligence is discussed in Figure 1. The flowchart represents the process of harmonic optimization using a PSO-based inverter system and graphically represented in Fig.1(a).

This is by beginning with initialization using reference parameters, which are input into the PSO tool herein as optimizer. The optimizer iteratively searches for switching angles to minimize harmonics as an outcome of this tool. A decision step checks whether optimized angles are realized or not. If not, the process loops back for further optimization through the next iteration. Once optimized angles are obtained, they are applied to a 2-level inverter, which drives the AC machine. Finally, the system performs FFT analysis to calculate Total Harmonic Distortion

(THD) and generate harmonic values, ensuring effective performance evaluation of the inverter–machine system.

c. PSO for Optimized Switching Angle

Figure 1 outlines the steps for an iterative optimization process. It commences with the initialization process, where the values of "particles" are created as initial angles. These particles showcase potential solutions in the search space. The core of the process is a loop that continues until a maximum number of iterations is achieved. Within this loop, the algorithm performs the following steps, and graphically represented in Fig.1(b):

i. Analyze objective: It evaluates the "basic error with harmonic penalty" for each particle. This step determines how good each solution is.
ii. Update personal and global bests: The algorithm preserves track of the best position of each particle. The best position exposed by any particle in the intact swarm (global best) and the one revealed thus far (personal best).
iii. Update PSO parameters: Each particle's location (P) and velocity (V) are modernized according to both its own and the swarm's optimal positions. This is a key characteristic of the PSO algorithm, guiding particles toward more optimal solutions.
iv. Enforce bounds and angle ordering: This step ensures that the updated particle positions remain within acceptable limits and follow a specific order, which is crucial for the problem being solved.
v. The loop repeats until the stopping condition (Max iteration?) is met. Once the process is complete, the algorithm outputs the final, optimized angles. This approach is typically used to find the best configuration or values for a complex problem by simulating the collective behavior of a swarm of particles.

Figure 1. (a)Harmonic Spectrum Optimization by swarm intelligence (b)swarm intelligence for optimized switching Angle

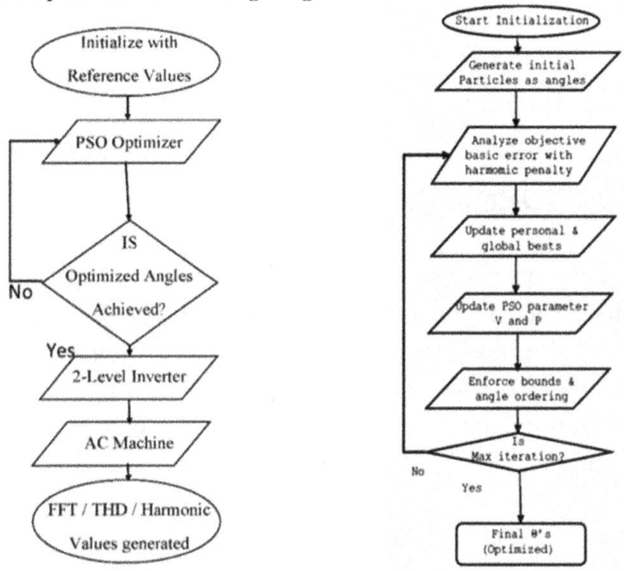

CLARKE TRANSFORMATION

The Clarke transformation updates 3-phase system to two-phase indicative orthogonal system termed in mathematical expression as,

$$i_\alpha = \frac{2}{3} i_a - \frac{1}{3}\left(i_b - i_c\right) \tag{4}$$

$$i_\beta = \frac{2}{\sqrt{3}}\left(i_b - i_c\right) \tag{5}$$

$$i_0 = \frac{2}{3}\left(i_a + i_b + i_c\right) \tag{6}$$

Utilizing i_α, i_β components surrounded by an orthogonal reference frame along with i_0 as homo-polar element, it's commonly observed that the homo-polar component is frequently negligible or absent in many applications. In this context, the space vector indicates initial three-phase input signal once lacking the homo-polar component. Now, let's consider a specific scenario where 'i' is combined. In these situations, 'i_a', 'i_b', 'ic' are converted to 'i_α' and 'i_β' using following mathematical treatment.

$$i_{\propto} = i_a \quad (7)$$

$$i_{\beta} = \frac{1}{\sqrt{3}} i_a + \frac{2}{\sqrt{3}} i_b \quad (8)$$

$$i_a + i_b + i_c = 0 \quad (9)$$

PARK'S TRANSFORMATION

In balanced 2-phase orthogonal system, such conversion shifts the vectors obsessed by an orthogonal rotating reference frame. These paired phase values serve the purpose of producing the reference vector and determining the angle of rotation within the d-q frame, subsequently influencing the modulation of the inverter output. The Park transformation equations, responsible for converting 3-phase (abc) quantities into 2-phase (d-q) quantities, are as follows.

$$V_d = \frac{2}{3}\left[V_a \sin(\omega t) + V_b \sin\left(\omega t - \frac{2\pi}{3}\right) + V_c \sin\left(\omega t + \frac{2\pi}{3}\right)\right] \quad (10)$$

$$V_q = \frac{2}{3}\left[V_a \sin(\omega t) + V_b cos\left(\omega t - \frac{2\pi}{3}\right) + V_c \cos\left(\omega t + \frac{2\pi}{3}\right)\right] \quad (11)$$

$$V_0 = \frac{2}{3}\left(V_a + V_b + V_c\right) \quad (12)$$

V_a, V_b, and V_c represent three sinusoidal voltages with identical amplitudes and frequencies but exhibiting a 120-degree phase difference, as shown in Figure 2 (Sharma et al., 2024; Anand et al., 2022). The space vector's magnitude remains mostly constant over time. However, the angle of the space vector develops over time, gradually increasing the rotational frequency of the vector to match that of the sinusoidal waveforms. In a 3-phase, six-step inverter unit, when output voltages are transformed in space vector and then plotted on a complex plane or trajectory, the resultant space vector increasingly assumes one of six distinct angles with respect to time progress.

Figure 2. Three phase voltage source inverter

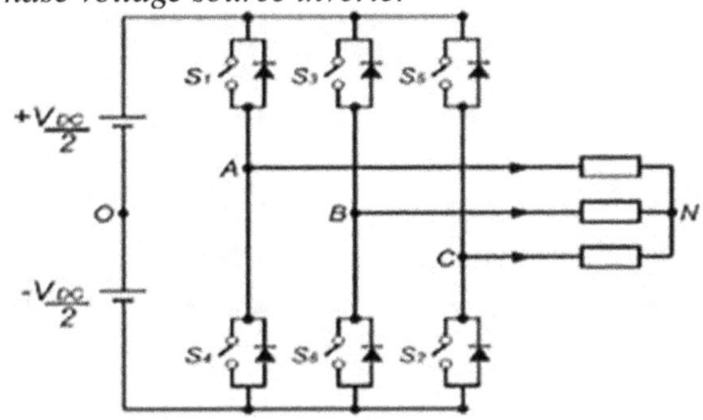

The primary objective of SVPWM is to create precise PWM signals to construct a vector with a specified angle. During operation, the lower transistor is deactivated when the upper transistor is activated. The ON/OFF stages of the lower power device are synchronized with those of the higher power device. The eight switch configurations of a 3-phase inverter, ranging from 000 to 111, form the fundamental basis of SVPWM. The inverter output voltages V_{ao}, V_{bo}, and V_{co} indicate independent pole voltages concerning the mid-terminal within DC source labeled as "o" in Figure 2. Among these, two are null states, while the remaining six represent active switching states. An eight-switching configuration for a 3-phase VSI is shown in Figure 3.

Figure 3. 8-switching configuration for 3- phase VSI

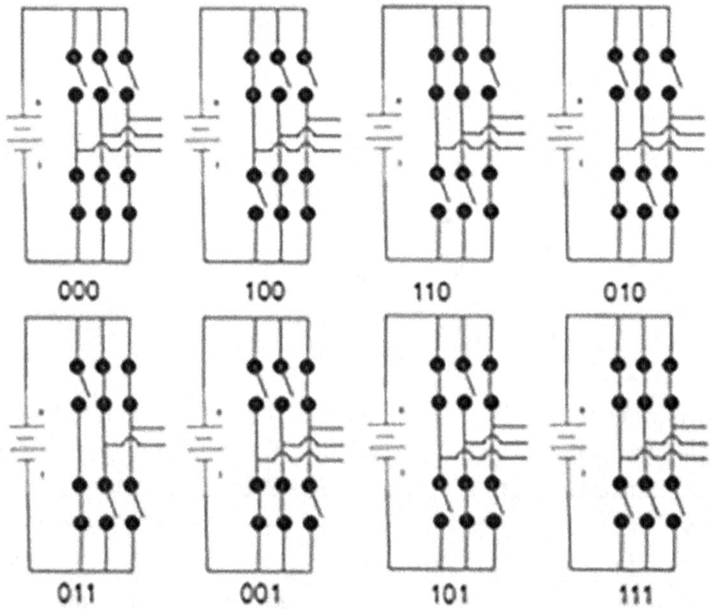

Figure 4. Switching vectors and sectors in hexagon

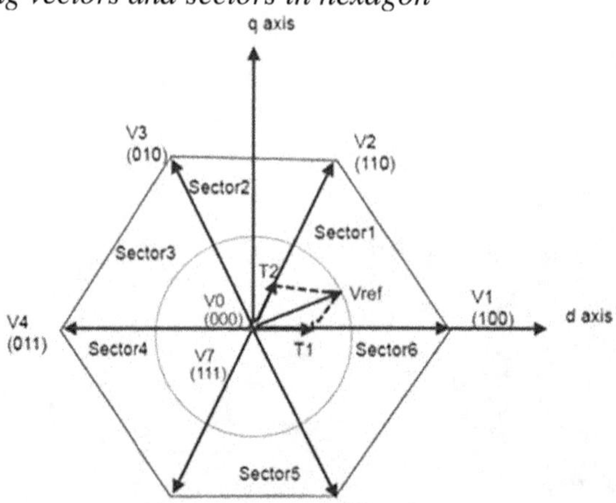

We use orthogonal coordinates to represent the 3-phase 2-level inverter in the phase diagram, as shown in Figure 4. To produce eight space vectors, there are eight possible inverter states. The complex vector expressions are,

$$\overline{V(k)} = \left\{\frac{2}{3}V_{dc}e^{j(k-1)\pi/3}\right\} if\, k = 1-6 \tag{13}$$

$$\overline{V(k)} = \{0\} if\, k = 0,7 \tag{14}$$

There are six 60-degree sectors of equal size dividing the entire area. Two active vectors encircle each sector. $V(k)$ and $V(0)$ are voltage vectors as shown in hexagon. The eight active and inactive state vectors are graphically represented in Figures 4 and 5. The SV-PWM switching arrangements for each sector are shown in Figure 6.

Figure 5. Components of available voltage space vector in (d, q) notations

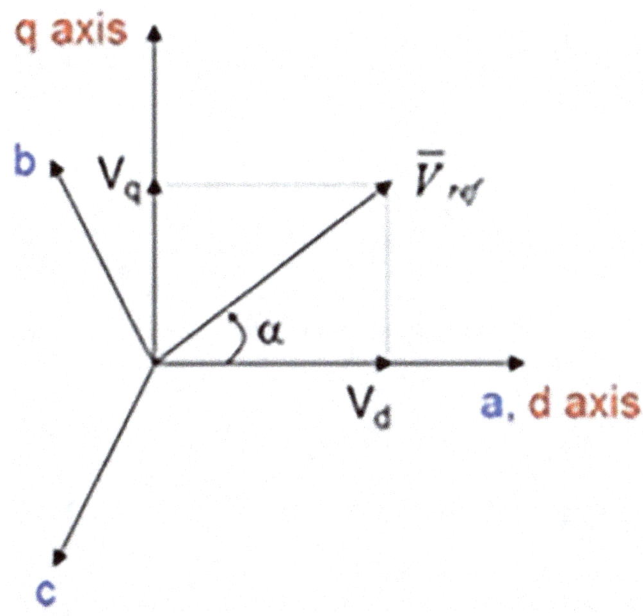

The system utilizes six active vectors to power loads, while the zero(0)vectors, V_0 as well as V_7 at origin, supply zero(0) voltage. With eight fundamental space vectors (V_0 to V_7), the SVPWM technique employs an 8-switching pattern to mimic the related reference vector. Consider the reference vector positioned within sector 1. T_1 signifies the application of active vector 100 (+ - -), while T_2 represents the application of active vector 110 (+ + -). The rotational angle of the reference vector is denoted by alpha (α). To determine the value of the reference vector, appropriate durations of applying T_1 and T_2 are calculated. T_s represents the total duration during which active vectors and null vectors (T_z) are employed. The calculations involving T_1, T_2, T_s, and T_z are subsequently utilized to derive V_d, V_q, V_{ref}, and the angle (α).

$$V_d = V_{an} - V_{bn}\cos(60) - V_{cn}\cos 60 \quad (15)$$

$$V_d = V_{an} - \frac{1}{2}V_{bn} - \frac{1}{2}V_{cn} \quad (16)$$

$$V_q = 0 + V_{bn}\cos(30) - V_{cn}\cos(30) \quad (17)$$

$$V_q = 0 + \frac{\sqrt{3}}{2}V_{bn} - \frac{\sqrt{3}}{2}V_{cn} \quad (18)$$

$$\begin{bmatrix} V_d \\ V_q \end{bmatrix} = \frac{2}{3}\begin{bmatrix} 1 & -\frac{1}{2} & -\frac{1}{2} \\ 0 & \frac{\sqrt{3}}{2} & -\frac{\sqrt{3}}{2} \end{bmatrix}\begin{bmatrix} V_{an} \\ V_{bn} \\ V_{cn} \end{bmatrix} \quad (19)$$

$$\overline{V_{ref}} = \sqrt{V_d^2 + V_q^2} \quad (20)$$

$$\alpha = \tan^{-1}\left(\frac{V_q}{V_d}\right) = \omega t = 2\pi f t \quad (21)$$

Figure 6. SVPWM switching patterns at every sector

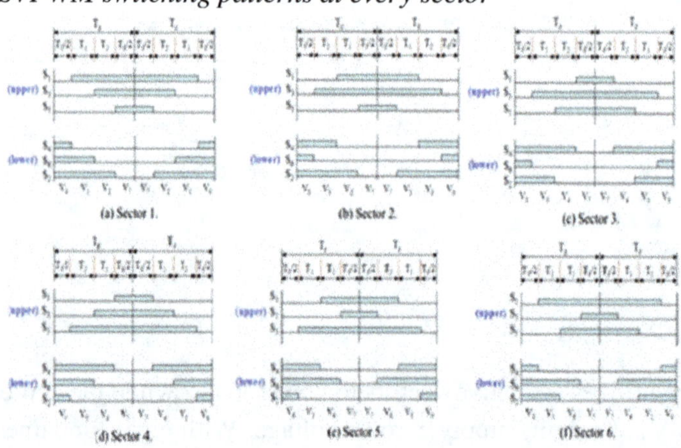

High-power nonlinear and time-varying loads lead to unwanted effects like reactive power demand and harmonic pollution. These nonlinear loads draw significantly distorted currents, particularly the high-power single-phase rectifier loads, leading to distorted outputs. To counter these impacts, a setup involving a dual inverter system (J. Hong et al., 2015; V. T. Somasekhar et al., 2005) connected to a DC supply powering the permanent magnet synchronous motor is utilized. This configuration resolves concerns linked to zero sequence currents within the open-ended winding drive system. These currents could otherwise generate circulating currents in the winding, place excessive strain on power semiconductor devices, and cause significant losses. To counteract this, the reference voltage is synthesized using a current approach that employs zero sequence switching combinations, effectively suppressing zero sequence voltage. The implementation of Nonlinear Loads drive unified SVPWM assists in defeating absolutely 0-sequence current(I) observed in open-ended winding. Additionally, the application of an alternative inverter PWM leads to decreased total harmonic distortion in the power electronic drive.

DESIGN OF UNIFIED SVPWM

The Unified SVPWM design incorporates two separate DC voltage sources, V_{dc1} and V_{dc2}, in the stated inverter setup used here. The ratio between these isolated DC voltages is denoted as $k=V_{dc1}/V_{dc2}$. The connection between the outputs of the dual inverters, V_{inv1} & V_{inv2}, and the reference voltage vector V_s is described as:

$$V_s = V_{inv1} - V_{inv2} \tag{22}$$

In the traditional SVPWM synthesis, two nonzero active voltage sources are used. To replicate this in the stated system, a perfect tuned unified SVPWM is suggested. Each inverter in this algorithm supplies an active voltage (V_x), characterized by their respective time lengths (t_x, t_y) are as follows.

$$t_x = \frac{\sqrt{3}\,V_{sm}}{V_{dc1}} T_s \sin\left(\pi/3 - \theta\right) \tag{23}$$

$$t_y = \frac{\sqrt{3}\,V_{sm}}{V_{dc2}} T_s \sin(\theta) \tag{24}$$

After obtaining t_x and t_y, their relationship with T_s for the time lengths for every inverter is determined and indicates the relationship. For the V_s seen in the first region, complementary voltages aren't needed, hence both t1y and t2x are bound to be set at zero. Whereas region 2 observing as, t_{1x} as limited to T_s, hence, INV2 generates the analogous voltage vector V_{2x}, and t_{2x} supposed to be calculated.

$$V_x t_{x=} V_{1X} T_X + V_{2x} t_x \tag{25}$$

Then the equation for t_{2x} is given by,

$$t_{2x} = (V_x t_x - V_{1x} T_x)/V_{2x} = k(t_x - T_x) \tag{26}$$

Similarly at region 3, where t_{2y} is constrained to $T_{s,}$ INV1 generates a complementary voltage vector, and t_{1y} is computed.

$$V_y t_y = V_{2Y} T_s + V_{1y} t_{1y} \tag{27}$$

Thus, the expression for t_{1y} is shown as,

$$t_{1y} = \frac{(V_y t_y + V_{2y} T_s)}{V_{1y}} + (t_y - T_s)/k \tag{28}$$

Voltage vectors V_{1x} as well as V_{2y} are selected from V_1, V_3, and V_5 based on the reference voltage vector's sector location. Similarly, compensatory voltage vectors V_{1y} and V_{2x} are selected from V_2, V_4, and V_6 to minimize transferring frequency within a sampling period. V_0 functions as the zero vector for both inverters. The vectors V_n assigned by INV_1 and INV_2 are represented with numbers n (n=1,2,3,4,5,6) and n', respectively. In Figure 4, black dotted circles delineate boundaries of various regions defined by the available two DC source voltages. These boundaries indicate the number of regions the reference voltage vector V_s traverses within a sector. The dotted circle implies the region with the peak linear modulation index, measuring $(V_{dc1} + V_{dc2})/p_3$. Voltage vector transferring patterns for each inverter across a sampling period within different areas, focusing on sector I: Voltage vector switching for INV_1, and Voltage vector switching also for INV_2, both inverters operate in switching mode within area 1. INV_1 operates in clinching mode while INV_2 operates in transferring mode in area 2, and INV_1 switches while INV_2 clamps in area 3. Irrespective of the voltage vector's placement, the dual inverter's switching time remains consistently four times longer compared to the dual-inverter dual-SV-PWM within a sampling period, which is only one-third of the former's duration. If the referred voltage vector V_s ruins small or within the smaller circle, both inverters solely operate in switching mode, moving solely within region 1 across all sectors. If the required voltage vector sits between the two spotted circles, regions 1 and 3 are traversed. Otherwise, in sector I, region 2 clamps the voltage vector V_1 chosen by INV1, while the remaining two regions receive only a fraction of it.

MATLAB IMPLEMENTATION OF PARTICLE SWARM OPTIMIZATION (PSO)

Selective Harmonic Elimination (SHE) switching-angle optimization

MATLAB script was written for Particle-Swarm-Optimization (PSO) algorithm to resolve a Selective Harmonic Elimination Pulse-Width Modulation (SHE-PWM) problem. The goal is to catch the optimal switching angles (θ) for a two-level inverter. These angles are preferred to eliminate specific low-order harmonics while also completing a desired fundamental voltage level. SHE-PWM is a technique utilized in power electronics, principally for voltage source inverters. Instead of expending a regular triangular carrier wave, it involves calculating and applying specific, non-uniformly spaced switching angles per quarter cycle. This method gives engineers specific control over the inverter's output voltage waveform, permitting them to (a) Eliminate specific harmonics: By solving a set of non-linear equations, certain low-order harmonics (like the 5th, 7th, etc.) can be completely removed from the output voltage spectrum and (b) Control the fundamental voltage: The amplitude of the fundamental frequency component can be controlled to a desired level.

Parameters of PSO selection as:

- **m**: The **modulation index**, a crucial parameter that determines the desired amplitude of the fundamental output voltage.
- N: The **switching angles** per quarter cycle number. This value is equal to the number of observed harmonics which can be eliminated, and add one. The script sets N = 3, meaning it can eliminate two harmonics while controlling the fundamental.
- **harmonics**: The specific low-order harmonics to be eliminated (5th, 7th, 11th, 13th). The script has N=3, but this list has four entries. This means the script is attempting to eliminate four harmonics with only three switching angles, which isn't possible in a typical SHE-PWM setup. Usually, the number of angles, N, governs the number of harmonics that can be eliminated (N-1), or the number of harmonics plus the fundamental (N). This is a potential point of confusion or a bug in the provided script.
- **theta_min** and **theta_max**: The bounds for the switching angles, which are constrained to be between 0 and π/2 (0 to 90 degrees) due to the quarter-wave symmetry.

Fitness Function for Objectives in SHE

This function, also known as the objective function, defines what the PSO algorithm is trying to minimize. A lower fitness value is better. The script's fitness function J is a weighted sum of two error terms:

1. Fundamental Voltage Error (err1): Calculates the squared difference between the actual fundamental voltage (V1) and the desired modulation index (m).
2. Harmonic Voltage Error (errh): Calculates the sum of the squared amplitudes of the harmonics that are targeted for elimination. The script weights this term 100*errh to prioritize harmonic elimination over achieving the precise fundamental voltage. This ensures the algorithm focuses on getting rid of the undesirable harmonics.

OPEN-END DESIGNED WINDING PMSM DRIVE TOPOLOGIES FED BY 2 INVERTERS

Figures 7 & 8 illustrate the setup of the dual inverter-fed open-end winding PMSM-drive. In this configuration, the neutral point of the motor windings remains open, and two inverters supply power to the windings from opposite ends. The circuit design for the dual inverter with isolated DC connection can be streamlined, requiring only a single DC power supply. However, it's important to note that despite this configuration, there remains the possibility for the zero-sequence current to flow through the DC bus.

Figure 7. A double inverter-fed open-end winding PMSM with an isolated DC

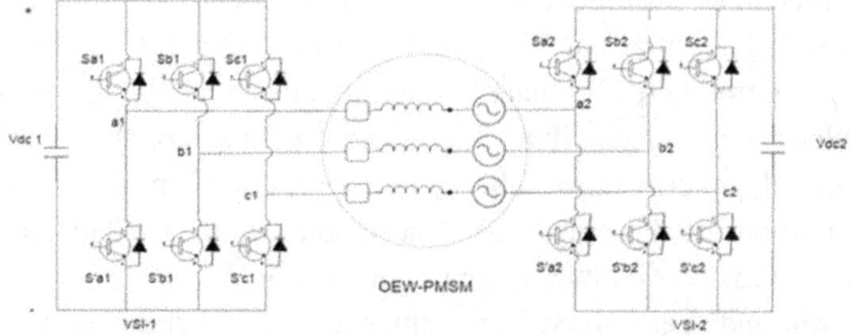

Figure 8. The dual inverter-fed open-end winding PMSM with a shared DC

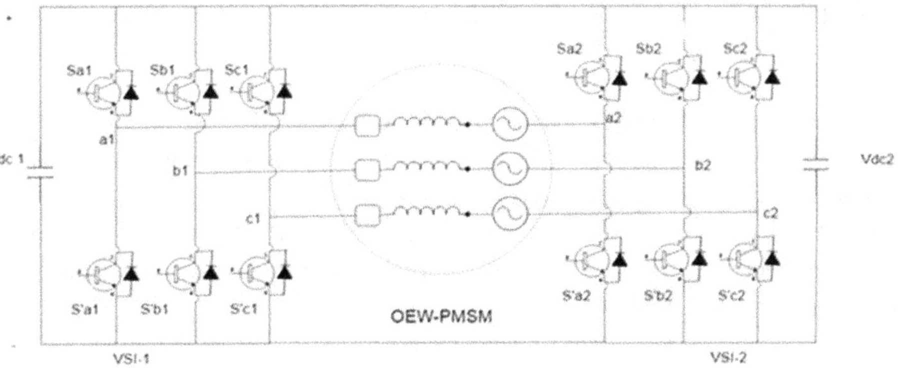

Figure 9. Open-end winding PMSM-an equivalent model

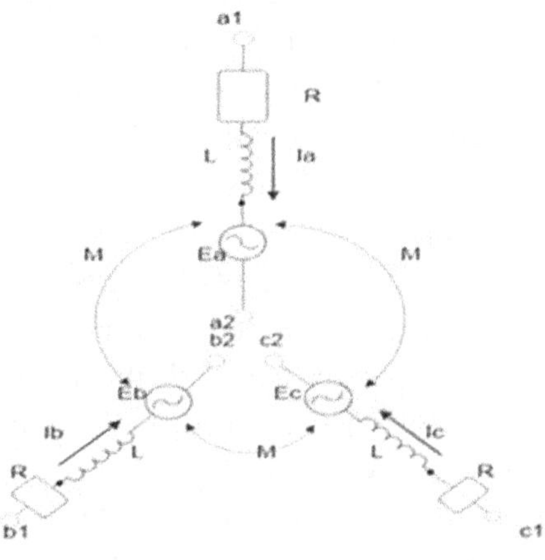

The model of an open-end winding PMSM comprises a series connection of resistance, inductance, and electromotive force (EMF) for each phase. An analogous circuit for a non-salient open-end PMSM winding is presented in Figure 9, and the model for each phase mirrors that of a star-connected motor. Specifically, the resistance, self-inductance, and mutual inductance across the three-phase windings are all identical. Consequently, the voltage equations for the open-ended PMSM can be formulated as:

$$\begin{bmatrix} U_a \\ U_b \\ U_c \end{bmatrix} = \begin{bmatrix} R & 0 & 0 \\ 0 & R & 0 \\ 0 & 0 & R \end{bmatrix} \begin{bmatrix} i_a \\ i_b \\ i_c \end{bmatrix} + \begin{bmatrix} L_m & M & M \\ M & L_m & M \\ M & M & L_m \end{bmatrix} \frac{d}{dt} \begin{bmatrix} i_a \\ i_b \\ i_c \end{bmatrix} + \begin{bmatrix} e_a \\ e_b \\ e_c \end{bmatrix} \tag{29}$$

Where the 3-ϕ voltages, currents, as well as back EMFs, uk, ik, and ek (k=a, b, and c), are present, respectively. Resistance of Stator, self & mutual inductances are represented by letters R, Ls, and M, respectively.

The novel OEW-PMSM model, employs Clarke and Park transformations to produce perfect sinusoidal EMFs in the available rotor reference frame and has a mathematical representation as:

$$\begin{bmatrix} U_d \\ U_q \\ U_0 \end{bmatrix} = \begin{bmatrix} R & \omega L & 0 \\ -\omega L & R & 0 \\ 0 & 0 & R \end{bmatrix} \begin{bmatrix} i_d \\ i_q \\ i_0 \end{bmatrix} + \begin{bmatrix} L & 0 & 0 \\ 0 & L & 0 \\ 0 & 0 & L \end{bmatrix} \frac{d}{dt} \begin{bmatrix} i_d \\ i_q \\ i_0 \end{bmatrix} + \begin{bmatrix} 0 \\ \omega\varphi_0 \\ e_c \end{bmatrix} \tag{30}$$

Here, ω_L represents the motor's electric angular speed, f is the flux linkage, and ud, uq, and u0 denote d-, q-, and zero-sequence voltages and currents. In nonsalient motors, the d-axis inductance equals the q-axis inductance (L=Ls-M). Additionally, the zero-sequence inductance, L0, is calculated as $L_0=L_s+2M$. In ideal conditions with perfect sinusoidal EMFs, the zero-sequence EMF (e_0) is zero. Kirchhoff's law typically restricts zero-sequence current flow in star-connected winding motors due to the summation of three-phase currents being zero. However, in open-end winding motors powered by a dual inverter with a shared DC connection, the constraint $i_a+i_b+i_c=0$ doesn't naturally apply. With the zero-sequence equation, the rapid responsiveness of zero-sequence current is due to the small value of L_0. Thus, the zero-sequence voltage (u_0) can easily trigger the zero-sequence current. The electromagnetic torque for a nonsalient Open-End Winding Permanent Magnet Synchronous Motor (OEW-PMSM) is expressed as:"

$$T_s = \frac{3}{2} n_p \psi_f i_q \tag{31}$$

Here, n_p represents the motor's pole pairs. The control strategy involves regulating the d-axis current to zero ($i_d=0$ control), allowing the q-axis current to produce torque, aiming for maximum torque per ampere. The open-ended winding PMSM is typically powered by a shared DC-link dual inverter configuration. This setup consists of dual 2-level voltage source inverters known as V_{SI1} and V_{SI2}. The voltage vectors for these inverters are defined as:

ZERO SEQUENCE ELEMENT BASED MODEL OF PMSG

The fluctuation in the stator flux can act as a portrayal of the back emf within the stator winding of an open winding PMSG (Chen, M, et al., 2016). Additionally, the existence of winding resistance fallouts in a voltage reduction during the flow of current through it. Therefore, the phase voltage equations for the open winding PMSG can be expressed as follows:

$$\begin{bmatrix} U_a \\ U_b \\ U_c \end{bmatrix} = \frac{d}{dt}\begin{bmatrix} i_d \\ i_q \\ i_0 \end{bmatrix} + \begin{bmatrix} \psi_a \\ \psi_b \\ \psi_c \end{bmatrix} - \begin{bmatrix} R & 0 & 0 \\ 0 & R & 0 \\ 0 & 0 & R \end{bmatrix}\begin{bmatrix} i_a \\ i_b \\ i_c \end{bmatrix} \tag{32}$$

Within a DC bus applied to open winding system, zero-sequence circuit may be an essential part of the mathematical model for PMSG. Thus, careful consideration of the third harmonic back EMF holds significant importance. Assuming '1r' and '3r' as the amplitudes of the fundamental and predominant third harmonic components observed in rotor flux linkage, the expression for the flux linkage of the three-phase winding is as follows:

$$\begin{bmatrix} \psi_a \\ \psi_b \\ \psi_c \end{bmatrix} = -\begin{bmatrix} L_s & M_s & M_s \\ M_s & L_s & M_s \\ M_s & M_s & L_s \end{bmatrix}\begin{bmatrix} i_a \\ i_b \\ i_c \end{bmatrix} + \begin{bmatrix} \psi_{1r}\cos(\theta)r + \psi_{3r}\cos(3\theta)r \\ \psi_{1r}\cos\left(\theta_r - \frac{2}{3}\pi\right)r + \psi_{3r}\cos(3\theta)r \\ \psi_{1r}\cos\left(\theta_r + \frac{2}{3}\pi\right)r + \psi_{3r}\cos(3\theta)r \end{bmatrix} \tag{33}$$

LS and MS stand for self-inductance and mutual-inductance, respectively, and r represents the angle of the rotor.

$$T_{abc\to dq0} = \frac{2}{3}\begin{bmatrix} \cos(\theta_r) & \cos(\theta_r - 120^0) & \cos(\theta_r + 120^0) \\ -\sin(\theta_r) & -\sin(\theta_r - 120^0) & -\sin(\theta_r + 120^0) \\ 1/2 & 1/2 & 1/2 \end{bmatrix} \tag{34}$$

$$\begin{bmatrix} \psi_d \\ \psi_q \\ \psi_0 \end{bmatrix} = T_{abc\to dq0}\begin{bmatrix} \psi_a \\ \psi_b \\ \psi_c \end{bmatrix} - \begin{bmatrix} L_d & 0 & 0 \\ 0 & L_q & 0 \\ 0 & 0 & L_0 \end{bmatrix}\begin{bmatrix} i_d \\ i_q \\ i_0 \end{bmatrix} + \begin{bmatrix} \psi_{1r} \\ 0 \\ \psi_{3r} Cos(3\theta)r \end{bmatrix} \tag{35}$$

Where, Ld, Lq, L_0 are inductance based component on dq0 axis.

$$T_s = \frac{3}{2}n_p\left[L_q i_q i_d + (\psi_{1r} L_d i_d) i_q 6\psi_{3r} sin(3\theta)\, i_\theta\right] \tag{36}$$

Figure 10. Positions of space vectors for a three-level dual inverter configuration

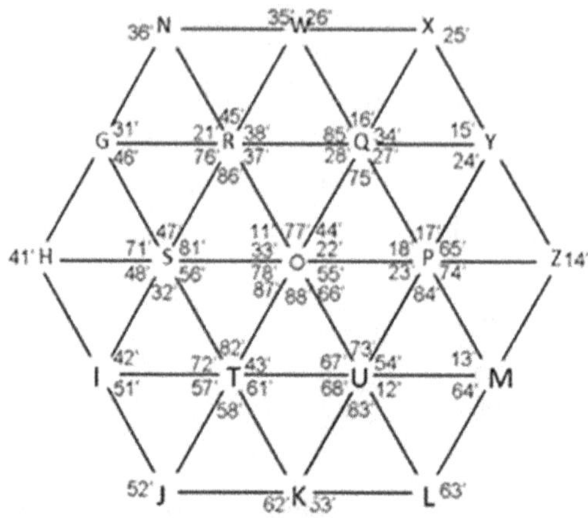

The three-level space_vector arrangement is made up of a central hexagon called “PQRSTU” that is centred on the letter “O” as seen in Figure 10 and six smaller hexagons called “OQYZMU,” “ORWXYP,” “OSGNWQ,” “OTIHGR,” “OUKJIS,” and “OPMLKT” that are centred on the letters “P,” “Q,” “R,” “S,” “T,” and “U,” respectively. Any one of the two inverter legs can independently reach level '0' or 'Vdc/2'. You can determine the voltage across a specific phase winding by,

$$V_{AA'} = V_{AO} V_{A'O} \tag{37}$$

$$V_{BB'} = V_{BO} V_{B'O} \tag{38}$$

$$V_{CC'} = V_{CO} V_{C'O} \tag{39}$$

One of the levels, ‘0’, ’-Vdc/2’, and +Vdc/2’ are attained by phase windings. The overall impact of the voltages in 3- windings is given by,

$$V_{sr} = V_{AA'} + V_{BB'} e^{j2\pi/3} + V_{CC'} e^{-j2\pi/3} \tag{40}$$

The equation demonstrates how different switching amalgamations evolve the locations of voltage space vectors. The collective voltage space vector is the totality of those produced by two 2-level inverters, Inverter-1 and Inverter-2.

$$V_{sr} = V_{s1} + V_{s2} \tag{41}$$

The 2 inverters in this drive are run off of the same DC-link voltage, as shown in Table 1. A DC-link voltage of $V_{dc}/2$ powers both inverters 1 and 2, as shown in Table 2.

Table 1. Inverter states- Individual

Inv_1_State	Switch_ON_state	Inv_2_State	Switch_ON_state
1 _ (+ - -)	S6,S1,S2	1` _ (+ - -)	S60,S10,S20
2 _ (+ + -)	S1,S2,S3	2` _ (+ + -)	S10,S20,S30
3 _ (- + -)	S2,S3,S4	3` _ (- + -)	S20,S30,S40
4 _ (- + +)	S3,S4,S5	4` _ (- + +)	S30,S40,S50
5 _ (- - +)	S4,S5,S6	5` _ (- - +)	S40,S50,S60
6 _ (+ - +)	S5,S6,S1	6` _ (+ - +)	S50,S60,S10
7 _ (+ + +)	S1,S3,S5	7` _ (+ + +)	S10,S30,S50
8 _ (- + -)	S2,S4,S6	8` _ (- + -)	S20,S40,S60

Table 2. Inverters Pole-Voltage & Motor Phase Voltages

Inv_1_Pole_Voltage	Motor_Phase_Voltage	Inv_2_Pole_Voltage
$\frac{V_{dc}}{2}$	0	$\frac{V_{dc}}{2}$
$\frac{V_{dc}}{2}$	$\frac{V_{dc}}{1}$	$-\frac{V_{dc}}{2}$
$-\frac{V_{dc}}{2}$	0	$-\frac{V_{dc}}{2}$
$-\frac{V_{dc}}{2}$	$-\frac{V_{dc}}{1}$	$\frac{V_{dc}}{2}$

The research expands the space-vector-based PWM approach to accommodate the dual-inverter configuration commonly found in a two-level inverter system. This method involves calculating switch timings for the active inverter, determined merely by instantaneous reference voltages.

These voltages dictate the fixing states of the respective inverters (either inverter 1 or inverter 2). During each complete 360-degree cycle, the inverters take turns serving as clamping and switching inverters based on the Sub-Hexagonal Centre (SHC), as described in Tables 3 & 4. The magnitudes of the three instantaneous phase reference voltages serve as the sole criteria used to determine the location of the Nearest Sub-Hexagonal Centre (NSHC).

Table 3. Inverter States throughout First Cycle

N.S.H.C.	A	B	C	D	E
Inv-1	Fixed to1 (+ - -)	Fixed to2 (+ + -)	Fixed to3 (- + -)	Fixed to4 (- + +)	Fixed to5 (- - +)
Inv-2	Switching mode during the entire second cycle				

Table 4. Inverter States throughout Second Cycle

N.S.H.C.	A	B	C	D	E
Inv-1	Switching mode during the entire second cycle				
Inv-2	Fixed to4 (+ - -)	Fixed to5 (+ + -)	Fixed to6 (- + -)	Fixed to1 (- + +)	Fixed to2 (- - +)

SIMULATION AND RESULTS

Experimental tests were carried out on a dual 2-level inverter system employing an open winding PMSM to evaluate the efficiency of the suggested unified SVPWM algorithm. The table contains the parameters concerning the PMSM. Over the course of the experiments, the combined voltages of the two DC links, V_{dc1} and V_{dc2}, were consistently maintained at 140 V. The Alternate inverter PWM switching strategy in Simulink is given in Figure 11.

Figure 11. Alternate inverter pwm switching strategy

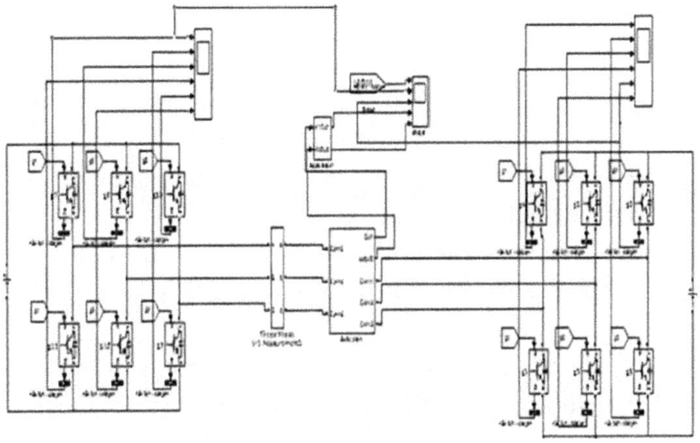

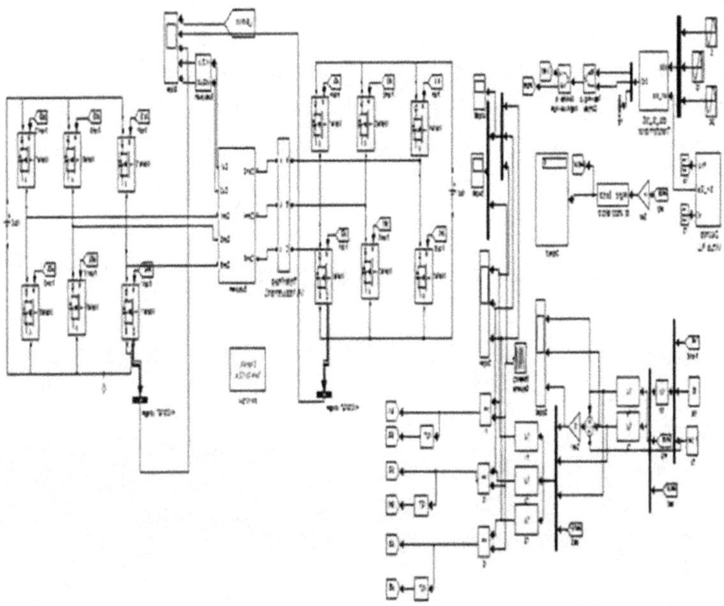

Tests were conducted on an open-end winding PMSM using a dual two-level inverter system to assess the suggested unified SVPWM algorithm's effectiveness.

Case 1: Alternate inverter PWM at 500 r/min Vs Unified SVPWM at 500 r/min at k= 70:70=1

The results demonstrate that

- At 500 rpm, INV1 and INV2 represent the motor's torque and three-phase current. Their waveforms exhibit three-phase pole voltages in alternate inverter PWM (as depicted in Figure 12).
- The chosen voltage vector in different sectors of sector I is displayed in enlarged views. Results for the alternate inverter PWM at k=1 show effective performance in both algorithms.
- Nevertheless, there is observable distortion in current and torque outputs in the specified method, even with a slight difference in the two DC voltages for k = 1.
- As seen in the graphical representation, the performance remains slightly stable.
- Current harmonics are observed as satisfactory for unified SVPWM inverter
- Harmonics are under limit as per IEEE standard and hence no issue to operate the unified SVPWM in its as-is mode as represented in Figure 13.

Figure 12. A case study on alternate inverter PWM and unified SVPWM for motor speed = 500 rev/m

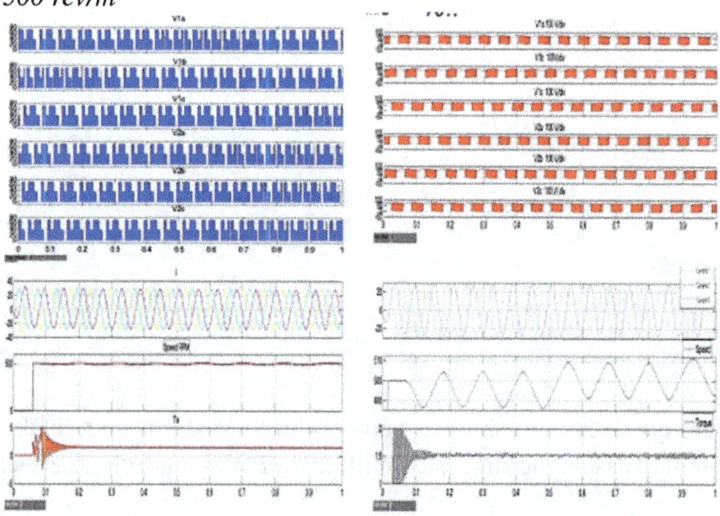

Figure 13. Current harmonic profiles at 500 rpm

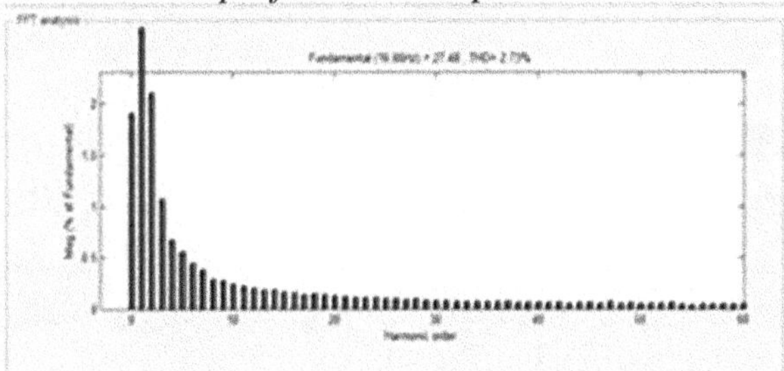

Case 2: Alternate inverter PWM at 1000 r/min Vs Unified SVPWM at 1000 r/min at k= 94:46

The results in case 2 are presented in simulation results, as graphically shown in Figure 14.

- Both inverters support producing the appropriate motor output torque for which the required three-phase current is required.
- Waveforms shown in the figure also visualize three-phase pole voltages. The voltage vectors that were chosen in various sectors are displayed in enlarged perspectives.
- The motor speed is maintained at 1000 rpm.
- Results display effective performance for both PWM switching methods in the inverter.
- The unified SVPWM technique exhibits controllable distortion. Minor variation in the DC voltages similarly affects current and torque outputs (k = 94:46).
- As seen in the graph, the performance remains stable no matter how large the voltage differential is.
- Current harmonics are also within the limit as per IEEE and IE standards.
- The unified SVPWM design appears to outperform alternate PWM inverters, suggesting its superiority as shown in Figure 15.

Figure 14. A case study on alternate inverter PWM and unified SVPWM for motor speed = 1000 rev/m

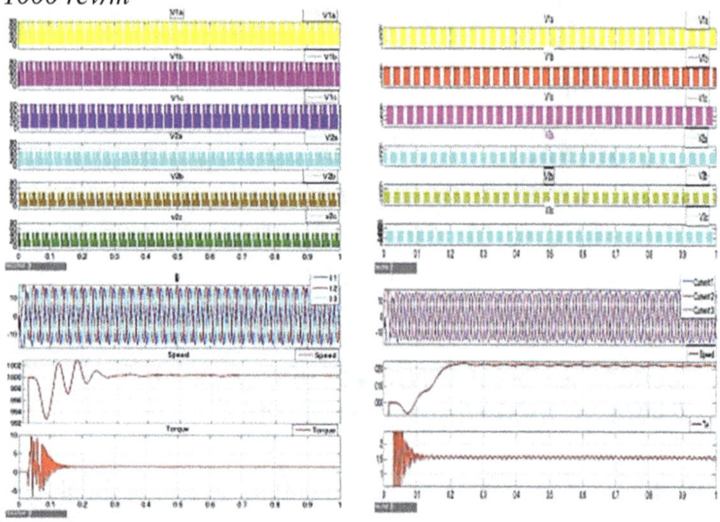

Figure 15. Current harmonic profiles at 1000 rpm

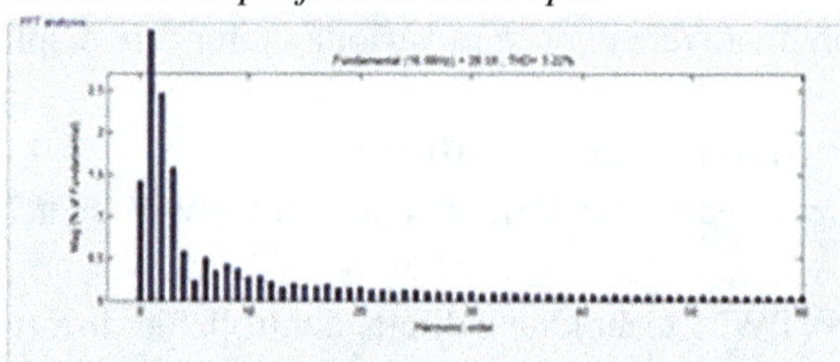

Case 3: Alternate inverter PWM at 1500 r/min Vs Unified SVPWM at 1500 r/min at k= 105:35

The concluding remarks for case 3 are

- The results demonstrate that both inverters produce resultant torque and 3-ϕ currents, respectively.
- Three-phase pole voltages may be seen in Figure 16. The voltage vectors that were chosen in various sectors are displayed in enlarged perspectives.

- At the operating speed of 1500 rpm, the output of the alternative inverter PWM switching technique is displayed in Figure 16.
- As seen in Figure 16, the performance remains stable no matter how large the voltage differential is.
- The unified SVPWM algorithm performs well across various speeds and allows adjustable ratios for two DC voltage supplies.

Figure 16. A case study on alternate inverter PWM and unified SVPWM for motor speed = 1500 rev/m

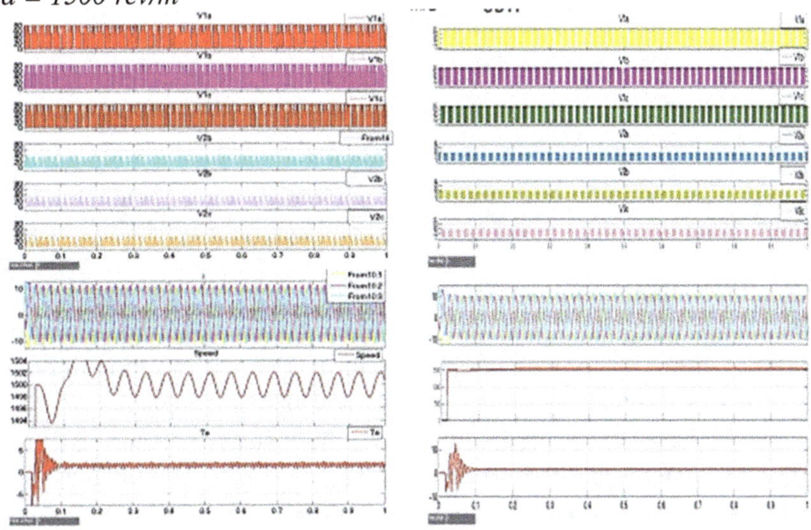

Harmonics in Unified SVPWM under different DC link voltages

Comparative analysis of harmonics in Unified SVPWM under different DC link voltages is shown in Figure 17. In any of the case studies, the current harmonic level does not exceed 2.73, whereas a minimum of 2.66 is observed for k=70/70.

Figure 17. Harmonics in Unified SVPWM under different DC link voltages

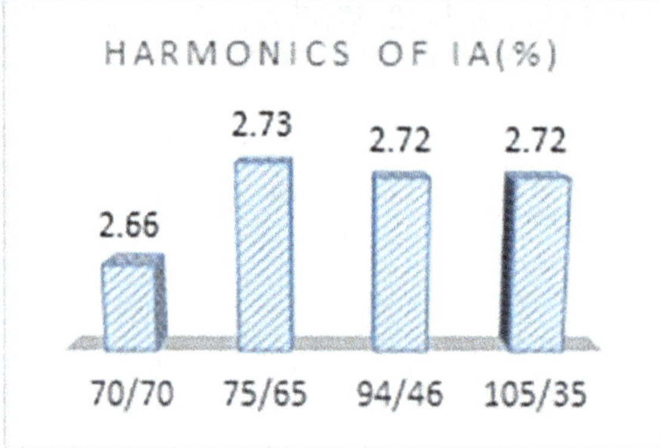

RESULTS

The optimizer iteratively searches for switching angles to minimize harmonics as an outcome of this tool. A decision step checks whether optimized angles are realized or not. If not, the process loops back for further optimization through the next iteration. Once optimized angles are obtained, they are applied to a two-level inverter, which drives the AC machine. Finally, the system performs FFT analysis to calculate Total Harmonic Distortion (THD) and generate harmonic values, ensuring effective performance evaluation of the inverter machine system. Simulink Model was planned for two Level Inverter with SPWM & PSO along with 3 phase AC machine (3 kW / 415 V L-L) using Unipolar Sinusoidal PWM (US SPWM). PSO (Particle Swarm Optimization) is used offline to optimize switching angles. A step by step build guide for a full simulation in MATLAB/Simulink for the purpose of Quick design choices & numerical targets as enlisted below. It is necessary to validate any intelligent tool. Herein, particle swarm intelligence was also tested and verified with real test data. All the coefficients used in the swarm intelligence equation were tuned properly along with the random variable. Final results were satisfactory and showed an increase in efficiency and a reduction of THD. After the PSO loop completes, the script outputs the final g_best (global best) set of switching angles in both radians and degrees. These are the optimal angles found by the algorithm that best satisfy the fitness function's objectives.

Simulink Model was planned for 2-Level Inverter with SPWM & PSO along with 3 phase AC machine (3 kW / 415 V L-L) using Unipolar Sinusoidal PWM (US SPWM). PSO (Particle Swarm Optimization) is used offline to optimize switching angles. A step-by-step build guide for a full simulation in MATLAB/Simulink for the purpose of Quick design choices & numerical targets as enlisted below.

- Machine: 3 kW, 4 pole, 50 Hz, 415 V (Line to line, RMS)
- Phase RMS voltage: V_phase_rms = 415 / $\sqrt{3}$ = 239.6 V
- Phase peak voltage: V_phase_pk = 239.6 × $\sqrt{2} \approx$ 339 V
- Desired modulation approach: Unipolar SPWM (US SPWM) with optional SVPWM
- Vdc (DC link): choose Vdc ≈ 750 V
- Switching frequency: fsw = 8 kHz (typical); carrier period Tcar = 125 μs
- Simulink solver sample time (recommended): Ts = 5e 6 s (25 samples per carrier).
- PWM dead time: 2–4 μs (for realistic IGBT/MOSFET gating)
- Filter (optional): Lf per phase = 200 μH (tune per motor); Cf = 10 μF
- DC link capacitance: Cdc = 2000 μF, ESR appropriate for ripple
- Motor thermal inertia (electromechanical): J = 0.089 kg·m^2 (example), B = 0.001 N·m·s

Note: The numerical motor parameters below are example values to let the Simulink model run quickly. Replace them with actual nameplate or datasheet values for precise results.

Table 5. PSO Parameter List

Parameter	Value
Swarm or population size	30
Max possible iterations	200
Inertia (w)	0.7-0.9
Cognitive coefficient	1.4-1.5
Social coefficient	1.4-1.5

Swarm intelligence use it's own format of PSO. It uses the parameter, and the parameter values are tabulated in Table 6 (Patel, C., et.al., 2012).

These parameters are used to find the switching angles required in a novel inverter. The outcome of the PSO that switches angles is tabulated in Table 7 (Sekhar & Srinivas et al., 2013).

PSO offline script (high level outline)

i) Define decision variables (e.g., carrier phase offsets φa, φb, φc or N switching angles θ1..θN).
ii) Form cost function: compute time-domain stepping waveform for one fundamental period using candidate parameters, compute FFT → THD metric (or weighted harmonic targets).
iii) PSO parameters: population = 30–60, maxIter = 200, inertia w = 0.7→0.4, cognitive c1 = 1.5, social c2 = 1.5 (as described in Table 5).
iv) Constraints: keep volt/phase fundamental amplitude within ±1% of target; enforce angle ordering.
v) Run PSO for each operating point (speed/torque) and save optimizedParams.

Table 6. Optimized Switching Angles (Example for N=3, m=0.85)

Angle (rad)	Angle (deg)
0.247	14.21°
0.714	40.87°
1.188	68.03°

Test Results:

a) No-load startup: step speed from 0 to 1440 rpm, log current, torque, speed response.
b) Rated load steady state: apply rated torque (~14 N·m for 3 kW at 1440 rpm), measure THD of line voltage and phase currents.
c) Torque step: apply a sudden torque step ±50% rated and measure current control settling.
d) PSO comparison sweep: for several speeds (0%, 25%, 50%, 75%, 100%) run baseline US SPWM, PSO optimized US SPWM, and SVPWM to compare THD and torque ripple.
e) Regeneration scenario: apply negative torque and validate DC link voltage rise capturing (add braking chopper if needed).

Table 7. Harmonic Spectrum Before vs After Optimization

Harmonic Order	Without SHE (SPWM)	With PSO-SHE	Improvement (%)
5th	12.5%	0.5%	96%
7th	8.1%	0.4%	95%
11th	4.9%	0.2%	96%
13th	3.2%	0.1%	97%
THD	18%	3%	83%

All the swarm intelligence is used to find the tuned switching angle, and accordingly, the inverter is utilized for improving efficiency.

With this, the harmonic spectrum before and after swarm optimization is tabulated in Table 8 (V. T. Somasekhar et.al., 2005). This indicates that swarm intelligence is helpful and the improvement is seen as 83% minimizing in THD and 95-97% in harmonic spectrum. This will remove the need for filters required for especially third harmonic. It is also represented in Figure 18 (Akanksha et.al., 2023), where THD reduction may be seen.

Figure 18. Harmonic Spectrum Before vs After Optimization

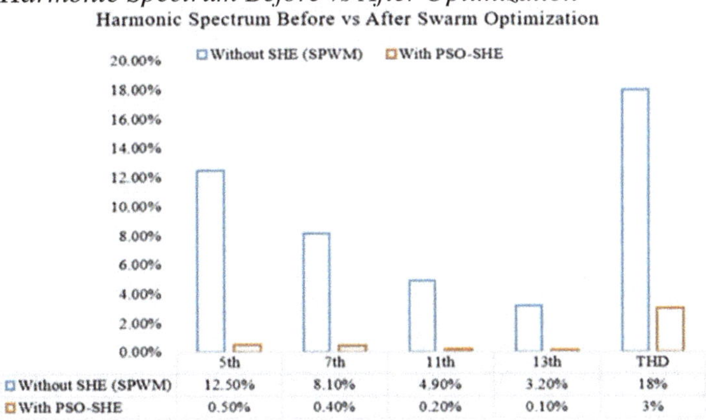

Table 8. Inverter Output Performance

Metric	Value
Fundamental Voltage	0.85 pu
Target Harmonics Removed	5, 7, 11, 13
THD (%)	~3%

CONCLUSION

This research work adopted the swarm optimization technique and evaluated PWM and Alternate Inverter PWM methodologies. Simulink models were developed and assessed using the MATLAB/SIMULINK environment to compare two different approaches. The total harmonic distortion (THD) was computed to compare simulation results for nonlinear loads. The findings indicate that Unified SVPWM demonstrates greater effectiveness in reducing harmonics within nonlinear loads. Different switching frequencies were analyzed with the help of PSO to examine current distortion, revealing that Unified Space Vector modulation significantly reduces THD more effectively across a wide speed range compared to Alternate Inverter pulse width modulation. Using swarm technology, the frequency spectrum improved by 95-97% and the THD reduction was by 83%. Swarm intelligence helps in removing the 3rd, 5th, 7th, and 11th harmonics. Overall, THD was reduced from 18% to 3% which highlights the importance of swarm intelligence. Concluding remarks accepted herein as a swarm intelligence tool if selected and used properly by tuning its social and cognitive coefficients, an increase in efficiency and a decrease in THD losses are seen as proven in the results.

REFERENCES

Anand, S., Kulkarni, S. V., Gupta, S., & Parthasarathy, V. (2022, December). Battery management system (BMS) for electric vehicle applications. In 2022 fourth international conference on emerging research in electronics, computer science and technology (ICERECT) (pp. 1-6). IEEE. DOI: 10.1109/ICERECT56837.2022.10060137

Awasthi, B., Gupta, S., Kumar, S., & Tripathi, A. (2024, January). Business management and planning using queuing model with heterogeneous customers for electrical applications. In 2024 2nd International Conference on Intelligent Data Communication Technologies and Internet of Things (IDCIoT) (pp. 279-283). IEEE. DOI: 10.1109/IDCIoT59759.2024.10467673

Chen, M., & Sun, D. (2017). A Unified Space Vector Pulse Width Modulation for Dual Two-level Inverter System. *IEEE Transactions on Power Electronics*, *32*(2), 889–893. DOI: 10.1109/TPEL.2016.2585223

Fotis, G., Gupta, S., Varshney, T., Dhar, S. L., & Gulzar, M. M. (Eds.). (2025). *Modern Computing Technologies for EV Efficiency and Sustainable Energy Integration*. IGI global., DOI: 10.4018/979-8-3373-2382-4

Gupta, S. (2025). *The Future of Machine Learning with Automation in Engineering*. Nova Science Publishers., DOI: 10.52305/YFIX2832

Gupta, S., & Tripathi, R. K. (2014). Two-area power system stability improvement using a robust controller-based CSC-STATCOM. *Acta Polytechnica Hungarica*, *11*(7), 135–155. DOI: 10.12700/APH.11.07.2014.07.9

Kapoor, G., Gakhar, P., Verma, P., Sonwane, P., & Rathore, B. (2021). A Double-Ended Scheme for Fault Identification and Phase-Type Categorization in Fixed Series Compensated Transmission Line during Distorted Signals. *2021 6th IEEE International Conference on Recent Advances and Innovations in Engineering (ICRAIE)*, 1–4. DOI: 10.1109/ICRAIE52900.2021.9703834

Kim, J.-S., & Sul, S.-K. (1996). A Novel Voltage Modulation Technique of the Space Vector PWM. *IEEJ Transactions on Industry Applications*, *116*(8), 820–825. DOI: 10.1541/ieejias.116.820

Kumari, A., Gopal, Y., Dhaked, D. K., & Panda, K. P. Y.N. Vijaya Kumar, A single source five-level switched-capacitor based multilevel inverter with reduced device count, *e-Prime - Advances in Electrical Engineering, Electronics and Energy*, Volume 5,2023,100235,ISSN 2772-6711.

Kumari, A., Gopal, Y., Dhaked, D. K., Panda, K. P., & Kumar, Y. N. V. (2023). A single source five-level switched-capacitor based multilevel inverter with reduced device count. *E-Prime - Advances in Electrical Engineering. Electronics and Energy*, *5*, 100235. DOI: 10.1016/j.prime.2023.100235

Lee, Y.-H., Kim, D.-H., & Hyun, D.-S. (n.d.). Carrier based SVPWM method for multi-level system with reduced HDF [harmonic distortion factor]. *Conference Record of the 2000 IEEE Industry Applications Conference. Thirty-Fifth IAS Annual Meeting and World Conference on Industrial Applications of Electrical Energy (Cat. No.00CH37129)*, 1996–2003. DOI: 10.1109/IAS.2000.882151

Loh, P. C., Holmes, D. G., & Lipo, T. A. (2005). Implementation and control of distributed PWM cascaded multilevel inverters with minimal harmonic distortion and common-mode voltage. *IEEE Transactions on Power Electronics*, *20*(1), 90–99. DOI: 10.1109/TPEL.2004.839830

Meena, G., Singh, V. P., Dixit, A., Mathur, A., & Bhatt, A. (2024, April). Steady-state models of ac microgrid for load flow solutions. In 2024 1st International Conference on Innovative Sustainable Technologies for Energy, Mechatronics, and Smart Systems (ISTEMS) (pp. 1-6). IEEE. DOI: 10.1109/ISTEMS60181.2024.10560356

Nabae, A., Takahashi, I., & Akagi, H. (1981). A New Neutral-Point-Clamped PWM Inverter. *IEEE Transactions on Industry Applications*, *IA-17*(5), 518–523. DOI: 10.1109/TIA.1981.4503992

Patel, C., P. P., R., Dey, A., Ramchand, R., Gopakumar, K., & Kazmierkowski, M. P. (2012). Fast Direct Torque Control of an Open-End Induction Motor Drive Using 12-Sided Polygonal Voltage Space Vectors. *IEEE Transactions on Power Electronics*, *27*(1), 400–410. DOI: 10.1109/TPEL.2011.2159516

Pradabane, S., Narasimharaju, B. L., & Srikanth, N. V. (2014). Two-quadrant clamping inverter scheme for three-level open-end winding induction motor drive. *2014 IEEE International Conference on Power Electronics, Drives and Energy Systems (PEDES)*, 1–4. DOI: 10.1109/PEDES.2014.7041988

Pradabane, S., Srikanth, N. V., & Narasimharaju, B. L. (2014). A new alternate Fixedbias inverter SVPWM scheme for open-end winding induction motor drive. *International Review of Electrical Engineering*, *9*(1).

Rao, M. S., Saraswathy, S., Neela, V., Arun, M. R., Gupta, S., & Taqui, S. N. (2023, November). Efficient Model Training in Decentralized Systems with Federated Learning. In *2023 International Conference on Communication, Security and Artificial Intelligence (ICCSAI)* (pp. 900-904). IEEE. DOI: 10.1109/ICCSAI59793.2023.10421469

Rufer, A., Veenstra, M., & Gopakumar, K. (1999). Asymmetric Multilevel Converter for High Resolution Voltage Phasor Generation. *Proceedings of the 1999 European Conference on Power Electronics and Applications (EPE 1999).*

Sekhar, K. R., & Srinivas, S. (2013). Discontinuous Decoupled PWMs for Reduced Current Ripple in a Dual Two-Level Inverter Fed Open-End Winding Induction Motor Drive. *IEEE Transactions on Power Electronics, 28*(5), 2493–2502. DOI: 10.1109/TPEL.2012.2215344

Semwal, S., Kant, S., Bandyopadhyay, K., Asim, M., Garg, R., & Goyal, S. (2024, October). Implicit Control of DC-DC Converter based DC Motor for the Charging of Electric Vehicle using Fuzzy Logic Controller. In 2024 5th IEEE Global Conference for Advancement in Technology (GCAT) (pp. 1-6). IEEE. DOI: 10.1109/GCAT62922.2024.10923979

Sen, P., Sahoo, A. K., Panda, K. P., & Jha, V. (2023). Single-phase switched-capacitor boost multilevel inverter interfacing solar photovoltaic system. *E-Prime - Advances in Electrical Engineering. Electronics and Energy, 6*, 100350. DOI: 10.1016/j.prime.2023.100350

Sharma, R. K., Saran, V. B., Kanaujia, S., & Gupta, S. (2024, May). Evaluation of mppt controller performance using fuzzy logic design with wireless sensor node. In *2024 International Conference on Smart Systems for applications in Electrical Sciences (ICSSES)* (pp. 1-6). IEEE. DOI: 10.1109/ICSSES62373.2024.10561365

Shukla, R. D., Tripathi, R. K., & Gupta, S. (2010, November). Power electronics applications in wind energy conversion system: A review. In *2010 International Conference on Power, Control and Embedded Systems* (pp. 1-6). IEEE. DOI: 10.1109/ICPCES.2010.5698663

Somasekhar, V. T., Srinivas, S., & Gopakkumar, K. A space vector based PWM switching scheme for the reduction of common-mode voltages for a dual inverter fed open-end winding induction motor drive, *Power Electronics. Specialists Conference*, 2005. PESC '05. IEEE 36th, 2005, pp. 816-821. DOI: 10.1109/PESC.2005.1581721

Somasekhar, V. T., Srinivas, S., & Kumar, K. K. (2008). Effect of Zero-Vector Placement in a Dual-Inverter Fed Open-End Winding Induction Motor Drive With Alternate Sub-Hexagonal Center PWM Switching Scheme. *IEEE Transactions on Power Electronics, 23*(3), 1584–1591. DOI: 10.1109/TPEL.2008.921170

Sonwane, P., Yadav, M., Bunker, N., Sonwane, S., & Shirsath, V. (2024). Economical Design Perspective for Aircraft by Optimizing Airfoil S1223. *E-Prime - Advances in Electrical Engineering. Electronics and Energy, 8*, 100531. DOI: 10.1016/j.prime.2024.100531

Srinivas, S., & Ramachandra Sekhar, K. (2013). Theoretical and Experimental Analysis for Current in a Dual-Inverter-Fed Open-End Winding Induction Motor Drive With Reduced Switching PWM. *IEEE Transactions on Industrial Electronics, 60*(10), 4318–4328. DOI: 10.1109/TIE.2012.2209615

Srinivas, S., & Somasekhar, V. T. (2008). Space-vector-based PWM switching strategies for a three-level dual-inverter-fed open-end winding induction motor drive and their comparative evaluation. *IET Electric Power Applications*, 2(1), 19–31. DOI: 10.1049/iet-epa:20070134

Stemmler, H. (1994). High-power industrial drives. *Proceedings of the IEEE, 82*(8), 1266–1286. DOI: 10.1109/5.301688

Thandlam, A. K., Vignesh, T., Saravanan, A., Subramani, P., Marimuthu, R., & Gupta, S. (2024, April). Next-gen vehicle safety: The futuristic approach to auto-stop in modern vehicles. In 2024 10th International Conference on Communication and Signal Processing (ICCSP) (pp. 265-270). IEEE. DOI: 10.1109/ICCSP60870.2024.10543666

Tyagi, M., Gupta, S., & Khan, M. I. (2024, May). Reduction of voltage instability in complex power networks through the utilization of D-STATCOM. In 2024 4th International Conference on Advance Computing and Innovative Technologies in Engineering (ICACITE) (pp. 1662-1666). IEEE. DOI: 10.1109/ICACITE60783.2024.10616856

Welchko, B. A. (2005). A double-ended inverter system for the combined propulsion and energy management functions in hybrid vehicles with energy storage. *31st Annual Conference of IEEE Industrial Electronics Society, 2005. IECON 2005.*, 6 pp. DOI: 10.1109/IECON.2005.1569110

Chapter 8
A Hybrid Q-Learning Decision Support Framework for Human-AI Interactive Energy Management in Industry 5.0

Vishal Sharma
https://orcid.org/0009-0009-0228-6384
National Institute of Technology, Delhi, India

Surya Karan Singh
https://orcid.org/0000-0001-9260-9216
National Institute of Technology, Delhi, India

Amit Kumar Singh
https://orcid.org/0000-0001-6711-2654
National Institute of Technology, Delhi, India

ABSTRACT

Industry 5.0 revolutionizes industrial energy management by fusing human–AI collaboration within cyber-physical ecosystems to enable adaptive, resilient, and sustainable operations. This proposal introduces a Hybrid Q-learning + Rule-based Energy Management System (EMS) that unites reinforcement learning's optimization capabilities with rule-based security to harmonize distributed energy resources in a grid-connected hybrid renewable system. Experimental validation on a realistic

DOI: 10.4018/979-8-3373-7041-5.ch008

microgrid demonstrates superior performance—Hybrid Q+Rule EMS achieves an operational profit of $0.47, while the conventional rule-based EMS incurs a $1.44 cost. The improvement stems from a 37% reduction in grid imports, $1.67 grid export revenue, and 100.4% renewable penetration. Wind power utilization rises to 89.6% (vs. 75%), with smart battery arbitrage transforming the system from cost-saving to profit-generating. This hybrid framework establishes a new paradigm for intelligent, sustainable, and economically resilient industrial energy management.

INTRODUCTION

Industrial industry today is found in a very pivotal transition phase which is characterized with the transformative movement of Industry 4.0 to Industry 5.0 (Romanova and Kuzmin, 2024; Mourtzis, 2024). Though Industry 4.0 introduced a paradigm of automation, data exchange, and cyber-physical systems, the concept of operational efficiency and productivity was at the center, whereas Industry 5.0 is a core correction in the direction of a human-focused approach to the sector of industrial activity, resilience, and sustainability (Awasthi et al., 2024; Thandlam et al., 2024). The new paradigm does not understand the factory floor as a fully autonomous system but as an exploratory ecosystem where the intelligence, creativity, and contextual reasoning of humans combine with the artificial intelligence (AI) computational power, accuracy, and predictability (Gupta, 2025; Johri et al., 2025). In associated energy-intensive industrial settings, this change requires an appropriate level of redesign of Energy Management Systems (EMS), which require designs that are not only adaptive and self-optimising, but also interpretable, credible, and capable of smooth interaction with human decision-makers, including the integration of human values and operational skills directly into the control loop. The need to achieve intelligent energy management has triggered broad research on the intersection of artificial intelligence and industrial control systems, and the Human-AI symbiosis approach has been recognized as an essential mechanism of optimization of efficiency and sustainability in various industries. Human expertise in the use of wireless sensor networks coupled with AI has been shown to make energy harvesting and routing solutions better to achieve a longer lifespan of the network and greater reliability (Mishra and Nandy, 2026). Similarly, in the domestic setting, Human-in-the-Loop learning algorithms automatically optimise the environmental setups depending on the behaviour of occupants with considerable energy conservation without affecting comfort (Verma et al., 2019; Rao et al., 2023). The adaptive control of HVAC and lighting systems by incorporating AI-driven building management has recorded 15-30% efficiency savings (Mane, 2026; Olatunde et al., 2024), and the deep integration of renewable energy sources, being AI-oriented, maintains the operational stability of vital infrastructure, including health facilities (Ukoba et al., 2024). Similar improvements have demonstrated parallel hybrid learning and reinforcement learning models to be highly effective in multi-objective energy optimisation in complex dynamic environments. As an instance, deep Q-network techniques have been used on hybrid electric vehicles, with a fuel efficiency of 5.31 L/100km and emissions reduction at the same time (Sharma et al., 2024; Zhang et al., 2022). Elsewhere in energy networks, Q-learning-based optimisers have prompted significant operational cost and carbon footprint reductions (Tu et al., 2024; Li et al., 2023). Algorithms like Q-Managed can be used to robustly explore Pareto-optimal solutions in the trade-off of competing energy goals (Fotis et al., 2025; Oliveira et al., 2021). In addition, hybrid RL models have also been shown to optimise revenue and energy cost of production in industries simultaneously, which highlights their relevance in manufacturing settings (Lu et al., 2024). In an effort to alleviate the lack of transparency of most RL models, an increasing number of scientific papers are urging a shift to explicable Decision Support Systems directing the implementation of RL with Bayesian reasoning, fuzzy logic, and edge computation to provide transparent, real-time operational guidance in industrial applications (Cochran et al., 2022; Kostopoulos et al., 2024; Almtrf, 2025), Pradheeba et al., 2025). Taken together, these studies point to the fact that the integration of advanced machine learning with human opinion and explainable AI tenets strengthens the resilience, adaptability, and reliability of modern energy infrastructure (Kostopoulos et al., 2024; Almtrf, 2025; Pradheeba et al., 2025; Sayed, 2025).

In spite of these improvements, there is a significant gap in the literature with no formal measures of or architectural modeling of human-AI partnership in the industrial energy management systems. Recently, it highlights the need of Human-in-the-Loop Decision Empowerment (HiL-DE) metrics to measure how much human feedback can be used to increase model interpretability and trust without impacting the quality of would like algorithm performance (Sele & Chugunova, 2023; Natarajan et al., 2024; Natarajan et al., 2025), and the Human-in-the-Learning-Loop (HILL) paradigm explores the means to formalise continuous human feedback on decision-making adaptation and ethical decision-making (Sele & Chugunova, 2023; So, 2020). However, a detailed construct that effectively incorporates adaptive reinforcement learning and a systematic empirically measurable layer of empowerment of industrial energy management is not yet fully realised. At the same time, Industry 5.0 studies are progressively focusing on human-centric production, resource optimization, and sustainability with the help of smart decision-making, and researchers state that Industry 5.0 reflects a transition into collaborative intelligence and responsible innovation (Anand et al., 2022; Ficzere, 2024).

To address this gap, the current research presents a new Hybrid -Q + Rule -Based Energy Management System that harmonises the exploratory, long-term optimisation of reinforcement learning and operational safety and interpretability of a deterministic rule-based system, in a designed flow that implements human control. The main among the contributions was that the work developed the Hybrid plus Rule + EMS which hybridises the adaptive RL with the more traditional deterministic rule-based security to enable stable operation; the work also developed the multi-objective optimisation formulation that aims at simultaneously ensuring economic efficiency, utilisation of renewable sources, and battery lifespan; it was also demonstrated in the work, with a rigour and working battery life of 24 hrs, that the framework is able to turn the system into a profit-making consumer rather than an entity causing costs and then a waste, and realise a net

SYSTEM MODELLING AND PROBLEM FORMULATION

The current section outlines a strict mathematical background of the building blocks of a grid-connected industrial microgrid, directly defining operation constraints and energy-management objective function. The hybrid includes an arrangement of natural-gas-fed fuel cell, stochastic renewable generation, battery storage, and bi-directional grid interface (Ali et al., 2024; Patel et al., 2024).

System Architecture and Power Flow Equations

The microgrid is operated in a grid-connected mode, with the alternating-current (AC) bus needing a constant power balance (Gupta & Tripathi, 2014; Mannan et al., 2024). The basic equation of power balance at a given discrete time point at an instant (t) is given by:

$$P_{load}(t) + P_{batt,c}(t) + P_{grid,exp}(t) = P_{pv}(t) + P_{wind}(t) + P_{fc}(t) + P_{batt,d}(t) + P_{grid,imp}(t) \quad (1)$$

Variable Definitions:

- $\left(P_{load}(t)\right)$: AC load demand on buses overall [kW]
- $\left(P_{pv}(t)\right)$: Power developed by the photovoltaic (PV) array [kW]
- $\left(P_{wind}(t)\right)$: Wind turbine power output [kW]
- $\left(P_{fc}(t)\right)$: Fuel cell power output [kW]
- $\left(P_{batt,c}(t)\right)$: Battery charging power [kW]
- $\left(P_{batt,d}(t)\right)$: Battery discharging power [kW]
- $\left(P_{grid,imp}(t)\right)$: Power imported from grid [kW]
- $\left(P_{grid,exp}(t)\right)$: Power exported to grid [kW]

Complementary Constraints:

$$P_{\text{batt,c}}(t) \cdot P_{\text{batt,d}}(t) = 0 \quad \forall t \tag{2}$$

$$P_{\text{grid,imp}}(t) \cdot P_{\text{grid,exp}}(t) = 0 \quad \forall t \tag{3}$$

$$P_{\text{batt,c}}(t), P_{\text{batt,d}}(t), P_{\text{grid,imp}}(t), P_{\text{grid,exp}}(t) \geq 0 \quad \forall t \tag{4}$$

These restrictions do not allow charging and discharging of the battery as well as importing and exporting simultaneously off the grid

Component Mathematical Modelling

Photovoltaic (PV) System Model

The PV output power is obtained as a determination of incident solar irradiance and ambient temperature (Karthick et al., 2024; Abbas et al., 2024):

$$P_{\text{pv}}(t) = P_{\text{pv,rated}} \cdot \frac{G(t)}{G_{\text{STC}}} \cdot \left[1 + \alpha_p \cdot \left(T_c(t) - T_{c,\text{STC}}\right)\right] \cdot \eta_{\text{pv}} \cdot \eta_{\text{inv}} \tag{5}$$

Where:

- $\left(P_{\text{pv,rated}}\right)$: Rated PV capacity under Standard Test Conditions (STC) [kW]
- $(G(t))$: Solar irradiance incident on PV [W/m^2]
- $\left(G_{\text{STC}}\right)$: Reference irradiance at STC (1000 W/m^2)
- $\left(T_c(t)\right)$: PV cell temperature [°C]
- $\left(T_{c,\text{STC}}\right)$: PV cell temperature at STC (25°C)
- $\left(\alpha_p\right)$: Power temperature coefficient [%/°C] (typically -0.3 to -0.5 \%/°C)
- $\left(\eta_{\text{pv}}\right)$: PV array efficiency [%]
- $\left(\eta_{\text{inv}}\right)$: Inverter efficiency [%]

The cell temperature is calculated through the Nominal Operating Cell Temperature (NOCT) model (Bayrak (2025), Ignatev et al. (2025)):

$$T_c(t) = T_{\text{amb}}(t) + \left(\frac{\text{NOCT} - 20}{800}\right) \cdot G(t) \tag{6}$$

Where:

- $\left(T_{\text{amb}}(t)\right)$: Ambient temperature [°C]
- (NOCT): Nominal Operating Cell Temperature [°C] (typically 40-48°C)

Operational Constraint:

$$0 \leq P_{pv}(t) \leq P_{pv,rated} \quad \forall t \tag{7}$$

Wind Turbine System Model

The power output of the turbine is modeled as the characteristic curves as a relation of the power to the wind speed (Marzouk, 2025; Yu et al., 2025):

$$P_{WECS}(t) = \begin{cases} 0 & V(t) < V_{cut\text{-}in} \text{ or } V(t) > V_{cut\text{-}out} \\ P_{wind,rated} \cdot \left(\frac{V(t) - V_{cut-in}}{V_{rated} - V_{cut-in}} \right)^3 & V_{cut\text{-}in} \leq V(t) < V_{rated} \\ P_{wind,rated} & V_{rated} \leq V(t) \leq V_{cut\text{-}out} \end{cases} \tag{8}$$

Parameters:

- ($V(t)$): Wind speed at hub height [m/s]
- ($V_{cut\text{-}in}$): Cut-in wind speed [m/s] (typically 3-4 m/s)
- V_{rated}): Rated wind speed [m/s] (typically 11-15 m/s)
- ($V_{cut\text{-}out}$): Cut-out wind speed [m/s] (typically 20-25 m/s)
- ($P_{wind,rated}$): Rated wind turbine power [kW]

Operational Constraint:

$$0 \leq P_{wind}(t) \leq P_{wind,rated} \quad \forall t \tag{9}$$

Natural Gas Fuel Cell Model

The fuel cell uses natural gas (methane, CH_4) through a reforming process (Lyu et al., 2025; He et al., 2024). The electrical energy performance and fuel management are in the form as shown below:

Electrical Power Output:

$$P_{fc}(t) = \eta_{fc,elec} \cdot \dot{m}_{NG}(t) \cdot \text{LHV}_{NG} \tag{10}$$

Fuel Consumption Rate:

$$\dot{m}_{NG}(t) = \frac{P_{fc}(t)}{\eta_{fc,elec} \cdot \text{LHV}_{NG}} \tag{11}$$

Where:

- $(\eta_{\text{fc,elec}})$: Electrical efficiency of fuel cell [%]
- $(\dot{m}_{\text{NG}}(t))$: Mass flow rate of natural gas, including natural gas outlet. [kg/s]
- (LHV_{NG}): Reduced heating worth of natural gas (≈ 50,000 kJ/kg)

Reforming Process Chemical Equation:

$$CH_4 + 2H_2O \rightarrow CO_2 + 4H_2$$

(Steam Methane Reforming) (12)

$$2H_2 + O_2 \rightarrow 2H_2O$$

(Fuel Cell Electrochemical Reaction) (13)

Total Reaction:

$$CH_4 + 2O_2 \rightarrow CO_2 + 2H_2O \tag{14}$$

Operational Constraints:

$$P_{\text{fc,min}} \leq P_{\text{fc}}(t) \leq P_{\text{fc,rated}} \quad \forall t \tag{15}$$

$$|P_{\text{fc}}(t) - P_{\text{fc}}(t-1)| \leq \Delta P_{\text{fc,max}} \quad \forall t \tag{16}$$

Where:

- $(P_{\text{fc,min}})$: Minimum stable operating power [kW]
- $(P_{\text{fc,rated}})$: Rated fuel cell power [kW]
- $(\Delta P_{\text{fc,max}})$: Maximum ramping rate [kW/min]

Battery Energy Storage System (BESS) Model

State of charge (SOC) variable is used to capture the battery dynamics (Aljohani and Almutairi, 2024):

SOC Evolution Equation:

$$\text{SOC}(t+1) = \text{SOC}(t) + \frac{\eta_{\text{batt,c}} \cdot P_{\text{batt,c}}(t) \cdot \Delta t}{E_{\text{batt}}} - \frac{P_{\text{batt,d}}(t) \cdot \Delta t}{\eta_{\text{batt,d}} \cdot E_{\text{batt}}} \tag{17}$$

Where:

- $(\mathrm{SOC}(t))$: State of Charge at time t [%]
- $\left(\eta_{\mathrm{batt,c}}\right)$: Battery charging efficiency [%]
- $\left(\eta_{\mathrm{batt,d}}\right)$: Battery discharging efficiency [%]
- $\left(E_{\mathrm{batt}}\right)$: Total battery energy capacity [kWh]
- (Δt): Time step duration [hours]

Power and Energy Constraints:

$$0 \leq P_{\mathrm{batt,c}}(t) \leq P^{\max}_{\mathrm{batt,c}} \cdot u_{\mathrm{batt,c}}(t) \quad \forall t \ (18)$$
$$0 \leq P_{\mathrm{batt,d}}(t) \leq P^{\max}_{\mathrm{batt,d}} \cdot u_{\mathrm{batt,d}}(t) \quad \forall t \ (19)$$
$$u_{\mathrm{batt,c}}(t) + u_{\mathrm{batt,d}}(t) \leq 1 \quad \forall t \ (20)$$
$$\mathrm{SOC}_{\min} \leq \mathrm{SOC}(t) \leq \mathrm{SOC}_{\max} \quad \forall t \ (21)$$

Where:

- $\left(P^{\max}_{\mathrm{batt,c}}, P^{\max}_{\mathrm{batt,d}}\right)$: Maximum charge/discharge power [kW]
- $\left(u_{\mathrm{batt,c}}(t), u_{\mathrm{batt,d}}(t)\right)$: Binary charging/discharging status variables

Grid Connection Model

The contact with the external grid is represented by a two-way power stream, which is regulated by temporal deviations of prices (Dai et al., 2025; Tyagi et al., 2024).

Grid Power Balance:

$$P_{\mathrm{grid}}(t) = P_{\mathrm{grid,imp}}(t) - P_{\mathrm{grid,exp}}(t) \tag{22}$$

Power Flow Constraints:

$$0 \leq P_{\mathrm{grid,imp}}(t) \leq P^{\max}_{\mathrm{grid,imp}} \quad \forall t \tag{23}$$

$$0 \leq P_{\mathrm{grid,exp}}(t) \leq P^{\max}_{\mathrm{grid,exp}} \quad \forall t \tag{24}$$

Where:

- $\left(P^{\max}_{\mathrm{grid,imp}}, P^{\max}_{\mathrm{grid,exp}}\right)$: The Maximum import/export power limits [kW]

Problem Formulation and Objective Function

The energy management issue is characterized in the form of a finite-horizon optimal control problem, leading to the minimisation of the overall operational cost that occurs during the optimisation period T:

Objective Function

$$\min \sum_{t=1}^{T} \left[C_{fc}(t) + C_{grid}(t) + C_{batt}(t) \right] \tag{25}$$

in which the constituent cost elements are defined in the following way:

Fuel Cell Operating Cost:

$$C_{fc}(t) = \dot{m}_{NG}(t) \cdot c_{NG} \cdot \Delta t = \frac{P_{fc}(t)}{\eta_{fc,elec} \cdot LHV_{NG}} \cdot c_{NG} \cdot \Delta t \tag{26}$$

Grid Interaction Cost:

$$C_{grid}(t) = \left[P_{grid,imp}(t) \cdot \pi_{imp}(t) - P_{grid,exp}(t) \cdot \pi_{exp}(t) \right] \cdot \Delta t \tag{27}$$

Battery Degradation Cost:

$$C_{batt}(t) = \frac{c_{batt}^{capital}}{N_{cycles} \cdot E_{batt}} \cdot \left(\frac{|P_{batt}(t)| \cdot \Delta t}{2} \right) \tag{28}$$

Parameters:

- $\left(c_{NG} \right)$: Natural gas price [$/kg]
- $\left(\pi_{imp}(t) \right)$: Time-of-use grid import price [$/kWh]
- $\left(\pi_{exp}(t) \right)$: Export feed-in tariff [$/kWh]
- $\left(c_{batt}^{capital} \right)$: Battery capital cost [$]
- $\left(N_{cycles} \right)$: Total cycle life of battery

System-Wide Constraints

The optimisation is constrained by the following system-wide:

Power Balance Constraint:

$$P_{load}(t) = P_{pv}(t) + P_{wind}(t) + P_{fc}(t) + P_{batt,d}(t) - P_{batt,c}(t) + P_{grid,imp}(t) - P_{grid,exp}(t) \quad \forall t \tag{29}$$

Renewable Utilization Constraints:

$$P_{\text{pv,used}}(t) \leq P_{\text{pv}}(t) \quad \forall t \tag{30}$$

$$P_{\text{wind,used}}(t) \leq P_{\text{wind}}(t) \quad \forall t \tag{31}$$

Equipment Operational Constraints:

$$P_{i,\min} \leq P_i(t) \leq P_{i,\max} \quad \forall t, \forall i \in \{\text{pv, wind, fc, batt,c, batt,d, grid,imp, grid,exp}\} \tag{32}$$

Ramping Constraints:

$$|P_{\text{fc}}(t) - P_{\text{fc}}(t-1)| \leq \Delta P_{\text{fc,max}} \quad \forall t \tag{33}$$

$$|P_{\text{batt}}(t) - P_{\text{batt}}(t-1)| \leq \Delta P_{\text{batt,max}} \quad \forall t \tag{34}$$

Problem Complexity Analysis

The given formulation is a mixed-integer nonlinear programming (MINLP) problem as it has the following properties: -Binary decision variables of battery charge/discharge. Constrained to binary decision variables. - Non-linear relations of battery efficiency. - Renewable generation models that are not linear. - Non-convex battery degradation cost model.

METHODOLOGY

The proposed research will present a new Hybrid Q-learning + Rule-based Energy Management System (Hybrid Q+Rule EMS) that can deal with the inherent dynamic, multi-objective optimisation opportunities of an Industry 5.0-aligned industrial microgrid. The overall goal is to coordinate long-term strategic planning and learning of Reinforcement Learning (RL) with the safety of operation, interpretability, and instant reliability of a rule-based system (Liao et al., 2023). Figure 1 defines the entire architectural process of the methodology proposed.

Its methodology designates three fundamental pillars: the Rule-based Supervisor, which will make sure that operational constraints are never compromised (Johansen et al., 2024; Shukla et al., 2010); the Q-learning Agent, which is able to learn an optimal policy to be applied in terms of economy and efficient dispatch; and the

Microgrid Environment, which will be the cyber-physical system. The interplay of these elements establishes a dynamic process of decisions, thus depicted in Figure 1.

Problem Formulation as a Markov Decision Process (MDP)

To simplify the learning process, the energy management problem is modelled as a Markov Decision Process (MDP) which is defined by the following the tuple $(\mathcal{S}, \mathcal{A}, \mathcal{P}, \mathcal{R}, \gamma)$, where:

- $\mathcal{S}$ is a finite set of states.
- $\mathcal{A}$ is a finite set of actions.
- $\mathcal{P}$ is the state transition probability matrix, $P\left(s_{t+1} s_t, a_t\right)$.
- $\mathcal{R}$ is the reward function, $R\left(s_t, a_t, s_{t+1}\right)$.
- γ is the discount factor, $\gamma \in [0,1]$.

State Space ($\mathcal{S}$)

The state space of a microgrid at a given time point t is meant to give a complete picture of the state of the microgrid system. It is defined as a vector:

$$s_t = \left[P_{\text{load}}(t), P_{\text{pv}}(t), P_{\text{wind}}(t), \text{SOC}(t), H(t), G_p(t)\right] \tag{35}$$

Where:

- $P_{\text{load}}(t)$ total AC bus load demand (kW).
- $P_{\text{pv}}(t)$ is the available power of photovoltaic (kW).
- $P_{\text{wind}}(t)$ is the available wind power (kW).
- $P_{\text{wind}}(t)$, $\text{SOC}(t)$ is the current State of Charge of the battery bank (%).
- $H(t)$ is the current hour of the day (0--23), providing temporal context for tariff structures and load patterns.
- $G_p(t)$ is the current grid electricity price (\$/kWh), which can be positive for import and negative for export tariffs.

Figure 1. Schematic flowchart of the proposed Hybrid Q-learning + Rule based EMS methodology, illustrating the integration of the Rule-based Supervisor, the Q-learning Agent, and the Microgrid Environment within a closed feedback loop.

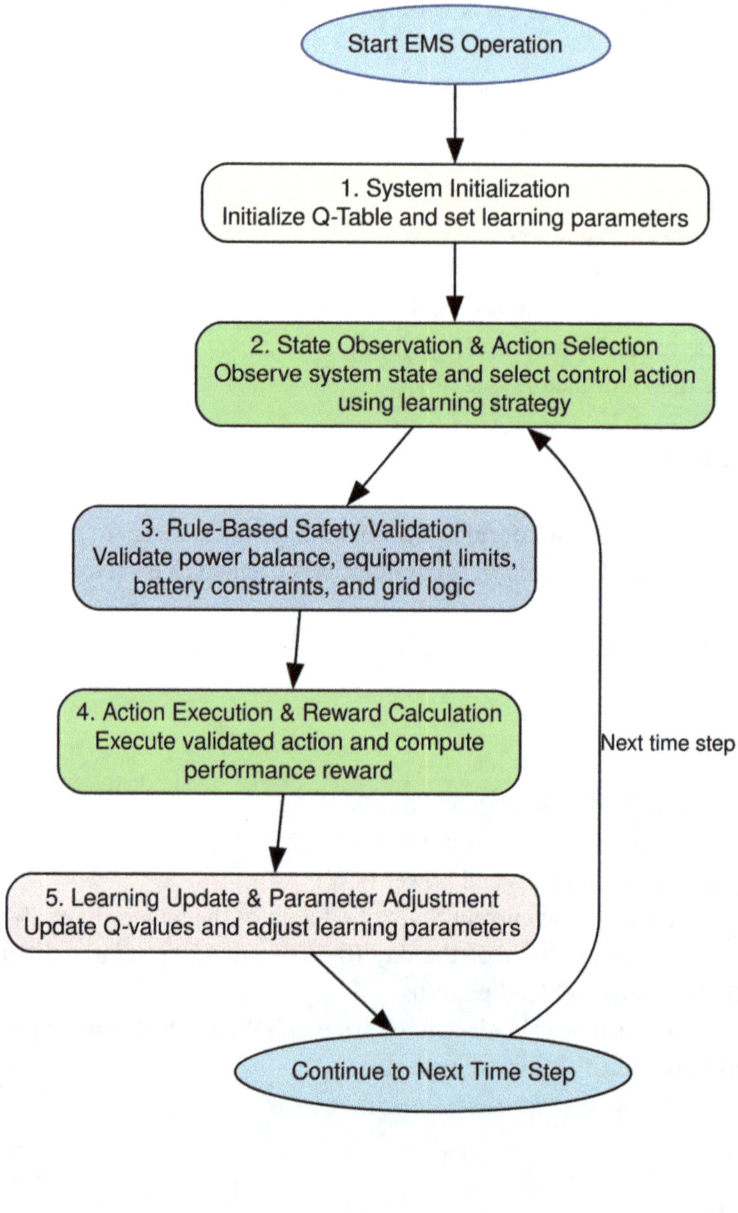

Action Space ($\mathscr{A}$)

Action space is the collection of all the potential decisions that the agent may make (Reyhan and Coleri, 2025; Han et al., 2025). In order to improve computational manageability and stability of the learning process, the action space is discretised. The main decision variable is the power setpoint of battery as well as the grid import/export. The action at time t is defined as:

$$a_t = \left[P_{\text{batt\_cmd}}(t), P_{\text{grid\_cmd}}(t)\right] \tag{36}$$

Where:

- $P_{\text{batt\_cmd}}(t)$ is the commanded battery power, discretized into levels (e.g., $-P_{\text{batt\_max}}, \ldots, 0, \ldots, +P_{\text{batt\_max}}$), where negative values indicate charging and positive values indicate discharging.
- $P_{\text{grid\_cmd}}(t)$ is the commanded grid power, discretized into levels (e.g., $-P_{\text{grid\_exp\_max}}, \ldots, 0, \ldots, +P_{\text{grid\_imp\_max}}$), where negative values indicate export and positive values indicate import.

Reward Function ($\mathscr{R}$)

The reward mechanism is a very important process of influencing the behaviour of an agent (Ibrahim et al., 2024; Ben-Porat et al., 2024). It is designed as an encapsulation of the multi-objective objectives of the EMS: cost reduction, maximisation of renewable use, and battery health maintenance. The instantaneous reward r_t at time t is a weighted sum of multiple components:

$$r_t = w_1 \cdot r_{\text{cost}}(t) + w_2 \cdot r_{\text{ren}}(t) + w_3 \cdot r_{\text{batt}}(t) + r_{\text{penalty}}(t) \tag{37}$$

1. **Economic Reward (r_{cost}):** This element directly punishes the cost of operation and the bonus of revenue:

$$r_{\text{cost}}(t) = -\left(C_{\text{fc}}(t) + C_{\text{grid\_imp}}(t) - R_{\text{grid\_exp}}(t)\right) \tag{38}$$

Where:

- $C_{\text{fc}}(t) = P_{\text{fc}}(t) \cdot c_{\text{fc}}$ is the fuel cell cost, with c_{fc} being the fuel cost ($/kWh).
- $C_{\text{grid\_imp}}(t) = \max\left(0, P_{\text{grid\_act}}(t)\right) \cdot G_p(t)$ is the cost of imported grid power.
- $R_{\text{grid\_exp}}(t) = \max\left(0, -P_{\text{grid\_act}}(t)\right) \cdot G_{p_{\text{exp}}}(t)$ is the revenue from exported grid power, where $G_{p_{\text{exp}}}(t)$ may be a different feed-in tariff.

2. **Renewable Utilization Reward** (r_{ren}): This term stimulates the maximum possible utilization of renewable energy sources and punishes reduced utilization:

$$r_{ren}(t) = -\left(P_{pv\_curt}(t) + P_{wind\_curt}(t)\right) \tag{39}$$

Where $P_{pv\_curt}(t)$ and $P_{wind\_curt}(t)$ are the curtailed (wasted) PV and wind power, calculated as the difference between available and utilized power.

3. **Battery Health Reward** (r_{batt}): This element will encourage long life by shunning extreme SOC actions and over cycling:

$$r_{batt}(t) = -\left(\alpha \cdot |P_{batt}(t)| + \beta \cdot I\{SOC(t) < SOC_{min}\} \cdot (SOC_{min} - SOC(t)) + \beta \cdot I\{SOC(t) > SOC_{max}\} \cdot (SOC(t) - SOC_{max})\right) \tag{40}$$

Where:

- The term $\alpha \cdot |P_{batt}(t)|$ imposes a small penalty proportional to the absolute power flow, discouraging unnecessary cycling.
- The terms with β apply a large penalty for violating the minimum (SOC_{min}) and maximum (SOC_{max}) SOC limits, scaled by the magnitude of the violation. $I\{\cdot\}$ is the indicator function.

4. **Constraint Penalty ($r_{penalty}$): ** When any hard system constraint is broken (i.e., load not served, severe over volt) a large, fixed negative reward is used to guarantee reliability of the system.

The weights w_1, w_2, w_3 are hyperparameters that balance the relative importance of these competing objectives.

Hybrid Q+Rule Control Architecture

The control loop works in the following way as it is illustrated in Figure 1:

Rule-Based Supervisor

This is the initial and absolute defence. For every proposed action a_t from the Q-learning agent, the Rule-based Supervisor performs a feasibility and safety check based on the current state s_t. These rules are deterministic and comprise:

- **Power Balance Verification:** The validation of the balance of power between the sum of all generation and storage power and the load is maintained bearing in mind the losses.
- **Equipment Limits:** Clamping $P_{batt\_cmd}(t)$ and $P_{grid\_cmd}(t)$ to their respective maximum and minimum power ratings.
- **Battery SOC Hard Limits:** Overriding the agent's command to charge if $SOC(t) \geq SOC_{max}$ or to discharge if $SOC(t) \leq SOC_{min}$, regardless of the proposed action.
- **Grid Transaction Logic:** Prohibiting simultaneous grid import and export.

The output of this stage is a feasible, safe action, $\boldsymbol{a}_{t_{feasible}}$, which is then executed in the environment.

Q-Learning Agent

The Q-learning agent has the responsibility of learning the long-term optimum policy. It uses a table-based Q -learning algorithm due to its simplicity and convergence properties in problems that may be described by a tractable state-action space.

1. Q-Value Update Rule:

The entire learning process revolves around the repetitive retraction of the Q-value function, $Q(s, a)$, which estimates the expected cumulative reward of taking action a in state s and thereafter, following the optimal policy. The update is performed after each transition (s_t, a_t, r_t, s_{t+1}) using the Bellman equation:

$$Q(s_t, a_t) \leftarrow Q(s_t, a_t) + \eta\left[r_t + \gamma \cdot \max_a Q(s_{t+1}, a) - Q(s_t, a_t)\right] \tag{41}$$

Where:

- η is the learning rate ($0 < \eta \leq 1$), controlling how much new information overrides old knowledge.
- γ is the discount factor ($0 \leq \gamma < 1$), determining the present value of future rewards. A value close to 1 makes the agent far-sighted.

2. Exploration-Exploitation Strategy (ϵ-Greedy):

To ensure the agent explores the state-action space sufficiently, an ϵ-greedy policy is used for action selection. In training, the agent will choose:

- A random action with probability ϵ (exploration).
- The greedy action, $\underset{a}{\mathrm{argmax}}\, Q(s_t, a)$, with probability $1 - \epsilon$ (exploitation).

An action selection is an ϵ is typically decayed over time from a high initial value (e.g., 1.0) to a small final value (e.g., 0.01), allowing the policy to converge towards the optimal one.

Training and Implementation Protocol

Figure 1 illustrates the training process, which consists of executing the microgrid simulation through thousands of episodes (e.g., through thousands of 24 -hours). During each time step during each episode:

- The agent observes the current state s_t.
- It selects an action a_t using the ϵ-greedy policy from its current Q-table.
- The action a_t is passed to the Rule-based Supervisor, which validates and modifies it to $a_{t_{\text{feasible}}}$.
- The feasible action is executed in the microgrid environment, leading to a new state s_{t+1} and a reward r_t.
- The Q-table is updated using the Q-learning update rule based on the experience (s_t, a_t, r_t, s_{t+1}).
- This is repeated until the episode is finished.

The system is implemented once it is complete and Q-values converge. In operation, exploration is switched off ($\epsilon = 0$), and the agent always chooses the best action, which has been presented in the learned Q-table, and the latter is then transmitted through the Rule-based Supervisor to obtain the last safety check before the action is realized. This protocol provides maximum and secure real time performance.

RESULT AND DISCUSSION

Industry 5.0 is the next transition in human history of industrial operations, as suggested by the name, as industry 4.0 implies the automation and exchange of data, whereas new industry 5.0 represents a human-focused, resilient, and sustainable model of the industrial processes. The main pillar of this vision is moving energy wisely in cyber-physical systems where distributed energy resources (DERs) should be synchronized not only to achieve reliability but also financial and environmental efficiency.

This chapter introduces and discusses the empirical confirmation of a new Hybrid Q -learning + Rule -based Energy Management System (Hybrid Q+Rule EMS), a framework that is developed to represent concepts of Industry 5.0. The system combines both the explore, long-term optimization of reinforcement learning (RL) and the deterministic and safety critical guarantees of rule-based control, and this results in a system of AI-enhanced ecosystems in which complex and multi-objective decision-making are executed within constraints posed by human operational understanding.

The experimental research is designed in the form of a strict comparative study with the familiar Rule-based EMS, as the counterpart. The model is evaluated on grid-connected industrial microgrids, and it is put through a 24-hour simulation under realistic and high-quality (high fidelity) environmental demands. The choice to incorporate performance metrics is based on the trio set by Industry 5.0 in terms of operational resiliency (power balance, resource adequacy), economic sustainability (net cost, revenue generation), and environmental stewardship (renewable penetration, curtailment reduction). Later paragraphs give a detailed analysis of how the Hybrid Q+Rule EMS manages within the complicated trade-offs in this world so that the microgrid can become a proactive value-generating component of the larger smart grid rather than a passive consumer of energy.

Experimental Setup and Industry 5.0-Aligned Microgrid Configuration

To authentically assess the proposed EMS, a high-fidelity model of an industrial microgrid was developed. This model represents a cyber-physical ecosystem, integrating physical energy assets (solar PV, wind turbine, fuel cell, battery storage, and grid connection) with a digital twin for control and optimization—a core tenet of Industry 5.0.

Environmental and Load Profile

The replication was carried out under renal weather conditions of a 24-hour time span, typical of semi arid industrial belt. The Gulf of Khambhat is situated along the west coast of India and is a bay on the Arabian Sea. It is north of Mumbai and the Diu Island (NASA (n.d.)) (The data is of the date 10th April 2025). These profiles shown in Figure 2 and (a) Solar Irradiance, (b) Wind Speed, (c) Temperature and (d) Relative Humidity sub-figure are the starting point of energy management challenge. The solar irradiance profile follows a predictable bell curve with its maximum insolation at noon whilst the wind speed profile follows a more stochastic pattern with moderate wind speeds especially during the night and in the early mornings. The accessible renewable energy amounted to 250.1kWh (PV: 124.5kWh: Wind: 125.6kWh). This was compared to a summative two-fazed industrial load demand of 254.2 kWh with high morning and evening peaks typical of normal industrial two shifts activities. The fact that renewable availability and load demand are brought to near-equilibrium is a purposeful design decision that creates a high-stakes, limited-resource environment in which the intelligence of the EMS is held to the utmost, as any potential error may have severe economic and operational impacts. This is not just a mere surplus/ deficit scenario but a complex energy arbitrage and maximization of value-stream.

System Architecture and Component Modeling

The design of the microgrid was developed with a focus on scalability and operating realism and incorporated a Photovoltaic (PV) system, a wind turbine, a fuel cell (FC), a battery energy storage system (BESS), and a grid connection in both directions. The operational structure and overall performance results acquired throughout the condition of a simulation are presented in a tabular form towards Table 1 and serves as the main guide towards the further comparative analysis.

Table 1. System Configuration and Overall Performance

Metric	Rule-based EMS (Jamal et al., 2024; Thirunavukkarasu et al., 2022)	Hybrid Q+Rule EMS
Total Load Demand	254.2 kWh	254.2 kWh
Total PV Available	124.5 kWh	124.5 kWh
Total Wind Available	125.6 kWh	125.6 kWh
PV Energy Used	122.9 kWh (98.7%)	121.9 kWh (97.9%)
Wind Energy Used	94.2 kWh (75.0%)	112.5 kWh (89.6%)
Fuel Cell Energy Used	38.5 kWh	39.9 kWh
Grid Import	4.3 kWh	2.7 kWh
Grid Export	0.0 kWh	16.7 kWh
Net Operational Cost	$1.44	**-$0.47**
Renewable Penetration	93.8%	100.4%
Min / Max Battery SOC	30.8% / 89.7%	30.0% / 90.0%

Figure 2. Input meteorological data profiles for the 24-hour simulation period (a) Solar Irradiance (b) Wind Speed (c) Temperature (d) Relative Humidity

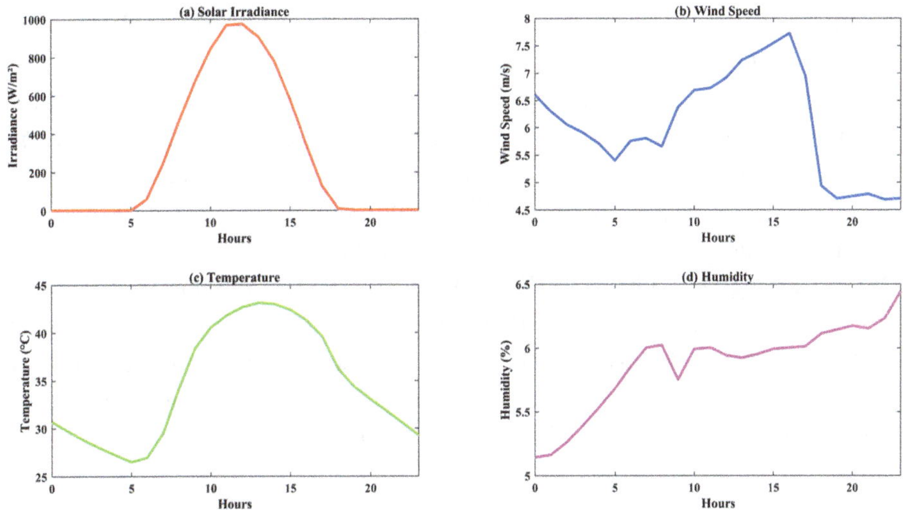

Comparative Analysis of Energy Management Strategies

Renewable Energy Utilization and Curtailment: From Myopic to Strategic Harvesting

One radical difference in the two energy management policies is reflected in the way they incorporate the idea of utilizing variable renewable energy, especially the wind energy. The rule-based EMS follows a predefined series or program of the so-called if-then logic. This would be effective in ensuring a basic power balance and in preventing faults in the system but this will be myopic; short sighted and it is not able to look into the future and see the state of energy and be able to adjust out of its set rules with regard to long term optimisation.

Table 1 and Figure 3b demonstrate that the rule-based controller used 75.0% (94.2 kWh) of the potential wind energy, thus, curtailing 31.4 kWh mainly at high wind periods (e.g. Hours 10-16) when solar energy is also high. This conservative rule-set, probably laying more emphasis on battery state-of-charge protection, or grid export avoidance, was not designed to exploit such excess energy, and will end up wasting zero-marginal-cost renewable resources.

Contrastingly, the hybrid Q +Rule EMS depicts what can be described as strategic harvesting. The system will optimize a policy by adjusting to the microgrid environment by incorporating Q -learning which is a model-free reinforcement learning algorithm. The Q-learning agent maximises a cumulative reward signal that is designed specifically to give consideration to both short-term expenses and long-term benefits (environmental costs) of utilisation of a renewable resource, battery life, and interaction with the grid.

This leads to the increased usage of wind energy beyond 89.6 percent (112.5 kWh), which is obtained in Table 1, a difference of 18.3 kWh compared to baseline. That policy behind, synthesised with the Q+Rule synergy, can be analysed with respect to comparative plots in Figure 3. Although Figure 3 a shows both strategies efficiently harness most of the available PV power, Figure 3b clearly shows that the hybrid strategy has a better ability to track and harness the power of the wind. Figures 3c and 3d draw another parallel between total renewable availability and the utilisation of the rule-based and hybrid EMS, respectively. Figure 3d confirms a statistically more matching availability and utilisation, with a significant percentage of the curtailment gap in Figure 3c being transformed into useful energy.

Figure 3. Renewable energy availability and utilization comparison between Rule-based and a novel Hybrid Q-learning + Rule based EMS (a) PV Power: Available vs. Utilized (b) Wind Power: Available vs. Utilized (c) Rule-based EMS: Total Renewable Availability vs. Usage (d) Hybrid Q+Rule EMS: Total Renewable Availability vs. Usage

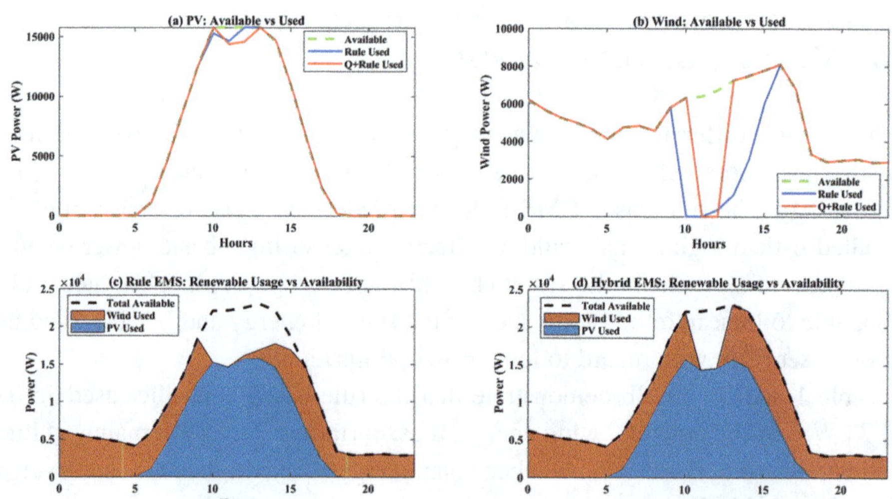

This integrated system of renewable management that made the system achieve a level of 100.4% renewable penetration (Table 1) implies that the microgrid was not only able to meet its full load of renewable throughout the simulation period but also gave back excess clean energy to the main grid, thus akin to the spirit of sustainability and collaboration that Industry 5.0 aims to facilitate.

Power Scheduling and Fuel Cell Operation: The Shift from Reactive to Proactive Dispatch

The impacts of scheduling power of dispatchable sources and controlling the power flow indicate the very essence of the respective EMSs intelligence. The source contributions and the balance of power as shown in Figure 4 gives a time story of this dynamic.

Figure 4b,c shows that the rule-based EMS has a reactive pattern in dispatch. Its control is primarily driven by instantaneous states of the system, the first one being predominantly net power balance (load- renewables) and the second one is the battery SOC. The system all depends on the fuel cell in times of peak demand (Hours 6-7 and 19-23), when the generation is low or zero through renewable sources, as shown in Figure 5a. Although this reactive dependence to the FC could guarantee reliability, it creates more fuel consumption and operational costs. The rule based logic perceives the FC as a major backup, which does not possess the strategic flexibility necessary to reduce its usage by better scheduling of battery.

On the other hand, the hybrid Q+Rule EMS applies proactive dispatch approach. It has Q-learning capabilities that make it have a visionary approach to decisions. Its functioning is analyzed (Figure 4d, e) and is more advanced, a sophisticated job:

- Pre-emptive Battery Deployment This is more back-up latent to a state of deficit; the hybrid EMS draws energy out of the battery during drought times, though it gets charged up strategically at earlier hours when the renewable production is high. This will ensure that the instantaneous load on the FC as well as the grid is minimized.
- Fuel Cell as a Strategic ally, but not a Crutch. The use of the FC by the hybrid EMS is more subtle. The context is critical although Table 1 shows that the total FC energy use (39.9 kWh versus 38.5 kWh) is slightly higher. This energy is often used in concert with other efforts in order to get the greatest economic advantage- such as to run the FC at a particular time to escape a high tariff to import to the grid or to charge the battery in order to store even cheaply produced renewable power, and hence increase the overall profit. Figure 5b supports the fact that the hybrid approach produces a high economic payoff on a unit of FC energy used. The FC is not only used to supply power but value optimisation.

This reactive to proactive dispatch initiation is indicative of a major Industry 5000 benefit, the ability not just to respond to the world, but also to predict and influence future conditions to the best possible results.

Figure 4. AC bus power balance and detailed generation source contributions for both EMS strategies (a) AC Bus: Delivered Power vs. Target and Actual Load (b) Rule-based EMS: Power Contribution from Sources (Area Plot) (c) Rule-based EMS: Power Contribution from Sources (Stacked Line Plot) (d) Novel Hybrid Q-learning + Rule based EMS: Power Contribution from Sources (Area Plot) (e) Novel Hybrid Q-learning + Rule based EMS: Power Contribution from Sources (Stacked Line Plot)

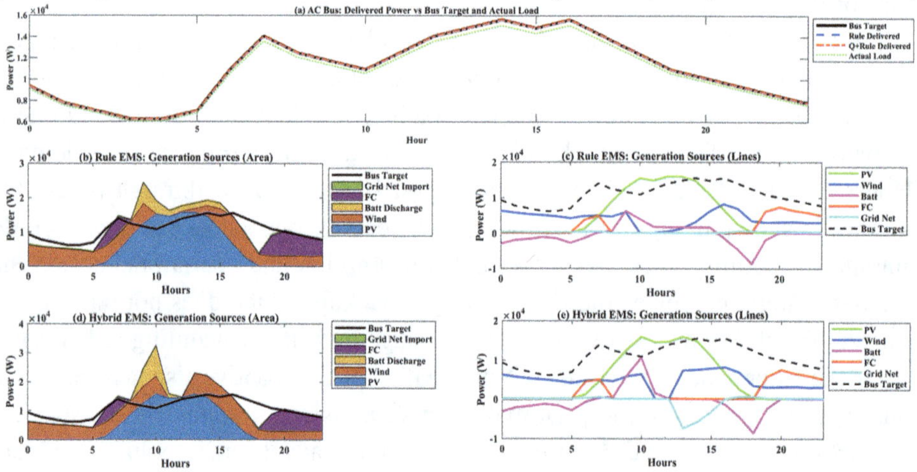

Figure 5. Fuel cell operational profile and performance metrics (a) Fuel Cell Power Output Comparison (b) Fuel Cell Performance and Cost Metrics

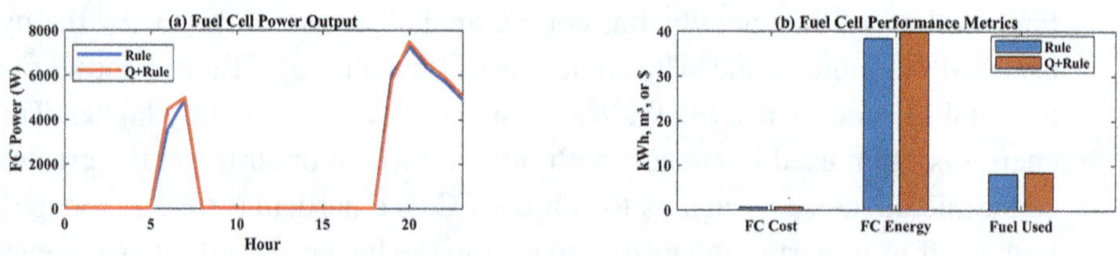

Battery State of Charge (SOC) and System Economics: The Nexus of Resilience and Profitability

State of Change (SOC) trajectory of a battery system is a diagnostic signature, which represents the strategic priorities of an Energy Management System (EMS). In particular, it represents a coded answer to the resilience, longevity, and economic arbitrage in the system. Figure 6a shows the evolutions of the SOC created by the two management strategies. Whether it was a SOC, it was limited to the prescribed envelope of safety between 30 and 90 percent and thus protect the health of the batteries and thus the resiliency of the system, which is a pre-condition in any industrial functionality.

However, the dynamics of the cycling trends was strikingly different. The Rule based EMS had a passive, buffer-based SOC profile where the battery was used to smooth the load and provide short-term backup. The resulting cycle count was therefore relatively low, and each cycle had been relatively shallow, provided by mere instantaneous power-balance triggers.

In contrast, the Hybrid Q+Rule EMS has been an active asset-based strategy. Its SOC curve (tracing the green line in Figure 6(b) displayed meaningful and deeper cycling. Strategic use of the battery was energy arbitrage: More vigorous charging was done when there was a high supply of renewable generation available cheaply, and less vigorous charging when the electricity price was high in the marketplace, or when there was high demand. This proactive cycling was the basis of economical change of the system.

The economic consequences as measured in Table 2 and plotted in Figures 6(e to f) are irrefutable.

Table 2 provides a breakdown of the cost of operations and revenues. The net cost under the Rule -based EMS per day was $1.44, necessitated by grid imports, which cost the system, at $0.77, and the cost of fuel cell (FC), which was $0.65; no revenue stream was created that would offset these two expenditures.

Table 2. Operational Cost Breakdown

Cost Component	Rule-based EMS	Hybrid Q+Rule EMS
Grid Purchase Cost	$0.77	$0.50
Grid Sales Revenue	$0.00	$1.67
FC Fuel Cost	$0.65	$0.67
Battery Cycling Cost	$0.025	$0.025
Net Operational Cost	$1.44	-$0.47

Figure 6. Battery performance, energy flow, and economic analysis for both EMS strategies (a) Battery State of Charge (SOC) Trajectories (b) Green SOC vs. Total SOC (c) Rule-based EMS: Total Energy Contributions by Source (d) Hybrid Q+Rule EMS: Total Energy Contributions by Source (e) Rule-based EMS: Operational Cost Breakdown (f) Hybrid Q+Rule EMS: Operational Cost Breakdown

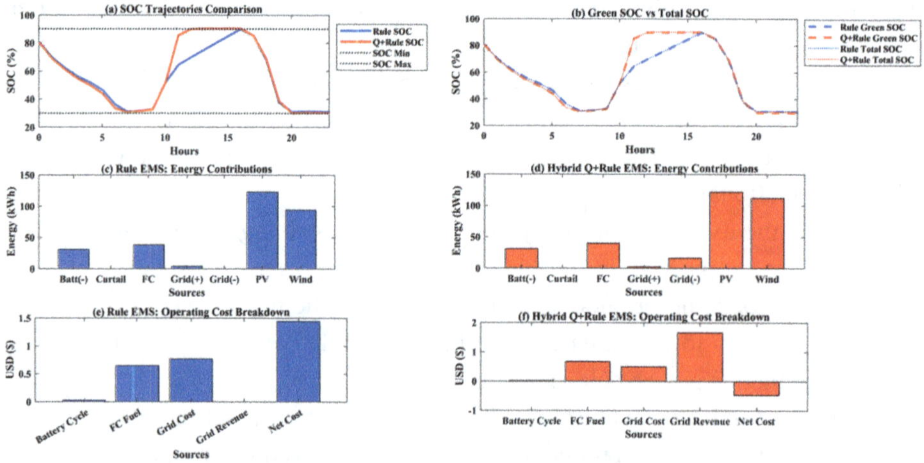

The net profit under Hybrid Q +Rule EMS was -0.47. This was achieved due to a two pronged strategy, well illustrated in cost breakdown of Figure 6f as compared to the Rule based breakdown in Figure 6e:

1. **Cost Reduction**: The importation costs on the grid have decreased by 37% (0.77) to 0.50 with the strategic control of the battery and the use of renewables.
2. **Revenue Generation**: The system earned a total of 16.7 -16.7 kWh of excess renewable energy that earned it revenue of 1.67. This stream marginally offset the increased FC fuel cost of 0.67 and the other operating costs resulting in a net profit.

There is the direct economic gain that is made upon the cognitive capabilities of the Hybrid Q+Rule framework which shift the operational paradigm of a microgrid to the value-generating one, which is one of the basic needs of the economic sustainability of the future industrial ecosystem. The combined energy input of each source as shown in Figure 6c and 6d of the two strategies, further contextualise the economic outcome as it indicates the redrawn energy mix.

Synthesis and Discussion: Forging a New Paradigm for Industrial Energy Management

The empiric evidence provided in the present requires indescribably few plates to come together to create a case that cannot be disputed stating the Hybrid Q-learning + Rule-based EMS as a fundamental technology in the Industry 5.0 world. Its capability to surpass traditional Rule based EMS includes much beyond incremental benefits: it constitutes a paradigm shift between automated control and cognitive optimisation.

The Synergy of AI and Human-Defined Operational Wisdom

The major innovation of the offered framework is not to replace the rule-based systems, but to enrich them. The rule-costed layer provides needed guardrails or safety, reliability, and equipment-integrity constraints that are never to be broken. This is the wisdom of operations that is condensed of the wisdom of man and this makes it likely to endure. The Q-learning layer does act as a cognitive layer that explores and exploits the solution space given these constraints, learns the complex, non-linear, relationships between the generation, consumption, storage, and market signals which would be difficult to encode by hand. The fact that this system is safe, intelligent, reliable, and adaptive is a resultant feature of this system.

Multi-Objective Optimization in a Dynamic Environment

The industrial energy environment is a complex optimisation challenge, with cost reduction, maximisation of the sub-objective of renewable penetration, minimisation of emissions, and reliability as common contradictory requirements. A rule-based EMS is an all-static organization that finds it hard to go through these compromises, and mostly it will go in a conservative, reliability-first position at the cost of the economics and sustainability.

The Hybrid Q+Rule EMS is doing well here. The system has the ability to organize the competing objectives in a natural manner by influencing the reward function of the Q-learning agent. Synthesized findings in Table 1, Table 2, Figure 5 and Figure 6 reveal that the Hybrid system meets the following concurrent goals of increased profitability, increased renewable penetration and reliability. It confirms that although these objectives are not counter-cyclical they can be synergistically worked to achieve this using smart dispensation.

Implications for the Sustainability and Viability of Industrial Microgrids

This means that having a net-negative work-in-progress cost has far reaching consequences. It significantly reduces the payback period of micro grid assets including batteries and renewable generators. The offered EMS will convert capital expenditures into payback assets, which will result in profitability generated by the establishment of a constant revenue stream, which can become the turning point in expediting the deployment of renewable-heavy microgrids in the industry. Such changes in the economy directly lead to global decarbonisation.

Also, reaching above 100% renewable penetration puts a path to energy-positive industries. The principle of Industry 5.0 vision is that factories do not just remain as an effective consumer in the system but become active and useful contributors to the local energy community and the bigger smart grid, and provide benefits to the overall system stability and sustainability.

Operational Implications and Future Directions

Conclusively, this empirical study in totality provides strong empirical support to Hybrid Q-learning + Rule based Energy Management System. The framework has been proved able to meet the metrics that are presented in Table 1, Table 2 and Figures 2-6:

- A 19.5 percent growth in the wind energy use (Figure 3b), significantly decreasing curtailment. 1.
- A shift in its economic performance in terms of a net cost that reduces to a net profit of 0.47 (Table, Figure 6e-f), countertestifying to its excellent performance.
- The achievement of 100.4% renewable penetration (Table 1), a new standard of sustainable industrial performance.
- Reactive to proactive dispatch (Figure 4, Figure 5a, Figure 6a-b), by using the battery storage and fuel cell capabilities more efficiently to ensure long-term value creation.

The Hybrid Q+Rule EMS is a new paradigm, which goes beyond the framework of traditional control measures and integrates within it a cognitive core based on learning into a safe, framework-based rule-driven scheme. This architecture can naturally be described: the rules provide it with a basis of transparent logic, and AI component optimises on this basis. It is flexible and it keeps on developing its policy since the conditions under which it operates are changing. In addition, it brings about economic change turning management of energy into a strategic asset. Therefore, this study provides a blueprint of a cognitively empowered, sustainable, and resilient energy system in industries which is relevant to the new Industry 5.0 paradigm.

CONCLUSION

This paper has clearly shown the transformative effectiveness of the Hybrid Q-learning + Rule based Energy Management System (EMS), and thus making it the new paradigm of human-centered, resilient and sustainable digit Industry 5.0. The hybrid structure was shown by rigorous comparative analysis in restricted, realistic microgrid operations to outperform a traditional rule-based EMS in all key performance measures by a significant margin and changed the system operation paradigm of a cost-sustaining organisation to a value-creating prosumer. The change is numerically proven by the fact that the system has reached a net operational profit of -0.47 which is hydrantly different than the cost that the rule based system makes-1.44. This was accomplished by a complex two-pronged approach which decreased imports into the grid by 37 percent and most crucially earned revenue of 1.67 as a result of tactical exporting of 16.7kWh of excess renewable energy to the grid. Moreover, the hybrid EMS was better in terms of environmental intelligence, with the roll of 89.6% of renewable curtailment, used to 75.0% of the available renewable power by the rule-based controller, enabling the impressive accomplishment of 100.4% of renewable penetration. As a result, the microgrid was able to cover its own demand of 254.2 0.99 kWh of clean energy not only but also provided a net excess of clean energy to the larger grid. The persuasive intelligence of the framework then receives a foresighted, proactive, dispatch pattern that is otherwise demonstrated by the battery state-of-charge (SOC) profiles and power voltage plots. The EMS arbitrarily cycled battery storage to arbitrately harness energy, pre-emptively used stored energy to slice the peak demand and to tune the fuel-cell operation to economic advantage instead of power balancing and carefully kept the system variables with a reasonable conformity to the safe operational limits. This ensures that there will be resilience without affecting economic or environmental performance. The power of the synergy between the exploratory and long-term optimisation abilities of the Q-learning algorithm and the safety-critical deterministic guardrails of the rule-based system resulted in a strong, adaptable, and understandable cognitive engine taking the form of in-run planning of the complicated, multi-objective trade-offs of a modular industrial microgrid within real-time. Therefore, the study confirms the legitimacy of the proposal that model-free reinforcement learning is being integrated into an ordered, rule-operationalized operational envelope can be not only an improvement, but a radical one. It offers a blueprint framework on how self-optimising, economically sustainable and resilient energy systems that learn, adapt, and spur transition to energy-positive smart industries can be developed. In the future, the study will be extended to larger multi-microgrid clusters, includes energy trading that is peer-to-peer and use of real-time carbon monitoring to directly optimize on the carbon neutrality goal with economic goals.

REFERENCES

Abbas, A., Rizoug, N., Trigui, R., Redondo-Iglesias, E., & Pelissier, S. (2024). Low-computational model to predict individual temperatures of cells within battery modules. *Batteries, 10*(3), 98. DOI: 10.3390/batteries10030098

Ali, S., Raza, A., Khan, I., Bhatti, M. K. L., & Malik, T. N. (2024). Energy management systems for microgrids. In *Microgrid* (pp. 65–90). CRC Press. DOI: 10.1201/9781003481836-5

Aljohani, T., & Almutairi, A. (2024). Modeling time-varying wide-scale distributed denial of service attacks on electric vehicle charging stations. *Ain Shams Engineering Journal, 15*(7), 102860. DOI: 10.1016/j.asej.2024.102860

Almtrf, A. (2025). Integrating explainable AI (XAI) into decision support systems: A framework for enhancing transparency and trust in managerial decision-making. *International Journal of Managerial Studies and Research, 13*(9), 9–22. DOI: 10.20431/2349-0349.1309002

Anand, S., Kulkarni, S. V., Gupta, S., & Parthasarathy, V. (2022, December). Battery management system (BMS) for electric vehicle applications. *In 2022 fourth international conference on emerging research in electronics, computer science and technology (ICERECT)* (pp. 1-6). IEEE. DOI: 10.1109/ICERECT56837.2022.10060137

Awasthi, B., Gupta, S., Kumar, S., & Tripathi, A. (2024, January). Business management and planning using queuing model with heterogeneous customers for electrical applications. *In 2024 2nd International Conference on Intelligent Data Communication Technologies and Internet of Things (IDCIoT)* (pp. 279-283). IEEE. DOI: 10.1109/IDCIoT59759.2024.10467673

Bayrak, F. (2025). Prediction of photovoltaic panel cell temperatures: Application of empirical and machine learning models. *Energy, 323*, 135764. DOI: 10.1016/j.energy.2025.135764

Ben-Porat, O., Mansour, Y., Moshkovitz, M., & Taitler, B. (2024, March). Principal-agent reward shaping in MDPs. *Proceedings of the AAAI Conference on Artificial Intelligence, 38*(9), 9502–9510. DOI: 10.1609/aaai.v38i9.28805

Cochran, D. S., Smith, J., Mark, B. G., & Rauch, E. (2022). Information model to advance explainable AI-based decision support systems in manufacturing system design. In Matt, D. T., Vidoni, R., Rauch, E., & Dallasega, P. (Eds.), *Managing and implementing the digital transformation* (Vol. 525). Springer. DOI: 10.1007/978-3-031-14317-5_5

Dai, J., Yuan, Q., Cai, H. H., Zhang, V., Hasanuzaman, M., & Selvaraj, J. (2025). Business-oriented optimization of EV-to-building energy flows: Predictive modeling and scenario evaluation. *Energy*, *333*, 137456. DOI: 10.1016/j.energy.2025.137456

de Oliveira, T. H. F., Medeiros, L. P. de S., Neto, A. D. D., & de Melo, J. D. (2021). Q-Managed: A new algorithm for multi-objective reinforcement learning. *Engineering Applications of Artificial Intelligence*, *9*, 100089.

Ficzere, P. (2024). Industry 5.0. Present or future? *Design of Machines and Structures*, *14*(2), 19–34. DOI: 10.32972/dms.2024.010

Fotis, G., Gupta, S., Varshney, T., Dhar, S. L., & Gulzar, M. M. (Eds.). (2025). *Modern Computing Technologies for EV Efficiency and Sustainable Energy Integration*. IGI Global., DOI: 10.4018/979-8-3373-2382-4

Gupta, S. (2025). *The Future of Machine Learning with Automation in Engineering*. Nova Science Publishers., DOI: 10.52305/YFIX2832

Gupta, S., & Tripathi, R. K. (2014). Two-area power system stability improvement using a robust controller-based CSC-STATCOM. *Acta Polytechnica Hungarica*, *11*(7), 135–155. DOI: 10.12700/APH.11.07.2014.07.9

Han, L., Zhang, H., & An, N. (2025). A continuous space path planning method for unmanned aerial vehicle based on particle swarm optimization-enhanced deep Q-network. *Drones (Basel)*, *9*(2), 122. DOI: 10.3390/drones9020122

He, L., Hu, X., Yin, G., Wang, G., Shao, X., & Liu, J. (2024). A current dynamics model and proportional–integral observer for state-of-charge estimation of lithium-ion battery. *Energy*, *288*, 129701. DOI: 10.1016/j.energy.2023.129701

Ibrahim, S., Mostafa, M., Jnadi, A., Salloum, H., & Osinenko, P. (2024). Comprehensive overview of reward engineering and shaping in advancing reinforcement learning applications. *IEEE Access : Practical Innovations, Open Solutions*, *12*, 175473–175500. DOI: 10.1109/ACCESS.2024.3504735

Ignatev, E., Deriugina, G., Suslov, K., & Balaban, G. (2025). Development of a hybrid model for medium-term wind farm power output forecasting. *Renewable Energy*, *249*, 123200. DOI: 10.1016/j.renene.2025.123200

Jamal, S., Pasupuleti, J., & Ekanayake, J. (2024). A rule-based energy management system for hybrid renewable energy sources with battery bank optimized by genetic algorithm optimization. *Scientific Reports*, *14*(1), 4865. DOI: 10.1038/s41598-024-54333-0 PMID: 38418902

Johansen, K. W., Teizer, J., & Schultz, C. (2024). Automated rule-based safety inspection and compliance checking of temporary guardrail systems in construction. *Automation in Construction, 168*, 105849. DOI: 10.1016/j.autcon.2024.105849

Johri, S., Selvakumar, P., Satyanarayana, P., Das, A., Gupta, S., & Bhattacharya, S. (2025). Urbanization and Energy: The Nexus of Global Challenges. In Geopolitical Landscapes of Renewable Energy and Urban Growth (pp. 453-478). *IGI Global Scientific Publishing*. DOI: 10.4018/979-8-3693-8814-3.ch016

Karthick, A., Shankar, R., & Dharmaraj, G. (2024). Energy forecasting of the building integrated photovoltaic system based on deep learning dragonfly-firefly algorithm. *Energy, 308*, 132926. DOI: 10.1016/j.energy.2024.132926

Kostopoulos, G., Davrazos, G., & Kotsiantis, S. (2024). Explainable artificial intelligence–based decision support systems: A recent review. *Electronics (Basel), 13*(14), 2842. DOI: 10.3390/electronics13142842

Li, P., Xue, Q., Zhang, Z., Chen, J., & Zhou, D. (2023). Multi-objective energy-efficient hybrid flow shop scheduling using Q-learning and GVNS driven NSGA-II. *Computers & Operations Research, 159*, 106360. DOI: 10.1016/j.cor.2023.106360

Liao, Y., Yu, G., Chen, P., Zhou, B., & Li, H. (2023). Integration of decision-making and motion planning for autonomous driving based on double-layer reinforcement learning framework. *IEEE Transactions on Vehicular Technology, 73*(3), 3142–3158. DOI: 10.1109/TVT.2023.3326548

Lu, R., Zhang, Y., Yang, T., Chen, Y., Wang, D., & Peng, X. (2024). A novel hybrid-action-based deep reinforcement learning for industrial energy management. *IEEE Transactions on Industrial Informatics, 20*(10), 1–15. DOI: 10.1109/TII.2024.3424529

Lyu, Z., Liu, Y., Sciazko, A., Komatsu, Y., Tao, J., Nakamura, A., Hara, T., Sun, K., Shikazono, N., & Han, M. (2025). Co-generation of electricity and chemicals from methane using direct internal reforming solid oxide fuel cells. *Advanced Energy Materials, 15*(10), 2403869. DOI: 10.1002/aenm.202403869

Mane, D. (2026). AI-driven optimization techniques for energy-efficient building systems. In *AI-Driven Approaches for Fully Automated Smart Engineering* (pp. 149–166). IGI Global.

Mannan, M., Mansor, M., Reza, M. S., Roslan, M. F., Ker, P. J., & Hannan, M. A. (2024). Recent development of grid-connected microgrid scheduling controllers for sustainable energy: A bibliometric analysis and future directions. *IEEE Access : Practical Innovations, Open Solutions, 12*, 90606–90628. DOI: 10.1109/ACCESS.2024.3419443

Marzouk, O. A. (2025). Wind speed Weibull model identification in Oman, and computed normalized annual energy production (NAEP) from wind turbines based on weather station data. *Engineering Reports, 7*(3), e70089. DOI: 10.1002/eng2.70089

Mishra, A., & Nandy, M. (2026). Human-AI collaboration in optimizing energy harvesting for sustainable wireless sensor networks. In *Multidisciplinary Advancements in Human-AI Augmentation* (pp. 225–242). IGI Global.

Mourtzis, D. (2024). Challenges and opportunities of the transition from Industry 4.0 to Industry 5.0. In *Manufacturing from Industry 4.0 to Industry 5.0* (pp. 97–131).

NASA Langley Research Center. (n.d.). *NASA POWER Data Access Viewer*. Retrieved October 25, 2025, from https://power.larc.nasa.gov/data-access-viewer/

Natarajan, S., Mathur, S., Sidheekh, S., Stammer, W., & Kersting, K. (2024). Human-in-the-loop or AI-in-the-loop? Automate or collaborate? *arXiv*.

Natarajan, S., Mathur, S., Sidheekh, S., Stammer, W., & Kersting, K. (2025). Human-in-the-loop or AI-in-the-loop? Automate or collaborate? *Proceedings of the AAAI Conference on Artificial Intelligence, 39*(27), 28594–28600. DOI: 10.1609/aaai.v39i27.35083

Olatunde, T. M., Okwandu, A. C., Akande, D. O., & Sikhakhane, Z. Q. (2024). Review of energy-efficient HVAC technologies for sustainable buildings. *International Journal of Science and Technology Research Archive, 6*(02), 12–20. DOI: 10.53771/ijstra.2024.6.2.0039

Patel, V., Giri, V. K., & Kumar, A. (2024). Efficient power management strategies for AC/DC microgrids with multiple voltage buses for sustainable renewable energy integration. *Energy Informatics, 7*(1), 97. DOI: 10.1186/s42162-024-00377-5

Pradheeba, P., Gopu, K., & Thanikasalam, A. (2025). *AI-enabled decision support systems for real-time industrial process automation and optimization.*

Rao, M. S., Saraswathy, S., Neela, V., Arun, M. R., Gupta, S., & Taqui, S. N. (2023, November). Efficient Model Training in Decentralized Systems with Federated Learning. *In 2023 International Conference on Communication, Security and Artificial Intelligence (ICCSAI)* (pp. 900-904). IEEE. DOI: 10.1109/ICCSAI59793.2023.10421469

Reyhan, B. G., & Coleri, S. (2025). Safe deep reinforcement learning for resource allocation with peak age of information violation guarantees. *IEEE Transactions on Communications*.

Romanova, O., & Kuzmin, E. (2024). From Industry 4.0 to Industry 5.0: Challenges and opportunities of evolutionary transition in Russia. In *The Future of Industry: Human-Centric Approaches in Digital Transformation* (pp. 55–72).

Sayed, E. (2025). Decision intelligence in the age of industrial transformation: A survey from Industry 4.0 to Industry 7.0. *Sustainable Machine Intelligence Journal*, *12*, 20–40. DOI: 10.61356/SMIJ.2025.12565

Sele, D., & Chugunova, M. (2023). Putting a human in the loop: Increasing uptake, but decreasing accuracy of automated decision-making. *PLoS One*, •••, 19. PMID: 38335162

Sharma, R. K., Saran, V. B., Kanaujia, S., & Gupta, S. (2024, May). Evaluation of mppt controller performance using fuzzy logic design with wireless sensor node. *In 2024 International Conference on Smart Systems for applications in Electrical Sciences (ICSSES)* (pp. 1-6). IEEE. DOI: 10.1109/ICSSES62373.2024.10561365

Shukla, R. D., Tripathi, R. K., & Gupta, S. (2010, November). Power electronics applications in wind energy conversion system: A review. *In 2010 International Conference on Power, Control and Embedded Systems (pp.* 1-6*)*. IEEE. DOI: 10.1109/ICPCES.2010.5698663

So, C. (2020). Human-in-the-loop design cycles: A process framework that integrates design sprints, agile processes, and machine learning with humans. *arXiv: Human–Computer Interaction*. DOI: 10.1007/978-3-030-50334-5_9

Thandlam, A. K., Vignesh, T., Saravanan, A., Subramani, P., Marimuthu, R., & Gupta, S. (2024, April). Next-gen vehicle safety: The futuristic approach to auto-stop in modern vehicles. *In 2024 10th International Conference on Communication and Signal Processing (ICCSP)* (pp. 265-270). IEEE. DOI: 10.1109/ICCSP60870.2024.10543666

Thirunavukkarasu, G. S., Seyedmahmoudian, M., Jamei, E., Horan, B., Mekhilef, S., & Stojcevski, A. (2022). Role of optimization techniques in microgrid energy management systems—A review. *Energy Strategy Reviews*, *43*, 100899. DOI: 10.1016/j.esr.2022.100899

Tu, N., Fan, Z., Pang, X., Yan, X., Wang, Y., Liu, Y., & Yang, D. (2024). A multi-objective scheduling method for hybrid integrated energy systems via Q-learning-based multi-population dung beetle optimizers. *Computers & Electrical Engineering*, *117*, 109223. DOI: 10.1016/j.compeleceng.2024.109223

Tyagi, M., Gupta, S., & Khan, M. I. (2024, May). Reduction of voltage instability in complex power networks through the utilization of D-STATCOM. *In 2024 4th International Conference on Advance Computing and Innovative Technologies in Engineering (ICACITE)* (pp. 1662-1666). IEEE. DOI: 10.1109/ICACITE60783.2024.10616856

Ukoba, K., Olatunji, K. O., Adeoye, E., Jen, T. C., & Madyira, D. M. (2024). Optimizing renewable energy systems through artificial intelligence: Review and future prospects. *Energy & Environment*, *35*(7), 3833–3879. DOI: 10.1177/0958305X241256293

Verma, M., Bhambri, S., Gupta, S., & Buduru, A. B. (2019). Making smart homes smarter: Optimizing energy consumption with human in the loop. *arXiv:1912.03298.*

Yu, C., Guo, Z., Pan, Z., Zhou, Y., Zhang, H., Chen, B., Tan, P., Guan, W., & Zhong, Z. (2025). Performance and stress analysis of flat-tubular solid oxide fuel cells fueled with methane and hydrogen. *Acta Mechanica Solida Sinica*, *38*(3), 402–414. DOI: 10.1007/s10338-024-00514-3

Zhang, S., Chen, J., & Tang, X. (2022). Multi-objective control and energy management strategy based on deep Q-network for parallel hybrid electric vehicles. *International Journal of Vehicle Performance*, *8*(4), 371–386. DOI: 10.1504/IJVP.2022.125943

Chapter 9

Optimal Placement and Sizing of Distributed Generation Using Random Drift Particle Swarm Optimization

Abubakarsadiq Ali Muhammad
Sharda University, India

Suman Lata
https://orcid.org/0000-0002-5806-7967
Sharda University, India

Tarun Varshney
https://orcid.org/0000-0002-4047-7351
Sharda University, India

Sandeep Gupta
https://orcid.org/0000-0002-3734-3723
Graphic Era University, Dehradun, India

Anand Shukla
https://orcid.org/0000-0003-2300-5376
Wollega University, Nekemte, Ethiopia

ABSTRACT

This paper presents the optimal placement and sizing of Distributed Generation (DG) using the Random Drift Particle Swarm Optimization (RDPSO) algorithm. The

DOI: 10.4018/979-8-3373-7041-5.ch009

study focuses on minimizing power losses, thus improving profiles of the voltage and maintaining stability of the system. The proposed methodology combines analysis of power flow through the Backward Forward Sweep (BFS) approach with RDPSO to ascertain optimal DG configurations. The proposed strategy is tested on the IEEE 33-bus system. The results obtained have shown a significant improvement in voltage stability and a reduction of power losses. Simulations with one, two, and three DGs highlight the scalability and effectiveness of the approach. The study underscores the potential of RDPSO in optimising DG placement, providing a workable way to enhance the dependability and efficiency of contemporary distribution networks.

INTRODUCTION

The adoption of DG into networks for power distribution has become a key strategy in modernizing and improving the performance of electricity grids. The term "DG" describes smaller power production units that are situated near the point of consumption and are frequently based on renewable resources like, solar panels, wind turbines, or small hydropower plants. Unlike centralized power plants, DG units generate electricity on a smaller scale, which can offer numerous advantages but also pose unique challenges.

The smooth operation of the power system hinges on meeting demand as efficiently and dependably as possible at every point in the power network. Conventional energy sources like fossil fuels, hydropower, nuclear power, etc., are used in traditional electric power generation systems to generate electricity. Utility generators with centralized control are the foundation of such conventional generating systems. Through a vast transmission and distribution network, these generators provide electricity to users who are geographically scattered. Large central-station plants are becoming less justified in the current climate because of declining traditional supplies, rising costs of transmission and distribution, deregulation trends, growing technology developments, and concerns for the environment. Many of these new problems can be solved by distributed generation. The placement of DG strategically can improve voltage stability, lower power losses, and increase overall dependability of the power system (Kayal & Chanda, 2013). If units of DG are not placed optimally, they have the potential to induce voltage variations and may compromise the stability of the network (Gupta et al., 2010; Iweh et al., 2021).

The optimal DG units sizing and DG units placement are complex tasks. Power flow characteristics of distribution network have become more complicated with the addition of DG, and traditional analytical methods may fall short in accurately modeling these changes. Furthermore, renewable DG sources are subject to variability, solar power output fluctuates with sunlight availability, and wind generation depends on wind speeds (Awasthi et al., 2024; Fotis et al., 2025). These uncertainties must be considered during planning to ensure a reliable and effective energy source.

To address these challenges, numerous optimization methods have been created. Conventional analytical techniques work well for simple networks, but they often don't scale well to larger, more complex systems. Metaheuristic algorithms, such as GA (Gupta & Tripathi, 2014) and PSO (M'dioud et al., 2024), have gained widespread popularity for DG placement problems due to their flexibility and robustness. PSO, driven by the social habits of flocks of birds, has proven to be particularly effective in handling nonlinear and multi-objective optimization problems (Gupta, 2025; Hassan et al., 2020)

More advanced versions of PSO, such as RDPSO (Sun et al., 2015), Chaotic PSO (Hassan, Sun, & Wang, 2020), NPSO (M'dioud et al., 2022), have been created to enhance the performance of the PSO algorithm. RDPSO is better able to explore the solution space and stay out of local optima. It introduces random fluctuations in particle movement. The RDPSO method adds stochastic drift to the standard PSO model, enhancing its performance in complex, multi-modal optimization landscapes encountered in DG placement. DG placement is framed as a multi-objective challenge where loss minimization is combined with other factors, like voltage stability or environmental impact. In various cases, such as with the WOA (H. P. C. et al., 2023), and MFO [(Elattar & Elsayed, 2020), ESGA (Huy et al., 2023), SOS (Truong et al., 2020). The algorithms capture complex relationships and non-linear constraints that often arise in DG systems.

For DG placement, a number of other techniques have been investigated. For example, an iterative approach that uses power flow analysis to determine which bus is most susceptible to voltage collapse at maximum loads is appropriate for DG installation. Hedayati et al., 2008 aimed to improve voltage stability margins, maximize loading, increase power transfer capacity, and decrease power losses. Studies done by (Tyagi et al., 2024; Rao et al., 2023) specifically targeted voltage profile improvements as a key outcome. These improvements are crucial in radial networks, where voltage drops can have significant downstream effects.

DG placement can bolster network resilience during large-scale disturbances, as confirmed by Kizito et al. (2020), by strategically integrating DG into weak areas of the network, power systems gain an additional layer of reliability. This trend aligns with the increasing need for resilient power grids in the face of natural disasters or other unpredictable disruptions.

Studies demonstrate that renewable-based DG units not only enhance grid stability but also significantly decrease emissions (Gupta et al., 2021; Thandlam et al., 2024). This environmental benefit aligns with the growing focus on decarbonizing power systems. Furthermore, DG placement contributes to reduced transmission costs (Truong et al., 2020; Sharma et al., 2024) as it minimizes the need for long-distance power transmission and enables localized energy generation.

The main contributions of this paper are

- To determine the optimal placement of DG.
- To determine the optimal sizing of DG.
- To reduce both active power and reactive power losses.

The rest of the paper is organized as follows:

Section 2 describes the problem formulation, objective function, system constraints, and power flow equation. Section 3 illustrates the basics of conventional PSO and RDPSO algorithms. In Section 4, simulation results are discussed, and finally, conclusions are drawn in Section 5.

PROBLEM FORMULATION

This work intends to find the optimal placement and optimal sizing of DG, and also to reduce active power loss and reactive power loss. Therefore, we aim to inject DG units in the most efficient way possible to achieve this intended objective. This will involve the optimal sizing and placement of the DG under the different constraints associated with various standard operational limits.

The objective function for power losses minimization is given by the following Equations 1 and 2.

$$Min\sum_{1}^{N} P_{loss}, I = Min\sum_{I=1}^{I=N} R_1 * I_I^2 \text{ for I} \in [1,2,3,\ldots\ldots,\text{N}] \tag{1}$$

$$Min\sum_{1}^{N} Q_{loss}, I = Min\sum_{I=1}^{I=N} X_1 * I_I^2 \text{ for I} \in [1,2,3,\ldots\ldots,\text{N}] \tag{2}$$

$$MinTVD = Min\sum_{j=1}^{j=M} \left|V_{norminal} - V_j\right| \text{ for j} \in [1,2,3,\ldots,\text{M}] \tag{3}$$

Where

P_{loss}: active power loss for branch I's
R_i: line's resistance.
I_I: branch current.
Q_{loss}: branch I's reactive power loss.
X_I: line I's reactance.
V_j: magnitude of the voltage on load bus j.

$V_{norminal}$: reference voltage, which is set at 1.0 p.u.
N: total number of branches in the Network.
M: The network's total number nodes.

The system voltage, current constraints, and DG units restriction are given by Eqs. 4, 5, and 6.

The voltage of each node must be within

$$V_{min} \leq V_j \leq V_{max}; i\epsilon[1,2,\ldots,M] \tag{4}$$

Where;

V_{min}: voltage's lowest limit.
V_j: voltage of the j-th node.
V_{max}: voltage's maximum limit.
M: network's total number of nodes.

Every branch's current must be below the conductor's maximum permitted value.

$$I_I \leq I_{max}; \text{for I} \in [1,2,\ldots,N] \tag{5}$$

Where;

I_I: line I's current.
I_{max}: maximum amount of current that can flow.

DG units Restrictions: The inequality below is used to prevent reverse power flow

$$0 \leq \sum_{i=1}^{i=M} P_{DG,i} \leq \sum_{i=1}^{i=M} P_{Load,i}; for\ i \in [1,2,\ldots,M] \tag{6}$$

Where;

$P_{DG,i}$: quantity of active power injected into node i.
$P_{Load,i}$: load on node i.
M: network's total number of nodes.

Hence, these two inequalities below are used for subjecting active and reactive power insertion

$$P_{DG,min} \leq P_{DG,i} \leq P_{DG,max}\ ; i \in [1,2,\ldots,M] \quad (7)$$

$$Q_{DG,min} \leq Q_{DG,i} \leq Q_{DG,max}\ ; i \in [1,2,\ldots,M] \quad (8)$$

Where;

$P_{DG,min}$, $Q_{DG,min}$: minimum value of active power and reactive power that can be injected into node i, respectively.
$P_{DG,i}$, $Q_{DG,i}$: value active power and reactive power injected into node i, respectively.
$P_{DG,max}$, $Q_{DG,max}$: maximum value of active and reactive power that can be injected into node i, respectively.
M: total number of nodes in the network.

Power Balance Equation

For the electrical network to remain stable, the total active power loss in the network, the total active power of all injected DG units, and the substation must equal the total load linked to all buses. Equations 9 and 10 demonstrate how the same standards apply to the network's reactive power.

$$P_{sub} + \sum P_{DG} = P_{load} + \sum P_{loss} \quad (9)$$

$$Q_{sub} + \sum Q_{DG} = Q_{load} + \sum Q_{loss} \quad (10)$$

Where;

P_{sub}: substation's active power.
$\sum P_{DG}$: total active power of all dispersed generation units.
P_{load}: total active load connected.
$\sum P_{loss}$: overall active power loss of the network.
Q_{sub}: substation's reactive power.
$\sum Q_{DG}$: is the total reactive power injected by all dispersed generation units.
Q_{load} : is the associated total reactive load.
$\sum Q_{loss}$: is the network's total reactive power loss.

Equations 11 and 12 are used to replace the power loss, and the new power balance is as follows:

$$P_{sub} + \sum P_{DG} = P_{load} + \sum_{l=1}^{l=N} R_1 * I_l^2\ ; for\ l \in [1, 2, \ldots, N] \quad (11)$$

$$Q_{sub} + \sum Q_{DG} = Q_{load} + \sum_{I=1}^{I=N} X_1 * I_I^2 \; ; for\ l \in [1, 2, \ldots, N] \tag{12}$$

Backward-Forward Sweep (BFS) for power flow solving

These studies suggest the Backward-Forward Sweep (BFS) method to analyze the power flow (Ouali & Cherkaoui, 2020; Shukla et al., 2010). There are three primary steps in this method:

Step 1: Each node I's nodal current is calculated using

$$I_i^t = conj\left(P_{m-i} + j * Q_{m-i}\right) / V_{m-i}^t \; ; for\ i = 1,2 \ldots m \tag{13}$$

Where:

P_{m-i}: The active power of the $m-i$ node.
Q_{m-i}: The reactive power of the $m-i$ node.
V_{m-i}^t : The voltage of the $m-i$ node at iteration I.

Figure 1. Single Line Diagram

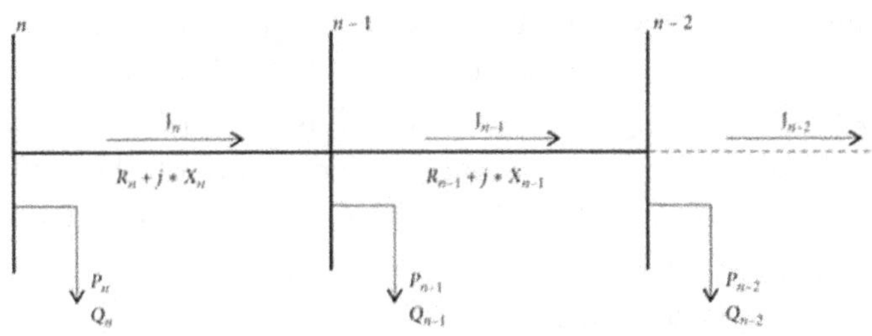

Step 2: the equation below is used at the last edge of the single line diagram to determine $J\,(i, i + 1)$. This stage is referred to as the backward sweep.

$$J_{m-1}^t = -\frac{conj\left(P_{m-i} + j * Q_{m-i}\right)}{V_{m-i}^t} + \sum_r J_{m-r}^t \; ; for\ r = 1,2 \ldots \tag{14}$$

Where:

$J(i,i+1)$ represents the electric current that flows between buses i and $i+1$.

$\sum_r J^t_{m-r}$ represents the amount of current flowing across the branches that emerge from bus i during iteration t.

Step 3: As the third step in the forward sweep, voltage for different nodes is calculated starting from the root node using the equation:

$$V^t_i = V^t_{i-1} - Z_i * J^t_i , for\ i = 2,3, \ldots m \quad (15)$$

where the series impedance of the branch "$i-1$, i" is shown by, Z-i.

Until the difference between each node's current and prior voltages is less than the tolerance limit, these steps are repeated. The inequality below determines this step.

$$Max\left(\left[V^{t+1}_i\right] - \left[V^t_i\right]\right) < \varepsilon\ ; for\ i = 1,2, \ldots\ldots m \quad (16)$$

PARTICLE SWARM OPTIMIZATION (PSO)

As an optimization tool, the PSO offers a population-based search process wherein particles, or people, alter their position (state) over time (Anand et al., 2022; Sudha et al., 2024). Particles fly over a multi-dimensional search space in a PSO system. Each particle in flight modifies its location based on its own experience (Pbest) and that of a neighboring particle (Gbest), using the best position that both the particle and its neighbor have experienced (Gupta et al., 2015). The idea of velocity might be used to illustrate this change. Each particle in PSO has a position (x) and velocity (v), according to the mechanics of PSO. The velocity controls the particle's motion throughout the search region, whilst the position indicates a potential arrangement for DG placement and sizing. The following equations are used to iteratively update each particle's position and velocity:

Velocity Update

$$Vi(t+1) = \omega Vi(t) + c1r1(Pbest - Xi(t)) + c2r2(Gbest - Xi(t))V_{i(t+1)} \quad (17)$$

where:

ω is called the inertia weight.
$c1$ represent cognitive coefficient.
$c2$ social acceleration coefficient.
$r1$ and $r2$ are random values between 0 and 1.

Pbest represents best position found by the particle.
Gbest represents best position found by the swarm.

Position Update:

$$Xi(t+1) = Xi(t) + Vi(t+1) \tag{18}$$

The key to PSO's success in optimization problem is striking a balance between exploitation (improving on well-known, effective solutions) and exploration (exploring new regions of the solution space). However, PSO may experience premature convergence in intricate optimization landscapes with numerous local optima.

RDPSO ALGORITHM

We suppose that the particle in the RDPSO moves in an external electric field like an electron in a metal conductor. Therefore, the particle's motion is the result of the superposition of the drift and thermal motions (Sinha et al., 2023). Whereas the drift motion primarily carries out the local search, the thermal motion carries out the particle's global search. As shown in the first paragraph of this chapter, the trajectory analysis shows that the particle's directional travel toward its local attractor, *P-i,n*, in the classical PSO represents the particle's local search. The directional movement towards, *P-i,n* is another definition of the particle's drift motion in the proposed RDPSO. The primary heritage of the RDPSO algorithm from the conventional PSO algorithm is that it embodies the amalgamation of the social and cognitive components of the canonical PSO (Sinha et al., 2024). The flowchart of RDPSO algorithm is shown in Figure 2.

Figure 2. Flowchart for the proposed RDPSO algorithm

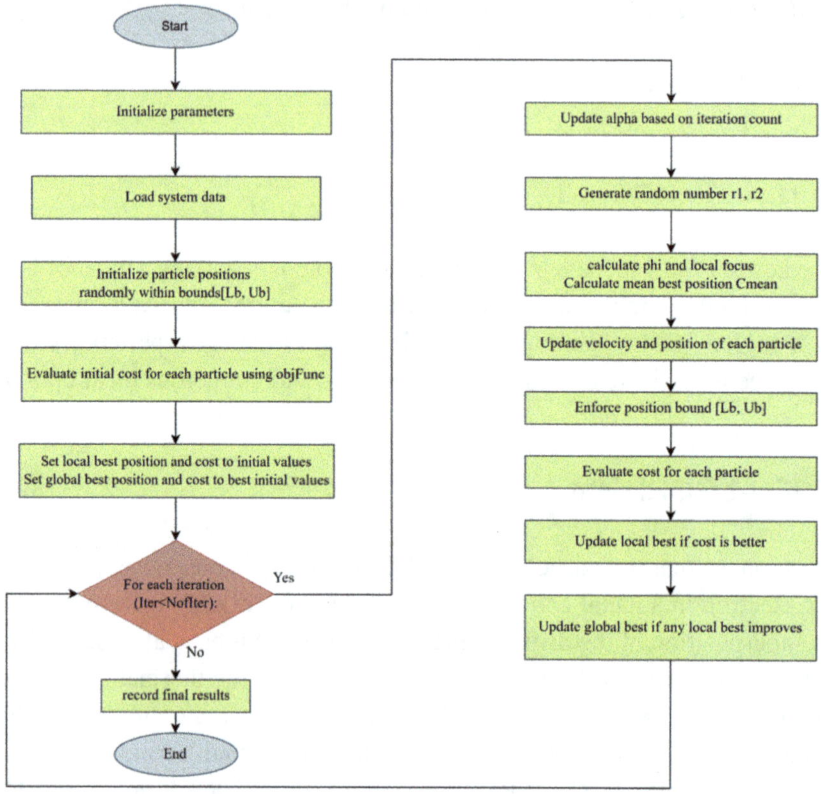

In this study, BFS approaches are used for power flow calculation, and RDPSO is used for the optimal placement and sizing of DG.

IEEE 33-Bus Test Network

This reference Baran & Wu, 1989) presents the features of the IEEE 33-bus network. The network architecture in its original configuration is shown in Figure 3. There are 37 branches and 33 buses in this electrical system's network. The network's five open switches are 33, 34, 35, 36, and 37.

Figure 3. IEEE 33 bus test system

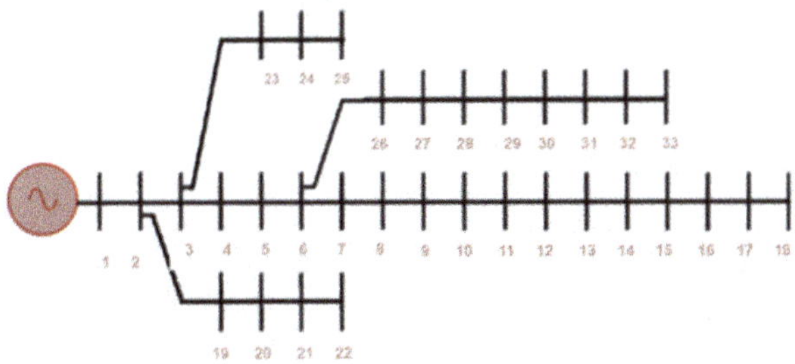

SIMULATION RESULTS AND DISCUSSIONS

The goals of this study are to minimize active and reactive power losses while identifying the best location and size for DG. All the simulations are carried out on the MATLAB 2018a platform. The voltage's upper and lower limits are assumed in this study to be 1.05*p.u.* and 0.95*p.u.*, respectively. Additionally, the electrical current should not exceed 300 A.

Figure 4. Voltage Profile at Various Buses for One DG Insertion

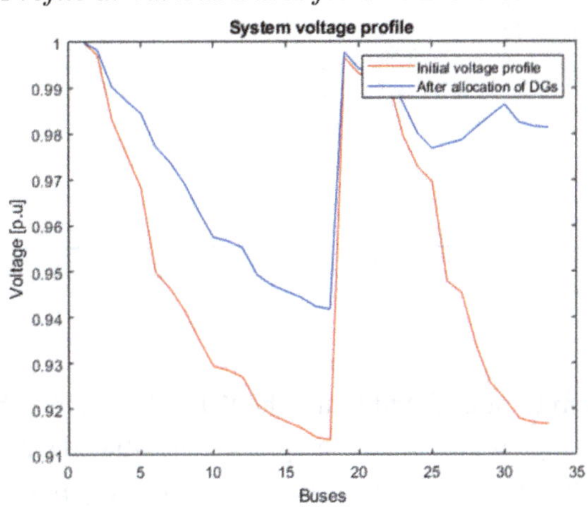

Figure 5. Voltage Profile at Various Buses for Two DGs Insertion

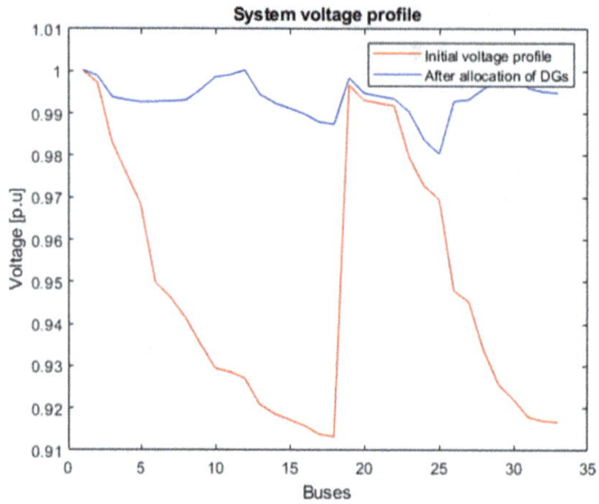

Figure 6. Voltage Profile at Various Buses for Three DGs Insertion

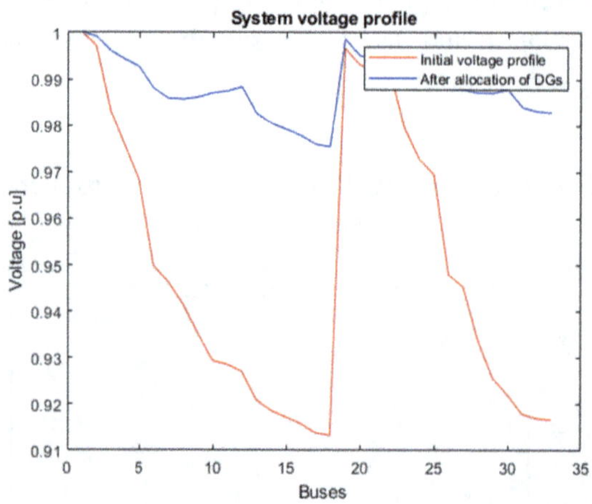

It is evident from Figures 5 and 6 that the IEEE 33 bus test shows a significant improvement in voltage profiles after integrating distributed generation units using RDPSO. Initially, voltage drops were observed, indicating potential inefficiencies and reliability issues. However, after DG units were integrated, voltage stabilization was observed, indicating a more balanced network. The strategic placement of DGs mitigated voltage drops, reducing power loss and enhancing system performance. The comparative simulation results for one, two, and three DGs integration are shown in Table 1.

Table 1. Simulation results for one, two, and three DGs integration

Nos. of DG	Size of DG(KW)	Location of DG	Active power loss (kW)	Reactive power loss (kVAR)
1	2408.56	6	61.5204	48.4215
2	789.6, 850.96	12,30	32.4006	22.7817
3	774.1, 701.2, 1006.9	30, 24, 12	68.5818	46.2648

The ideal size for the scenario involving a single DG unit is found to be 2408.56 kW, and it is installed at bus 6. The active and reactive power losses from this setup are 48.422 kVAR and 61.520 kW, respectively. Even if power losses are significantly lower than in the base scenario, this shows that a single DG unit might not be able to cover the entire network.

With the deployment of two DG units, sized at 789.6 kW and 850.96 kW and positioned at buses 12 and 30, there is a significant improvement. Active power loss drops to 32.401 kW, and reactive power loss decreases to 22.782 kVAR, showcasing the benefits of strategically distributing DGs. This setup effectively mitigates power losses and improves voltage profiles more efficiently than a single DG unit.

However, when three DG units are introduced, sized at 774.1 kW, 701.2 kW, and 1006.9 kW and placed at buses 30, 24, and 12, the active and reactive power losses are recorded at 68.582 kW and 46.265 kVAR. Interestingly, this configuration results in higher losses compared to the two-DG scenario, indicating potential inefficiencies in placement and power distribution.

The analysis suggests that simply increasing the number of DG units does not necessarily guarantee better outcomes. Instead, optimal placement and sizing play crucial roles in achieving minimal losses.

Table 2. Comparative study of three DGs integration

Method	DG scale (kW)	DG unit's placement (node)	ΔP (kW)	Loss reduction (kW)	Loss reduction %
Initial	-	-	202.686		
RDPSO	774.1, 701.2, 1006.9	30, 24, 12	68.5818	134.1045	66.16
NPSO (M'dioud et al., 2022)	1071, 759, 1100	24, 14, 30	69.3833	133.303	65.76
CSCA (Selim et al., 2020)	871, 1091.47, 954.08	13, 24, 30	71.94	130.7463	64.506

continued on following page

Table 2. Continued

Method	DG scale (kW)	DG unit's placement (node)	ΔP (kW)	Loss reduction (kW)	Loss reduction %
PSO (Karunarathne et al., 2020)	2420, 960, 1080	6, 11, 29	73.63	129.0653	63.67
ECOA (Pham et al., 2021)	737.6, 651.8, 1070.5	14, 25, 30	74.6	128.4263	63.36
TSO (Bhadoriya & Gupta, 2022)	771.54, 1103.65, 1064.57	14, 24, 30	72.79	129.8963	64.087
SM-PSO (Belbachir et al., 2021)	545.7, 768.6, 735.8	13, 24, 30	84.9083	117.778	58.1
ESFO (Nguyen, 2021)	1094.5, 1316.2, 804.6	30, 24, 13	72.2671	130.4192	64.35

The comparative studies for three DG integration while suing different optimization strategies, have been tabulated in Table 2. The RDPSO methodology for optimal DG placement performs better in reducing power losses and increasing efficiency in the IEEE 33-bus network. At 202.6863 kW, the baseline system with no DG units shows a significant active power loss. The introduction of different optimization methods significantly reduces these losses, but the effectiveness varies.

RDPSO achieves the most substantial loss reduction, with a decrease of 134.1045kW, corresponding to an impressive 66.16% reduction in active power loss. This method employs DG units of sizes 774.1 kW, 701.2 kW, and 1006.9 kW, placed optimally at nodes 30, 24, and 12. Compared to other techniques, RDPSO outperforms them all, showcasing its effectiveness in identifying the best DG configuration.

In contrast, the NPSO method, using DGs sized at 1071 kW, 759 kW, and 1100 kW at nodes 24, 14, and 30, achieves a loss reduction of 133.303 kW (65.76%). Similarly, CSCA yields a 64.506% reduction, with DG units strategically positioned at nodes 13, 24, and 30. Although these methods are efficient, their performance lags behind RDPSO.

PSO, ECOA, and TSO methods show further decline in performance, with loss reductions of 63.67%, 63.36%, and 64.087%, respectively. Notably, even though these methods use varying DG sizes and placements, their efficiency does not match that of RDPSO. The SM-PSO method, with a 58.1% reduction, is the least effective among the compared techniques.

Overall, RDPSO's ability to strategically place DG units for optimal loss reduction is evident. Its higher efficiency suggests that more precise and adaptable optimization strategies can significantly enhance power distribution system performance. The comparative analysis emphasizes the importance of advanced heuristic algorithms, such as RDPSO, for effectively reducing power losses and improving network reliability.

CONCLUSIONS

Based on the analysis and results obtained from simulation studies, it is evident that the proposed RDPSO algorithm provides superior performance in minimizing active power losses and improving the system's voltage profile. The study compared the effectiveness of various optimization techniques, including RDPSO, NPSO, CSCA, PSO, ECOA, TSO, SM-PSO, and ESFO, in reducing power losses.

The RDPSO approach demonstrated an outstanding active power loss reduction of 66.16%, significantly outperforming other methods. By strategically placing DG units with sizes of 774.1 kW, 701.2 kW, and 1006.9 kW at nodes 30, 24, and 12, RDPSO achieved the most efficient configuration for enhancing the system's performance. Other methods, such as NPSO, CSCA, and PSO, also achieved notable loss reductions but were less effective than RDPSO.

The results underscore the importance of employing advanced optimization techniques, such as RDPSO, to optimize DG placement in power distribution networks. The findings confirm that RDPSO not only minimizes power losses but also enhances the overall voltage stability of the system. The use of MATLAB 2018a provided an effective platform for simulating and analyzing the complex optimization problems associated with DG placement. Conclusively, RDPSO's efficacy for optimal DG integration makes it a viable option for upcoming upgrades to power distribution networks. The potential benefits of employing RDPSO in real-world systems are highlighted by the improved voltage profile and significant reduction in power losses.

REFERENCES

Anand, S., Kulkarni, S. V., Gupta, S., & Parthasarathy, V. (2022, December). Battery management system (BMS) for electric vehicle applications. *In 2022 fourth international conference on emerging research in electronics, computer science and technology (ICERECT)* (pp. 1-6). IEEE. https://doi.org/DOI: 10.1109/ICERECT56837.2022.10060137

Awasthi, B., Gupta, S., Kumar, S., & Tripathi, A. (2024, January). Business management and planning using queuing model with heterogeneous customers for electrical applications. *In 2024 2nd International Conference on Intelligent Data Communication Technologies and Internet of Things (IDCIoT)* (pp. 279-283). IEEE. https://doi.org/DOI: 10.1109/IDCIoT59759.2024.10467673

Baran, E., & Wu, F. F. (1989). Network reconfiguration in distribution systems for loss reduction and load balancing. *IEEE Transactions on Power Delivery*, *4*(2), 1401–1407. DOI: 10.1109/61.25627

Belbachir, N., Zellagui, M., Lasmari, A., El-Bayeh, C. Z., & Bekkouche, B. (2021). Optimal PV sources integration in distribution system and its impacts on overcurrent relay based time-current-voltage tripping characteristic. In *2021 12th International Symposium on Advanced Topics in Electrical Engineering (ATEE)* (pp. 1–7). IEEE. https://doi.org/DOI: 10.1109/ATEE52255.2021.9425155

Bhadoriya, J. S., & Gupta, A. R. (2022). A novel transient search optimization for optimal allocation of multiple distributed generators in the radial electrical distribution network. *International Journal of Emerging Electric Power Systems*, *23*(1), 23–45. DOI: 10.1515/ijeeps-2021-0001

Elattar, E. E., Elsayed, S. K., & Elsayed, S. K. (2020). Optimal location and sizing of distributed generators based on renewable energy sources using modified moth flame optimization technique. *IEEE Access : Practical Innovations, Open Solutions*, *8*, 109625–109638. DOI: 10.1109/ACCESS.2020.3001758

Fotis, G., Gupta, S., Varshney, T., Dhar, S. L., & Gulzar, M. M. (Eds.). (2025). *Modern Computing Technologies for EV Efficiency and Sustainable Energy Integration.* IGI Global., DOI: 10.4018/979-8-3373-2382-4

Gupta, S. (2025). *The Future of Machine Learning with Automation in Engineering.* Nova Science Publishers., DOI: 10.52305/YFIX2832

Gupta, S., & Tripathi, R. K. (2014). Two-area power system stability improvement using a robust controller-based CSC-STATCOM. *Acta Polytechnica Hungarica*, *11*(7), 135–155. DOI: 10.12700/APH.11.07.2014.07.9

Gupta, S., & Tripathi, R. K. (2015). Transient Stability Enhancement of Multimachine Power System Using Robust and Novel Controller Based CSC-STATCOM. *Advances in Power Electronics, 2015*(1), 626731. https://onlinelibrary.wiley.com/doi/10.1155/2015/626731. DOI: 10.1155/2015/626731

Gupta, S., Tripathi, R. K., & Shukla, R. D. (2010, November). Voltage stability improvement in power systems using facts controllers: State-of-the-art review. In *2010 International Conference on Power, Control and Embedded Systems* (pp. 1-8). IEEE. https://doi.org/DOI: 10.1109/ICPCES.2010.5698665

Gupta, Y., Doolla, S., Chatterjee, K., & Pal, B. C. (2021). Optimal DG allocation and volt-var dispatch for a droop-based microgrid. *IEEE Transactions on Smart Grid, 12*(1), 169–181. DOI: 10.1109/TSG.2020.3017952

Hassan, A. S., Sun, Y., & Wang, Z. (2020). Multi-objective for optimal placement and sizing DG units in reducing loss of power and enhancing voltage profile using BPSO-SLFA. *Energy Reports, 6*, 1581–1589. DOI: 10.1016/j.egyr.2020.06.013

Hedayati, H., Nabaviniaki, S. A., & Akbarimajd, A. (2008). A method for placement of DG units in distribution networks. *IEEE Transactions on Power Delivery, 23*(3), 1620–1628. DOI: 10.1109/TPWRD.2007.916106

Huy, T. H. B., Vo, D. N., Truong, K. H., & Van Tran, T. (2023). Optimal distributed generation placement in radial distribution networks using enhanced search group algorithm. *IEEE Access : Practical Innovations, Open Solutions, 11*, 103288–103305. DOI: 10.1109/ACCESS.2023.3316725

Iweh, C. D., Gyamfi, S., Tanyi, E., & Effah-Donyina, E. (2021). Distributed generation and renewable energy integration into the grid: Prerequisites, push factors, practical options, issues and merits. *Energies, 14*(17), 5375. DOI: 10.3390/en14175375

Karunarathne, E., Pasupuleti, J., Ekanayake, J., & Almeida, D. (2020). Optimal placement and sizing of DGs in distribution networks using MLPSO algorithm. *Energies, 13*(23), 6185. DOI: 10.3390/en13236185

Kayal, P., & Chanda, C. K. (2013). Placement of wind and solar based DGs in distribution system for power loss minimization and voltage stability improvement. *International Journal of Electrical Power & Energy Systems, 53*, 795–809. DOI: 10.1016/j.ijepes.2013.05.047

Kizito, R., Li, X., Sun, K., & Li, S. (2020). Optimal distributed generator placement in utility-based microgrids during a large-scale grid disturbance. *IEEE Access : Practical Innovations, Open Solutions, 8*, 21333–21344. DOI: 10.1109/ACCESS.2020.2968871

M'dioud, M., Bannari, R., & Elkafazi, I. (2022). Optimal DG allocations in the distribution networks using an improved PSO algorithm. [IJRER]. *International Journal of Renewable Energy Research, 12*(3).

M'dioud, M., Bannari, R., & Elkafazi, I. (2024). A novel PSO algorithm for DG insertion problem. *Energy Systems, 15*(1), 325–351. DOI: 10.1007/s12667-022-00546-z

Nguyen, T. T. (2021). Enhanced sunflower optimization for placement distributed generation in distribution system. *International Journal of Electrical and Computer Engineering, 11*(1), 107–113. DOI: 10.11591/ijece.v11i1.pp107-113

Ouali, S., & Cherkaoui, A. (2020). An improved backward/forward sweep power flow method based on a new network information organization for radial distribution systems. *Journal of Electrical and Computer Engineering, 2020*, 5643410. Advance online publication. DOI: 10.1155/2020/5643410

Pham, T. D., Nguyen, T. T., & Dinh, B. H. (2021). Find optimal capacity and location of distributed generation units in radial distribution networks by using enhanced coyote optimization algorithm. *Neural Computing & Applications, 33*(9), 4343–4371. DOI: 10.1007/s00521-020-05239-1

Rao, M. S., Saraswathy, S., Neela, V., Arun, M. R., Gupta, S., & Taqui, S. N. (2023, November). Efficient Model Training in Decentralized Systems with Federated Learning. *In 2023 International Conference on Communication, Security and Artificial Intelligence (ICCSAI)* (pp. 900-904). IEEE. https://doi.org/DOI: 10.1109/ICCSAI59793.2023.10421469

Selim, A., Kamel, S., & Jurado, F. (2020). Efficient optimization technique for multiple DG allocation in distribution networks. *Applied Soft Computing, 86*, 105938. DOI: 10.1016/j.asoc.2019.105938

Sharma, R. K., Saran, V. B., Kanaujia, S., & Gupta, S. (2024, May). Evaluation of mppt controller performance using fuzzy logic design with wireless sensor node. *In 2024 International Conference on Smart Systems for applications in Electrical Sciences (ICSSES)* (pp. 1-6). IEEE. https://doi.org/DOI: 10.1109/ICSSES62373.2024.10561365

Shukla, R. D., Tripathi, R. K., & Gupta, S. (2010, November). Power electronics applications in wind energy conversion system: A review. *In 2010 International Conference on Power, Control and Embedded Systems* (pp. 1-6). IEEE. https://doi.org/DOI: 10.1109/ICPCES.2010.5698663

Sinha, R., Kaur, N., Gupta, S., & Thakur, P. (2023, November). Diagnosis of Parkinson's Disease using Hybrid Ensemble Technique. *In 2023 International Conference on Ambient Intelligence, Knowledge Informatics and Industrial Electronics (AIKIIE)* (pp. 1-5). IEEE. https://doi.org/DOI: 10.1109/AIKIIE60097.2023.10390458

Sinha, R., Sinha, K. K., Patel, M., Gupta, S., & Priya, S. (2024, July). Detection of leukemia disease using convolutional neural network. *In 2024 5th International Conference on Image Processing and Capsule Networks (ICIPCN)* (pp. 451-456). IEEE. https://doi.org/DOI: 10.1109/ICIPCN63822.2024.00080

Sudha, G., Natha, A., Rajendiran, M., Vanitha, U., Kiran, V., Patil, H., & Rajaram, A. (2024). Optical oasis harnessing energy efficiency in wireline, optical, and wireless communications. *Journal of Environmental Protection and Ecology*, *25*(4), 1179–1187.

Sun, J., Wu, X., Palade, V., Fang, W., & Shi, Y. (2015). Random drift particle swarm optimization algorithm: Convergence analysis and parameter selection. *Machine Learning*, *101*(1–3), 345–376. DOI: 10.1007/s10994-015-5522-z

Thandlam, A. K., Vignesh, T., Saravanan, A., Subramani, P., Marimuthu, R., & Gupta, S. (2024, April). Next-gen vehicle safety: The futuristic approach to auto-stop in modern vehicles. *In 2024 10th International Conference on Communication and Signal Processing (ICCSP)* (pp. 265-270). IEEE. https://doi.org/DOI: 10.1109/ICCSP60870.2024.10543666

Truong, K. H., Nallagownden, P., Elamvazuthi, I., & Vo, D. N. (2020). A quasi-oppositional-chaotic symbiotic organisms search algorithm for optimal allocation of DG in radial distribution networks. *Applied Soft Computing*, *88*, 106067. DOI: 10.1016/j.asoc.2020.106067

Tyagi, M., Gupta, S., & Khan, M. I. (2024, May). Reduction of voltage instability in complex power networks through the utilization of D-STATCOM. *In 2024 4th International Conference on Advance Computing and Innovative Technologies in Engineering (ICACITE)* (pp. 1662-1666). IEEE. https://doi.org/DOI: 10.1109/ICACITE60783.2024.10616856

Chapter 10

Optimization and Control of Solar Energy System by Fuzzy Logic for Wireless Sensor Network

Sandeep Kanaujia
https://orcid.org/0000-0001-9109-5030
United University, India

Ajay Sharma
https://orcid.org/0000-0002-9112-6228
United College of Engineering and Research, Prayagraj, India

Rajev Kumar Sharma
https://orcid.org/0000-0003-2245-1390
United University, Prayagraj, India

ABSTRACT

The limited energy storage capacity of batteries in wireless sensor networks that is use for various different remote areas presents a significant constraint, limiting network lifespan. Energy harvesting (EH) systems offers a promising solution to overcome this limitation. While various energy sources exist, solar energy (clean and green energy) is widely utilized for energy harvesting due to its flexibility, maturity, and external power availability. The aim of this chapter is to extract effective solar energy from harvesting technology that uses solar energy with maximum power to give explanation for restricted cell life issue of wirelesssensor nodes. This research will provide enhanced and effective energy harvesting by solar method for wireless sensor nodes that combines maximum power point to track and approximate reasoning

DOI: 10.4018/979-8-3373-7041-5.ch010

methods together. The research goal is to enhance the total overall efficiency of the harvesting methodology, which is dependent on types of solar panel, fuzzy based logic controls, and maximum power point tracking efficiency, among other aspects.

INTRODUCTION

Wireless (remote) Sensor Networks (WSN) have gained popularity and widespread usage across various fields. These networks comprise sensor nodes tasked with monitoring diverse environmental conditions and transmitting data to a central sink for further processing (Kanaujia and Sharma, 2023; Sharma et al., 2024).

Sensor nodes within WSNs typically rely on batteries for power, which can be refillable or non-refillable. The use of non-refillable batteries presents challenges as it limits the lifespan of WSNs and can adversely affect their performance, posing a critical issue for network operation. Consequently, rechargeable batteries are preferred; however, replacing these batteries in remote or inaccessible areas where many WSN applications are deployed can be inconvenient and logistically challenging. Therefore, extending the lifespan of WSNs within finite power constraints remains a significant challenge (Akyildiz, 2002; Meena et al., 2024).

Energy harvesting (EH) technologies have a solution for battery power limitations and extend the lifespan of WSNs. Various EH sources, such as photovoltaic, airflow, vibration, electromagnetic-frequency (EF), and heat energy, can be harnessed for this purpose (Sah et al.,2020; Gupta et al., 2026). Among these, solar EH techniques are particularly favored due to the high-power density provided by solar cells, which can reach up to 15 mW/cm^3. This ample power density makes solar energy a viable option to meet the power requirements of WSNs (Chalasani et al., 2008; Gupta, 2025). The different energy densities of the type of harvesting sources are given below in Table 1.

Table 1. Energy densities of harvesting approach

S. N0	Types	Energy density
a.	photovoltaic cells	15 mW/cm^3
b.	Piezo material	330μW/ cm^3
c.	Kinetic energy harvester	116 μW/ cm^3
d.	Thermoelectric	40 μW/ cm^3
e.	Acoustic noise	960nW/ cm^3

The tracking of maximum power points algorithm is designed to boost the power output of a photoelectric solar cell system by continuously adjusting to changing conditions, as shown in Figure 1. Factors such as solar luminescence and degrees significantly affect the achievement of the photovoltaic system (Karami et al., 2017; Ali et al., 2020). As these environmental conditions fluctuate, tracking of maximum power points of the PV factor shifts, altering optimal operating voltage and current. The MPPT algorithm adjusts accordingly to promise the system consistently intervenes at its most efficient point, delivering maximum power to the load (Li et al., 2019; Mao et al., 2020; Sorte et al., 2022).

Figure 1. Diagrammatic representation of fuzzy logic control

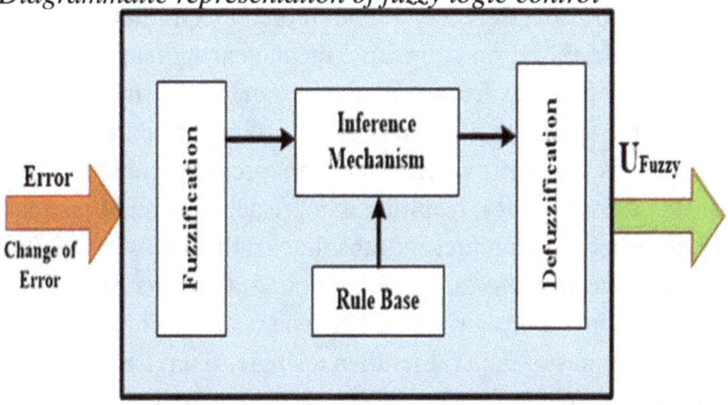

Fuzzy logic control has become increasingly popular over the last ten years due to its capacity to manage nonlinearity, process imprecise data, and operate without requiring a precise mathematical formulation (Malik et al., 2024; Bhatt et al., 2024). The rise of microcontrollers has significantly enhanced the application of fuzzy logic control. This approach consists of three main phases: fuzzification, where crisp inputs are transformed into fuzzy values; the inference system, which applies fuzzy rules to derive conclusions; and defuzzification, which is the inverse of fuzzification (Dash et al., 2015; Semwal et al., 2024). Fuzzification describes the way of transforming "crisp" mathematical feed into "fuzzy" semantic changeable, depending on membership level in specific sets. Benchmark for Fuzzy control is generally structured in an IF-THEN format (Nabulsi et al., 2012; Sutikno et al., 2021).

FUZZY LOGIC: SOLVING UNCERTAIN PROBLEMS IN MATH AND ENGINEERING

The word "fuzzy" describes things that are uncertain, unclear, or not well-defined. In real life, many situations cannot be classified as completely true or completely false. Fuzzy logic helps in these situations by allowing flexible reasoning (Fotis et al., 2025; Nishad et al., 2024). It considers uncertainty and allows variables to have multiple values rather than just true or false. This makes it possible to find better solutions to complex problems.

The concept behind fuzzy logic was first introduced in the 1920s. However, the terminology fuzzy logic was initially given by Lotfi Zadeh in 1965, a professor at the University of California, Berkeley. He monitored that traditional computer logic could not handle data based on subjective or unclear human concepts. Logic of approximate reasoning comes from mathematics and helps study vague or unclear concepts, often represented as fuzzy sets of data (Gunasekaran et al., 2024; Sinha et al., 2024). It solves problems where data is not precise, allowing a range of possibilities instead of exact values, resulting in more accurate and practical solutions.

In a Boolean system, high represents absolute truth and low represents absolute false. Fuzzy logic removes this strict distinction. Statements can be partially true and partially false, with values anywhere between 0 and 1. This allows decisions to be made based on ranges of data rather than a single, fixed value.

Objective to Use Logic of Fuzzy: Fuzzy logic is widely used in practical and commercial applications because it can handle uncertainty and provide flexible reasoning:

- Operates and manages machines and household or consumer devices.
- Provides acceptable reasoning, even if not perfectly accurate.
- Helps manage uncertainty in engineering and data analysis.

Components of a Fuzzy Logic System:

- Fuzzification Module – Converts precise input values into fuzzy sets. Inputs can be categorized as: HP: High Positive; MP: Moderate Positive; S: Slight; MN: Moderate Negative; HN: High Negative (Shankar et al., 2023).
- Knowledge Base – Contains expert-defined IF-THEN rules used for decision-making.
- Inference Engine – Determines how well the current fuzzy inputs match each rule and selects which rules to apply, combining them to form control actions.
- Defuzzification Module – Translates the fuzzy outputs from the inference engine into precise, real-world numerical values.

Fuzzy Set: The logic of Fuzzy is based on fuzzy set theory, given by Lotfi Zadeh in 1965. It generalizes classical set theory (crisp sets) by allowing partial membership. A fuzzy set is a collection of values that range between 0 and 1, instead of being limited to just 0 or 1. Fuzzy sets are usually denoted with a tilde (~). Mathematically, a fuzzy set $\tilde{A}$ is defined as a pair (U, M), where:

- U is the universe of discourse (all possible values, also denoted Ω or X).
- M is the membership function, giving each element a membership degree ranging from 0 to 1.

Partial membership allows an element to be a member of a set to some extent, making fuzzy sets ideal for modeling real-world uncertainty.

Advantages of Fuzzy Logic Systems:

- Can handle imprecise, noisy, or distorted inputs.
- Easy to construct and understand.
- Simple reasoning based on mathematical set theory.
- Efficient solutions for complex problems, similar to human decision-making.
- Requires little data and memory.
- Robust because precise inputs are not necessary.
- Can work even if feedback sensors fail.
- Easy to modify to improve system performance.
- Allows use of inexpensive sensors, reducing cost and complexity.
- Effective in AI, engineering, and data mining applications.

Fuzzy logic systems are built using four main parts:

- Rule Base – Contains IF-THEN rules from experts to guide decision-making using human-like language. Modern methods decreased the number of rules needed.
- Fuzzification – Changes exact inputs (like temperature, pressure, or RPM) into fuzzy sets.
- Inference Engine – Decides and establishes how well inputs synchronize with each rule and selects which rules to apply. Combined rules form the system's control actions.
- Defuzzification– Converts fuzzy outputs into crisp, usable values. Different methods are used to reduce errors.

Fuzzy Algorithm:

A fuzzy logic algorithm addresses challenges by analyzing all available data and selecting the most optimal decision. It allows computers to handle information that is neither completely true(1) nor completely false(0), similar to person reasoning. Examples include assessing situations like “a little dark” or “somewhat bright.”

Fuzzy algorithms are used in many fields, from control systems to artificial intelligence:

A control system is a configuration of physical components that is designed to regulate or manage another system to ensure it performs in a desired manner.

Why Use Fuzzy Logic in Control Systems:

- Traditional control requires precise models, which is often difficult in real-world situations.
- Fuzzy logic allows using human expertise to design controllers.
- Fuzzy control rules (IF-THEN) are practical for building controllers in complex systems.

HOW A SEH-WSN NODE WORKS

Insolation, a switching regulator, a power converter, a battery that can be re-charged, MPPT logic, and a wireless sensor node linked as a DC load make up the system shown in Figure 2. The solar panel gathers and transforms ambient light energy into an electrical form of energy. The assembled voltage is controlled by the DC-DC power converter before it is sent to the reusable battery (Himanshu et al., 2018; Tyagi et al., 2024). The following are the model components.

Figure 2. Module of solar energy harvesting system connected with wireless sensor node

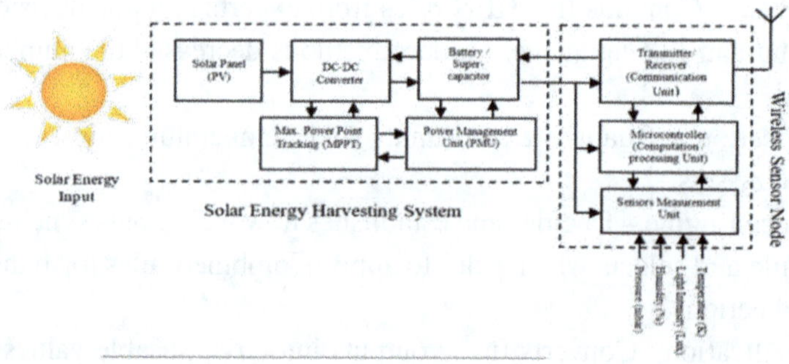

Solar Energy Harvesting (SEH) Blocks

- **Solar (PV)**

Through the PV effect, solar energy's electromagnetic radiation can be transformed into electrical power. Electron-holes are created when photons that have energy higher than semi-conductor's band-gap energy proportionate to the incident irradiation when they are exposed to sunlight. In essence, a photo cell is a p-n junction placed on a tiny semiconductor wafer (Anand et al., 2022; Shukla et al., 2010). They serve as the basic components of solar panels, which produce clean, renewable energy from sunshine for a range of uses, including lighting up establishments, technological gadgets, and residences.

- **DC-DC Converter**

A device that converts a direct current (DC) that flow in one direction only, input voltage to a different DC output voltage is known as a DC-DC converter (Gupta & Tripathi, 2014). It is used to effectively control voltage levels in a variety of applications, including battery charging, feasible energy devices, and electronic device power sources. Depending on the needs of the application, DC-DC converters can step up means boost, step down means buck, or reverse the voltage level. They are essential for maximizing power transfer and guaranteeing that various voltage sources and loads are compatible.

- **Battery/Super capacitor storage**

Rechargeable batteries and ultracapacitors are two methods of storing electrical energy in a system. This is known as battery or supercapacitor storage. These storage devices are vital parts of numerous applications, including as electric automobile, electronics easy to carry, and feasible energy systems. Whereas super capacitors store energy electrostatically and have varied properties, such as energy per unit volume, specific power, and cycle, that is, lifespan, batteries store energy chemically. They offer a way to stabilize power fluctuations, store extra energy for later use, and guarantee dependable system performance.

- **Maximum Power Point Tracking (MPPT)**

A method used in solar systems is called Tracking of peak Power (MPPT), to regularly adjust the set off point to the maximum power point (MPP) under switching external circumstances, such as heat degree and shine intensity. To optimize the amount of power generated by panels, by altering the electrical load or the photovoltaic array's electrical properties, MPPT algorithms dynamically track the MPP, guaranteeing effective energy conversion and optimizing the quantity of power collected from the solar cells group. To peak energy output and raise the general efficiency of solar energy systems, MPPT is necessary.

- **Power management unit(PMU)**

An integrated circuit or device called a Power Management Unit (PMU) is in charge of controlling the distribution and use of power in an electronic system. In order to ensure effective and dependable operation, it controls the flow of electrical power to various components or subsystems from different types, including batteries, solar panels, or external power supply. Features like power sequencing, voltage management, current limitation, and energy harvesting optimization are all possible with PMUs. They are essential for maximizing energy efficiency, extending the life of batteries, and shielding electronic equipment from damage brought on by overvoltage or overcurrent situations.

Wireless Sensor Network (WSN) Node

The important blocks of the sensor-based wireless network block are as follows:

- **Communication unit (Transmitter/Receiver)**

The transmitter and the receiver make up the communication unit. The transmitter is in charge of transforming sensor node data into electromagnetic signals that may be transmitted wirelessly. Subsequently, these signals are sent over the air to adjacent nodes or a central base station. In contrast, the transmitted signals are captured by the receiver and transformed back into digital data for processing. The communication unit makes it possible for sensor nodes to seamlessly exchange data, which makes data gathering, aggregation, and transmission in WSNs easier. (Sharma et al., 2024)

- **Computation/Processing Unit (i.e. Microcontroller)**

In most Wireless Sensor Networks (WSNs), an embedded system such as a microcontroller serves as the compute or processing unit. Within the sensor nodes, this microcontroller is in charge of carrying out the logic and algorithms required for data processing, analysis, and decision-making. It connects to a range of sensors to gather information, processes the data it receives, and manages the node's general operations. The microcontroller also controls sensor node functions, communication protocols, and power consumption. It functions as the sensor node's brain overall, enabling intelligent processing and effective functioning in WSNs.

- **Sensors Measurement Unit**

The Sensors Measurement Unit in a Wireless Sensor Network (WSN) is responsible for acquiring data from various sensors deployed in the environment. It interfaces with sensors to collect measurements such as temperature, humidity, pressure, and other environmental parameters. This unit usually includes analog-to-digital converters (ADCs) to convert cognate sensor signals into digital data that can be prepared by the on chip computer. Additionally, it may include signal conditioning circuits to improve accuracy of sensor and reliability by its measurements. The Sensors Measurement Unit plays a crucial role in gathering real-time data for monitoring and analysis in WSN applications.

SOLAR ENERGY HARVESTING (CAPTURE)

By giving the WSN node a DC power source, the solar energy gathering system plays a crucial part. Using solar cell panels, which directly transform light energy into direct current electrical energy, this voltage is obtained from ambient sunshine (Mathews et al., 2015). This DC volts is then controlled by the DC to DC converter in order to charge the rechargeable battery. Last but not least, the wireless sensor node is powered by the charged battery.

Solar energy stands out as one of the most environmentally friendly technologies available. Its intermittent availability aligns well with the requirements of Wireless sensor networks, thereby extending the network's lifespan (Harmandeep et al., 2020). To capitalize on this, we implemented a solar Energy Harvesting (EH) technique to enhance the network's longevity. Solar energy-based harvesting is a widely adopted method in remote sensor networks. The primary objective of utilizing solar energy is to either eliminate the need for batteries or reduce their usage, providing an alternative solution for energy resource management. Sun power can be collected using photoelectric solar cell conversion techniques (Bhuvaneswari et al., 2009), which convert sun energy into electric power (Li et al., 2015). Photovoltaic panels typically comprise various cells, which can be connected in succession or analogous configurations that meet specific power requirements by adjusting current and voltage levels. These cells are made with a variety of technologies, including thin films, polycrystalline silicon, amorphous silicon, and monocrystalline silicon. (Zamchiy et al., 2022). Fuzzy control is types of DC to DC converter control techniques as shown in Figure 3.

Based on the voltage and current recorded from the solar panel, the MPPT controller adjusts the duty cycle of the MOSFET in the DC to DC Buck converter. (Shiau et al., 2013). The wireless sensor node then runs on the battery voltage and is responsible for sensing, computing, and communicating. This makes it possible to use SEH-WSN nodes for autonomous monitoring and control of a variety of physical phenomena, including temperature exit, humidity in environment, pressure at those moments, and acceleration. By this, the solar energy harvester circuit's performance is crucial since low efficiency can result in insufficient battery recharge and a shorter wireless sensor network lifespan.

Figure 3. Block diagram for solar energy harvesting system by using MPPT control using Fuzzy logic control.

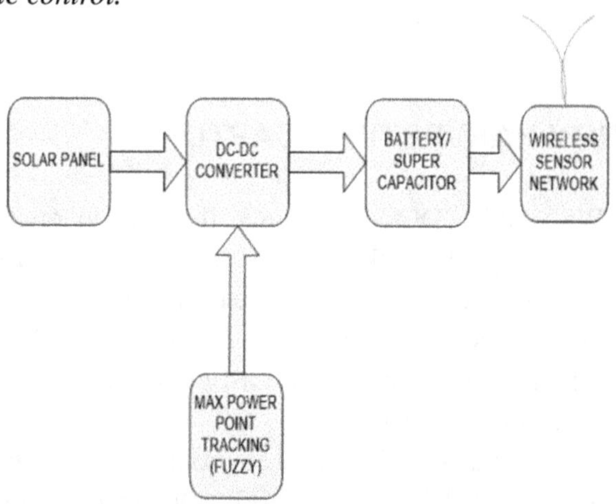

SOLAR PV PANEL MODELING

Sunlight comprises small packets of energy known as photons, which are detected in solar radiation. The light quality and its intensity depend on solar radiation (Sivakumar et al.,2016). An electron-hole pair is generated when a photon of light energy (hv>Eg) interacts with a solar cell. Light-generated current (IL) is the term used to describe this phenomenon. The solar panels' current and voltage (I-V) have the following relationship (Pandey et al., 2023).

$$\text{Isolarcellcurrent} = \text{IL} - \text{IS}\left[\exp\left(\frac{qV_{solar}}{\alpha kT}\right) - 1\right] \tag{1}$$

where T is the junction temperature in unit Kelvin, Is is the reverse saturation current, and k is Boltzmann's constant, α is the diode ideal factor, while q is the electron charge expressed in absolute terms. The electric produced by the solar panel is computed using Equation 1 as follows:

$$PSolarcellpower = Vsolar(IL + Is) - IsVsolar \times \exp(qV_{solar}/\alpha kT) \tag{2}$$

The electricity produced by the solar panel is computed using Equation 1 as follows:

$$\frac{I_L + I_S}{I_S} = \left(\frac{qV_{MPP}}{\alpha kT} + 1\right) * \exp\left(\frac{qV_{MPP}}{\alpha kT}\right) \tag{3}$$

Conversion efficiency(η), (solar energy into electrical energy),

$$\eta = \frac{FF * V_{OC} * I_{SC}}{P_m} \tag{4}$$

Where, FF = Fill Factor, Vac stands for open-circuit voltage, Isc for short-circuit current, and Pm for incident optical power, as shown in Figures 4 & 5 (Sharma et al., 2018).

Figure 4. Circuit of Solar Cell

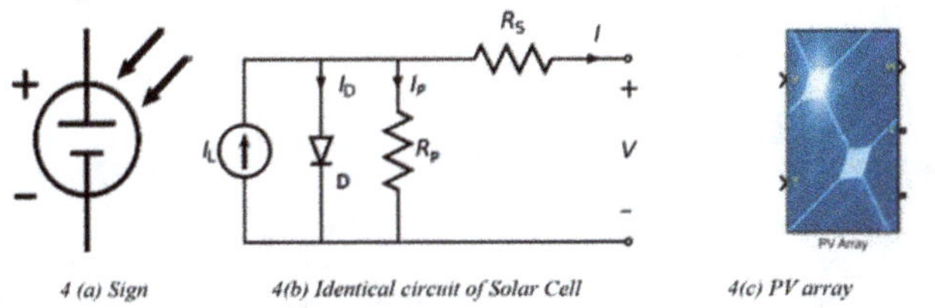

Figure 5. Current-Voltage curve and power curve of a solar cell

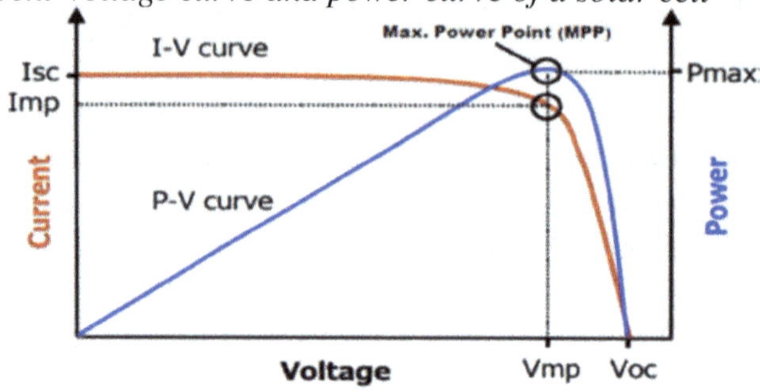

Effect of Temperature (T)

Figure 6a illustrates how a rise in solar panel temperature causes a fall in output current and vice versa. That means temperature changes and the output current have an inverse relationship. Likewise, Figure 6b illustrates how a rise in temperature causes a fall in output power and vice versa. As a result, changes in temperature directly correspond to the sensor node's output power.

Figure 6. Features of solar panels with temperature variations (degree-celcius)

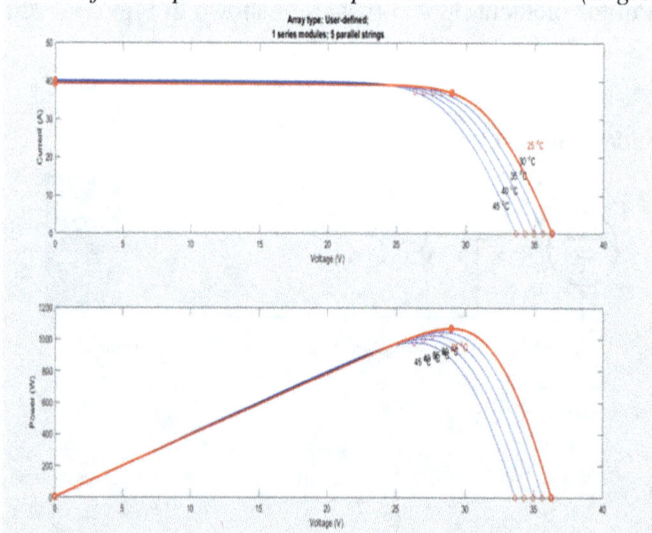

The MATLAB-Simulink program is used to model characteristics of the I-V (current-voltage) and P-V (power-voltage). The results are shown in Figure 6.

Solar Radiation (G) Effect

Solar model efficiency (η) is directly impacted by changes in solar radiation. An increase in solar radiation causes the solar cell efficiency (η) to grow. Figure 7 displays the current-voltage (I-V) relationship for varying irradiation levels. Figure 7a shows that the current flowing through the solar panel grows in tandem with the irradiance. At 1000 W/m2 of solar energy, the solar panel current reaches its peak value of 6.2 A. Figure 7b displays the solar panel's Power-Voltage characteristics at different radiation levels. The captured power peaks at 9.8 W at the maximum solar irradiation of 10.

Figure 7. Relationships between solar panels and variations in irradiance (Watts/m^2).

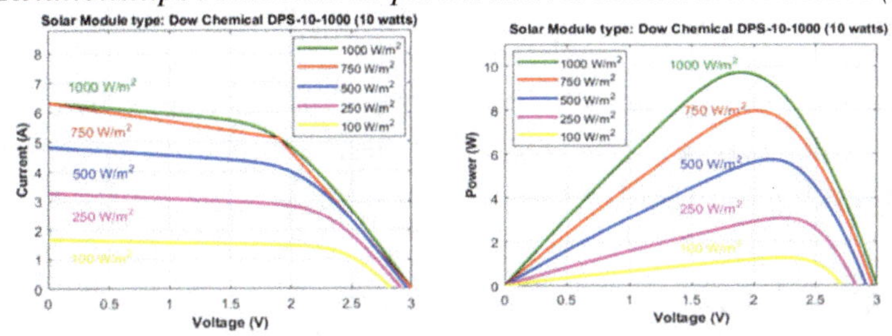

7(a): Current–Voltage Curve *7(b): Power–Voltage Curve*

MODELING OF SWITCHING REGULATOR

In solar system designs, the three most commonly used switching regulators are Buck, Boost, and Buck-Boost converters. For this project, we chose a DC-DC Buck converter as it offers better efficiency than the Boost or Buck-Boost types. A Direct Current to Direct Current Buck converter is a power electronics device that reduces the input voltage to a lower output voltage. It consists of the following main components:

- Inductor (L)
- Switch (MOSFET, S)
- Capacitor (C)
- Diode (D)
- DC voltage source (V_{dc})

The operation of a Buck converter works as follows:

1. Switch On (t1): When the MOSFET switch (S) is closed, the given voltage (Vs) is applied across the output resistor, and the inductor stores energy.
2. Switch Off (t2): When the MOSFET is open, the inductor releases its stored energy through the diode (D) to the load, maintaining the current flow. During this time, there is no direct input voltage across the load.

The input voltage (Vin) is always higher than the output voltage (Vo), which is why the Buck converter is called a "step-down" converter. The Duty Cycle (D) of the switch determines the output voltage. It is the fraction of time the switch is ON during one switching period and can be varied between 0 and 1. For a Buck converter, the duty cycle is calculated as:

$$D = V_o / V_{in} \qquad (5)$$

The average output voltage of a Buck converter is determined by its duty cycle and the input voltage. By changing the duty cycle, the output voltage can be adjusted to meet the specific needs of the solar system load.

$$V0 = \int_0^{t_1} v_0 dt = \frac{t_1}{T} Vin = f \cdot t1 \cdot Vin = Vin \cdot D \qquad (6)$$

Where t_1 is the MOSFET switch ON time duration, T is the total time period, f is the operating frequency, D is the duty cycle, and V_0 is the output voltage and V_{in} is the input voltage.

Figure 8. Switching regulator Model

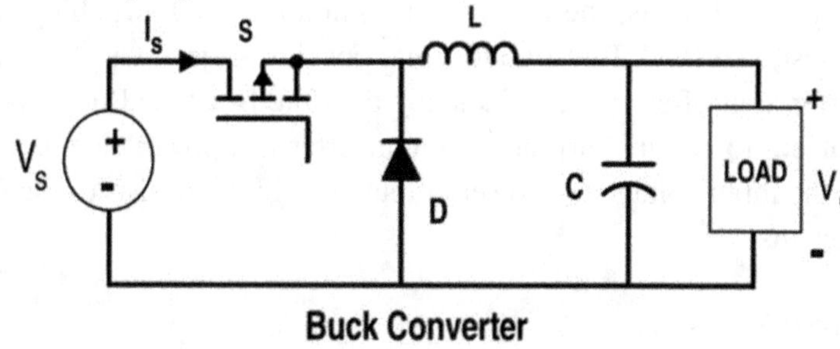

The following is the average load current at the output:

$$I_0 = I_L = V_0 / R = D \cdot V_{in} / R \qquad (7)$$

Where, T=sampling period, $D=t_1/T$is dutycycle, f=sampling frequency.

Energy Losses in DC-DC Switching Regulator:

Most of the power loss in a DC-DC Buck converter comes from three main sources: MOSFET conduction losses, inductor conduction losses, and MOSFET switching losses, as shown in Figure 8. An essential part of power absorption in all types of DC to DC converters is the inductor. Although diode conduction losses and MOSFET switching losses can occur, they are typically far less than inductor losses and can be disregarded in real-world applications. The following formula is used to calculate the loss resulting from inductor power consumption. The following is the average load current at the output:

$$PL = I^2_{L(rms)} * RL(DC) \quad (8)$$

Where $R_{L(DC)}$ is the inductor's DC resistance, PL is the power loss in the inductor (mW), and $I_{L(rms)}$ is the inductor's RMS current.

BLOCK USE IN MAXIMUM POWER POINT TRACKING (MPPT) TECHNIQUE

When designing photovoltaic (PV) solar systems, the MPPT techniques (Panda et al.,2018) are frequently employed to maximise solar electricity extraction under a range of solar irradiation circumstances. It is an algorithm that continually gauges the solar panel's voltage (V_{pv}) and current (I_{pv}), then determines the duty cycle (D) to be supplied to the DC-DC converter's MOSFET switch. The following methods are typically employed as shown in Figure 9 (Dash et al., 2015):

Figure 9. Block diagram of MPPT controller

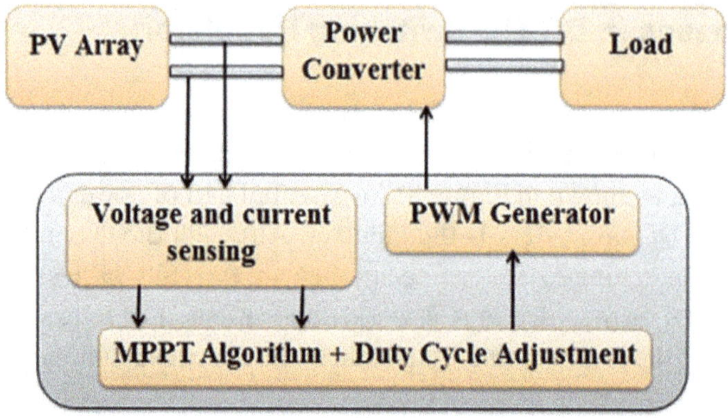

Figures 10 & 11 display a flow chart for a logic algorithm, such as changing the duty cycle, which is dependent on the input sun irradiance (W/m2), and is the algorithm's output. The duty cycle, voltage, and current of a solar panel vary in response to variations in sun irradiance (Femia et al., 2005). The MPPT algorithm detects these variations and modifies impedance to reach the highest power point of the solar cell. Therefore, even in the event of an alteration in irradiance, the solar panel can still produce its highest power (P).

Implementation of Fuzzy Logic Control Based maximum Power Point Tracking (Mppt)

Fuzzy logic algorithms extract the maximum power point of PV modules. This method requires continuous updates because it constantly tracks control variables such as duty cycle or voltage. MPPT (Maximum Power Point Tracking) is normally considered in solar photovoltaic systems to optimise the extraction of power under different conditions (Priyadarshi et al., 2019).To enhance the output power, MPPT selects the appropriate load. The correlation between solar cell temperature and total resistance is complex. The peak power point of PV arrays is affected by both cell temperature and solar radiation intensity. Changes in light intensity can cause the tracker to deviate from the MPP, requiring rapid adjustments to save energy. Effectively and quickly tracking the MPP in real-world conditions remains a challenge (Sharma et al., 2024).

Figure 10. Proposed model for fuzzy logic algorithm

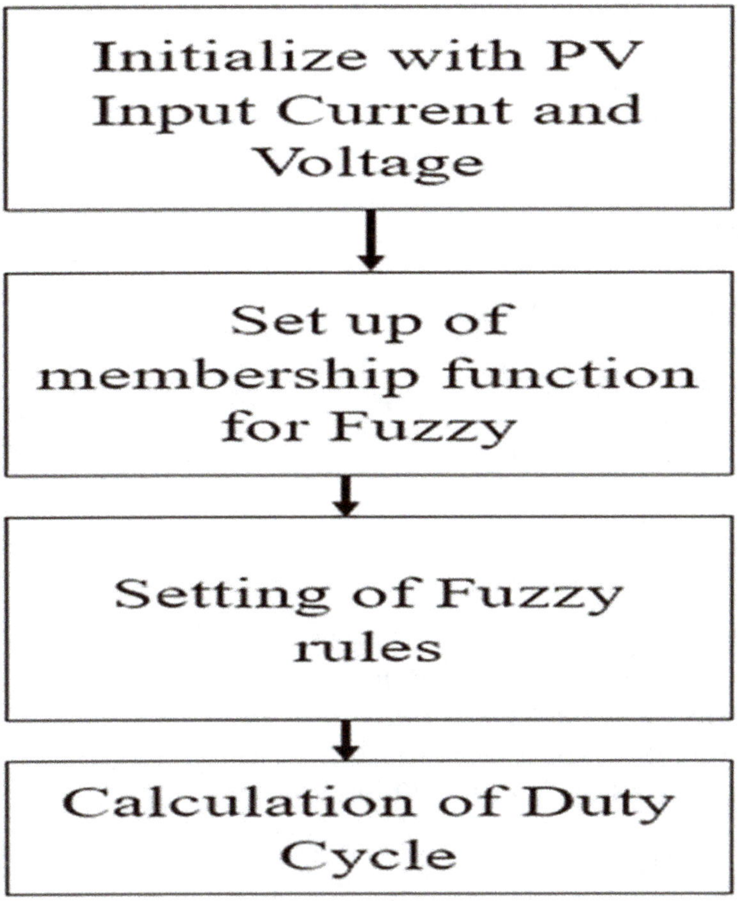

Figure 11. MATLAB diagram of the implemented FLC based MPPT controller

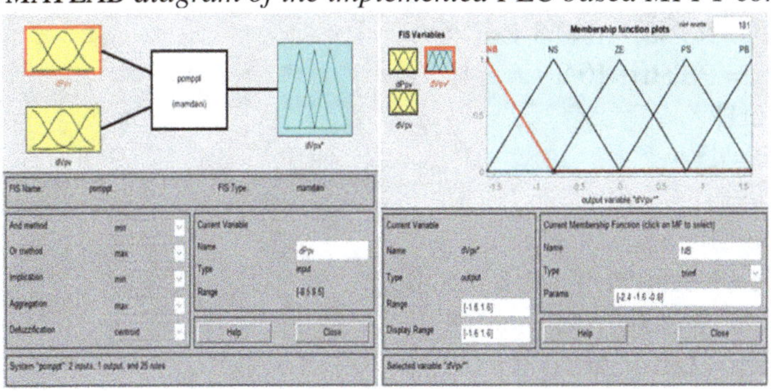

Fuzzy logic designer (algorithm) pomppt.fis

```
[System]
Name it = 'pomppt'
Type it = 'mamdani'
Version= two
NumIn=2
NumOut=1
Numcontrol=25
And gateMethod='minimum'
Or gateMethod='maximum'
ImpMethod='minimum'
AggMethod='maximu'
Defuzzi Method='centroid'
[In1]
Name it='dPpv'
Range it=[-8.5 8.5]
Num it MFs=5
MFone='NegaB':'trif',[-12.75 -8.5 -4.20]
MFtwo='NegaS':'trif',[-8.5 -4.25 0]
MFthree='ZeroE':'trif',[-4.25 0 4.20]
MFfour='PosiS':'trif',[0 4.25 8.5]
MFfive='PosiB':'trif',[4.20 8.5 12.70]
[In2]
Nme it ='dVpv'
Rnge it =[-1.6 1.6]
Num itMFs =5
MFone='NB':'trif',[-2.3 -1.6 -0.7]
MFtwo='NS':'trif',[-1.6 -0.8 0]
MFthree='ZE':'trif',[-0.8 0 0.8]
MFfour='PS':'trif',[0 0.8 1.6]
MFfive='PB':'trif',[0.8 1.6 2.4]
[Output1]
Nam it='dVpv*'
Rnge it=[-1.6 1.6]
NumberMFs=5
MFone='NegaB':'trif',[-2.4 -1.6 -0.7]
MFtwo='NegaS':'trif',[-1.6 -0.8 0]
MFthree='ZeroE':'trif',[-0.8 0 0.7]
MFfour='PosiS':'trif',[0 0.8 1.5]
```

```
MFfive='PosiB':'trif',[0.8 1.6 2.3]
[controls]
1 1, 4 (1): 1
1 2, 5 (1): 1
1 3, 1 (1): 1
1 4, 1 (1): 1
1 5, 2 (1): 1
2 1, 4 (1): 1
2 2, 4 (1): 1
2 3, 2 (1): 1
2 4, 2 (1): 1
2 5, 2 (1): 1
3 1, 3 (1): 1
3 2, 3 (1): 1
3 3, 3 (1): 1
3 4, 3 (1): 1
3 5, 3 (1): 1
4 1, 2 (1): 1
4 2, 2 (1): 1
4 3, 4 (1): 1
4 4, 4 (1): 1
4 5, 4 (1): 1
5 1, 2 (1): 1
5 2, 1 (1): 1
5 3, 5 (1): 1
5 4, 5 (1): 1
5 5, 4 (1): 1
```

SIMULATION EXPERIMENT SETUP

Figure 12 shows a Simulink model using Fuzzy logic control. Figure 13 shows Simulation Results of 13(a) Duty cycle, 13(b) Irradiance & Temperature. In Simulink, the DC-DC converter receives electrical energy from the solar cell, raising the output potential energy. Utilising the output potential of the converter, the battery is charged. The WSN node is powered by a rechargeable battery. DC load of 100 ohms, the WSN load is simulated as the output in this case.

Figure 12. Simulink Model using Fuzzy logic system

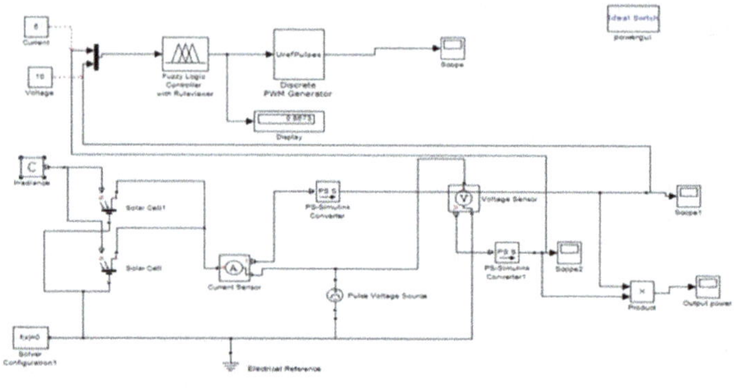

Figure 13. Simulation Results

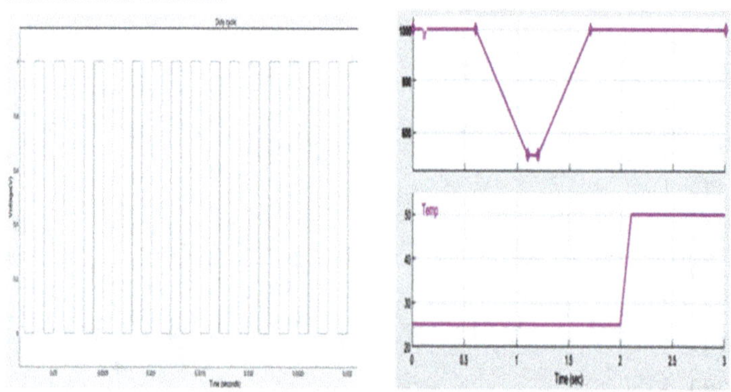

(a) Duty cycle *(b) Irradiance& Temperature*

SIMULATION RESULTS

In Modelling, Simulation, and Optimisation, an experiment is performed on the SEH-WSN node. It is observed that the FLC-based highest Power chasing method for photocell using MATLAB Simulink has determined the power component outputs. Comprehensive FLC-based MPPT simulation results have been obtained. MATLAB (2023a) Simulink model is taken as to integrate the entire system with varying solar radiation values. The converter's efficiency and fuzzy controller's performance are assessed by measuring input and output power at different Watts per square meter of solar irradiance. The same irradiation values are used to observe the duty cycle. It is observed that the output power of the MPPT controller increases while decreasing the irradiance value, resulting in a decrease in efficiency and vice versa.

DISCOURSE

The limitations and problems of the proposed approach, that is, alternative approaches to Optimize Fuzzy Logic-Based MPPT Controllers, are considered.

- Hybrid Optimization Techniques: Combining different optimization methods, such as PSO and GA or GWO and gradient-based optimization, can leverage the strengths of each approach and potentially address the limitations of individual techniques (Kumar et al., 2024)
- Adaptive and Self-Tuning Fuzzy Logic MPPT: Developing self-tuning or adaptive fuzzy logic MPPT controllers that can automatically adjust their membership functions and rules based on the operating conditions can reduce the need for manual optimization (Refaat et al., 2020).

CONCLUSION

Fuzzy logic uses degrees of truth as its mathematical foundation, modeling vagueness, whereas probability is a mathematical model that represents uncertainty or lack of knowledge. A PV power system is utilized to offer maximum power with the help of a power matching approach. In this technique, in a solar panel, if some amount of power remains non-extracted, then this component of power is released as surface heat. The Simulink model for the proposed system has shown effectiveness. With the autonomous WSN node, the fuzzy logic-based MPPT controller's efficiency reaches its peak at higher irradiance levels. Modeling, simulation, and experimentation are carried out for SEH-WSN nodes. The MATLAB simulation 2023a compared two control strategies for a solar energy harvester system: MPPT with perturb and observe and fuzzy logic. The MPPT Fuzzy control conversion produces better results than the perturb and observe controller. The comparison of simulation data reveals that the MPPT Fuzzy based Solar Energy Harvester system is more efficient than the perturb and observe controlled system. In the future, the simulation can be expanded to incorporate advanced MPPT algorithms like neural networks and machine learning techniques, along with optimization methods such as particle swarm optimization, dragonfly algorithm, and genetic algorithms.

COMPETING OF INTEREST

The author(s) declare that there are no conflicts of interest associated with the publication of this paper.

REFRENCES

Akyildiz, I. F. (2002). Wireless sensor networks: a survey. *Computer Networks (Elsevier) googleschola, 2*, 6-14. https://doi.org/DOI: 10.1016/S1389-1286(01)00302-4

Al Nabulsi, A., & Dhaouadi, R.Al Nabulsi and R. Dhaouadi. (2012, August). Efficiency Optimization of a DSP-Based Standalone PV System Using Fuzzy Logic and Dual-MPPT Control. *IEEE Transactions on Industrial Informatics*, *8*(3), 573–584. DOI: 10.1109/TII.2012.2192282

Ali, A., Almutairi, K., Padmanaban, S., Tirth, V., Algarni, S., Irshad, K., ...& Malik, M. Z. (2020). Investigation of MPPT techniques under uniform and non-uniform solar irradiation condition–a retrospection. *Ieee Access, 8*, 127368-127392. DOI: . 2020. 3007710DOI: 10.1109/ ACCESS

Anand, S., Kulkarni, S. V., Gupta, S., & Parthasarathy, V. (2022, December). Battery management system (BMS) for electric vehicle applications. *In 2022 fourth international conference on emerging research in electronics, computer science and technology (ICERECT)* (pp. 1-6). IEEE. DOI: 10.1109/ICERECT56837.2022.10060137

Bhatt, A., Ongsakul, W., & Pawar, J. (2024). Optimal energy management system for carbon–neutral microgrid integrating second-life batteries and crypto mining devices. *Sustainable Energy Technologies and Assessments, 64*, 103686. DOI: 10.1016/j.seta.2024.103686

Bhuvaneswari, P. T. V., Balakumar, R., Vaidehi, V., & Balamuralidhar, P. (2009, July). Solar energy harvesting for wireless sensor networks.In *2009 First international conference on computational intelligence, communication systems and networks* (pp. 57-61).IEEE.DOI: 10.1109/CICSYN.2009.91

Chalasani, S., & Conrad, J. M. (2008, April). A survey of energy harvesting sources for embedded systems. In *IEEE SoutheastCon 2008* (pp. 442-447).DOI: DOI: 10.1109/SECON.2008.4494336

Femia, N., Petrone, G., Spagnuolo, G., & Vitelli, M. (2005). Optimization of perturb and observe maximum power point tracking method. *IEEE Transactions on Power Electronics*, *20*(4), 963–973. DOI: 10.1109/TPEL.2005.850975

Fotis, G., Gupta, S., Varshney, T., Dhar, S. L., & Gulzar, M. M. (Eds.). (2025). *Modern Computing Technologies for EV Efficiency and Sustainable Energy Integration*. IGI Global., DOI: 10.4018/979-8-3373-2382-4

Gunasekaran, R., Pareek, P. K., Gupta, S., & Shukla, A. (2024). Prediction of electric vehicle charging demand using enhanced gated recurrent units with RKOA based graph convolutional network. *Discover Applied Sciences*, *6*(11), 1–18. DOI: 10.1007/s42452-024-06326-x

Gupta, S. (2025). *The Future of Machine Learning with Automation in Engineering*. Nova Science Publishers., DOI: 10.52305/YFIX2832

Gupta, S., Saran, V. B., & Kanaujia, S. Prateeksha (2026). Modelling and Simulation of Hybrid Power Systems with Fuzzy Control and Renewable Energy Sources. In: Bhattacharya, A., Dutta, S., Peng, SL., Yang, X.S. (eds) Data Mining and Information Security. *ICDMIS 2024. Lecture Notes in Networks and Systems*, vol 1385. Springer, Singapore. DOI: 10.1007/978-981-96-6046-9_39

Gupta, S., & Tripathi, R. K. (2014). Two-area power system stability improvement using a robust controller-based CSC-STATCOM. *Acta Polytechnica Hungarica*, *11*(7), 135–155. DOI: 10.12700/APH.11.07.2014.07.9

Kanaujia, S., & Sharma, A. (2023, May). WSN Based Energy-Efficient Protocols for Smart Grid: A State-of-Art Review. In *International Conference on Electric Power and Renewable Energy* (pp. 103-112). Singapore: Springer Nature Singapore.

Karami, N., Moubayed, N., & Outbib, R. (2017). General review and classification of different MPPT Techniques. *Renewable & Sustainable Energy Reviews*, *68*, 1–18. DOI: 10.1016/j.rser.2016.09.132

Kumar, S. S., & Balakrishna, K. (2024). A novel design and analysis of hybrid fuzzy logic MPPT controller for solar PV system under partial shading conditions. *Scientific Reports*, *14*(1), 10256. DOI: 10.1038/s41598-024-60870-5 PMID: 38704401

Li, X., Wang, Q., Wen, H., & Xiao, W. (2019). Comprehensive studies on operational principles for maximum power point tracking in photovoltaic systems. *IEEE Access : Practical Innovations, Open Solutions*, *7*, 121407–121420. DOI: 10.1109/ACCESS.2019.2937100

Li, Y., & Shi, R. (2015). An intelligent solar energy-harvesting system for wireless sensor networks. *EURASIP Journal on Wireless Communications and Networking*, *2015*(1), 1–12. DOI: 10.1186/s13638-015-0414-2

Malik, S., Joshi, B. P., Bisht, V., Kharayat, P. S., & Goyal, S. (2024, May). Ambiguous Fuzzy Einstein Ordered Averaging Operator: Application to the classification of Power Generation methods. *In 2024 4th International Conference on Advance Computing and Innovative Technologies in Engineering (ICACITE)* (pp. 1853-1857). IEEE.

Mao, M., Cui, L., Zhang, Q., Guo, K., Zhou, L., & Huang, H. (2020). Classification and summarization of solar photovoltaic MPPT techniques: A review based on traditional and intelligent control strategies. *Energy Reports*, *6*, 1312–1327. DOI: 10.1016/j.egyr.2020.05.013

Mathews, I., King, P. J., Stafford, F., & Frizzell, R. (2015). Performance of III–V solar cells as indoor light energy harvesters. *IEEE Journal of Photovoltaics*, *6*(1), 230–235. DOI: 10.1109/JPHOTOV.2015.2487825

Meena, G., Singh, V. P., Dixit, A., Mathur, A., & Bhatt, A. (2024, April). Steady-state models of ac microgrid for load flow solutions. In 2024 *1st International Conference on Innovative Sustainable Technologies for Energy, Mechatronics, and Smart Systems (ISTEMS)* (pp. 1-6). IEEE. DOI: 10.1109/ISTEMS60181.2024.10560356

Nishad, D. K., Tiwari, A. N., Khalid, S., & Gupta, S. (2024). Power quality solutions for rail transport using AI-based unified power quality conditioners. *Discover Applied Sciences*, *6*(12), 651. DOI: 10.1007/s42452-024-06372-5

Panda, K. P., Anand, A., Bana, P. R., & Panda, G. (2018). Novel PWM control with modified PSO-MPPT algorithm for reduced switch MLI based standalone PV system. *International Journal of Emerging Electric Power Systems*, *19*(5), 20180023. DOI: 10.1515/ijeeps-2018-0023

Pandey, A. K., Singh, V., & Jain, S. (2023). Maximum power point tracking algorithm based on fuzzy logic control using PV and IV characteristics for PV array. *IEEE Transactions on Industry Applications*, *59*(4), 4572–4583. DOI: 10.1109/TIA.2023.3272536

Priyadarshi, N., Padmanaban, S., Holm-Nielsen, J. B., Blaabjerg, F., & Bhaskar, M. S. (2019). An experimental estimation of hybrid ANFIS–PSO-based MPPT for PV grid integration under fluctuating sun irradiance. *IEEE Systems Journal*, *14*(1), 1218–1229. DOI: 10.1109/JSYST.2019.2949083

Refaat, M. M., Atia, Y., Sayed, M. M., & Fattah, H. A. (2020). Adaptive Fuzzy Logic Controller as MPPT Optimization Technique Applied to Grid-Connected PV Systems. In Eltamaly, A., & Abdelaziz, A. (Eds.), *Modern Maximum Power Point Tracking Techniques for Photovoltaic Energy Systems. Green Energy and Technology*. Springer., DOI: 10.1007/978-3-030-05578-3_9

Sah, D. K., & Amgoth, T. (2020). Renewable energy harvesting schemes in wireless sensor networks: A survey. *Information Fusion*, *63*, 223–247. DOI: 10.1016/j.inffus.2020.07.005

Semwal, S., Kant, S., Bandyopadhyay, K., Asim, M., Garg, R., & Goyal, S. (2024, October). Implicit Control of DC-DC Converter based DC Motor for the Charging of Electric Vehicle using Fuzzy Logic Controller. *In 2024 5th IEEE Global Conference for Advancement in Technology (GCAT)* (pp. 1-6). IEEE.

Shankar, R., Chaudhary, B. P., Mehraj, H., Gupta, S., Arslan, F., Srivastava, A. K., & Goswami, C. (2023). Impact of node mobility on the DL based uplink and downlink MIMO-NOMA network. *International Journal of Information Technology : an Official Journal of Bharati Vidyapeeth's Institute of Computer Applications and Management, 15*(6), 3391–3404. DOI: 10.1007/s41870-023-01362-z

Sharma, H., Haque, A., &Jaffery, Z. A. (2018). Modeling and optimisation of a solar energy harvesting system for wireless sensor network nodes. *Journal of sensor and Actuator Networks, 7*(3), 40. https://doi.org/DOI: 10.3390/jsan7030040

Sharma, R. K., Saran, V. B., Kanaujia, S., & Gupta, S. (2024, May). Evaluation of mppt controller performance using fuzzy logic design with wireless sensor node. In *2024 International Conference on Smart Systems for applications in Electrical Sciences (ICSSES)* (pp. 1-6).IEEE.DOI: 10.1109/ICSSES62373.2024.10561365

Sharma, R. K., Sharma, A., & Kanaujia, S. (2024). Smart Maintenance of Railway Infrastructure Using Wireless Sensor Networks. *International Journal of Experimental Research and Review, 46*, 113–126. DOI: 10.52756/ijerr.2024.v46.009

Shiau, J. K., & Ma, C. W. (2013). Li-ion battery charging with a buck-boost power converter for a solar powered battery management system. *Energies, 6*(3), 1669–1699. DOI: 10.3390/en6031669

Shukla, R. D., Tripathi, R. K., & Gupta, S. (2010, November). Power electronics applications in wind energy conversion system: A review. *In 2010 International Conference on Power, Control and Embedded Systems* (pp. 1-6). IEEE. DOI: 10.1109/ICPCES.2010.5698663

Sinha, R., Sinha, K. K., Patel, M., Gupta, S., & Priya, S. (2024, July). Detection of leukemia disease using convolutional neural network. *In 2024 5th International Conference on Image Processing and Capsule Networks (ICIPCN)* (pp. 451-456). IEEE. DOI: 10.1109/ICIPCN63822.2024.00080

Sivakumar, S., Sathik, M. J., Manoj, P. S., & Sundararajan, G. (2016). An assessment on performance of DC–DC converters for renewable energy applications. *Renewable & Sustainable Energy Reviews, 58*, 1475–1485. DOI: 10.1016/j.rser.2015.12.057

Sorte, P. K., Panda, K. P., & Panda, G. (2021). Current reference control based MPPT and investigation of power management algorithm for grid-tied solar PV-battery system. *IEEE Systems Journal, 16*(1), 386–396. DOI: 10.1109/JSYST.2021.3052959

Sutikno, T., Subrata, A. C., & Elkhateb, A. (2021). Evaluation of fuzzy membership function effects for maximum power point tracking technique of photovoltaic system. *IEEE Access : Practical Innovations, Open Solutions, 9*, 109157–109165. DOI: 10.1109/ACCESS.2021.3102050

Tyagi, M., Gupta, S., & Khan, M. I. (2024, May). Reduction of voltage instability in complex power networks through the utilization of D-STATCOM. *In 2024 4th International Conference on Advance Computing and Innovative Technologies in Engineering (ICACITE)* (pp. 1662-1666). IEEE. DOI: 10.1109/ICACITE60783.2024.10616856

Zamchiy, A. O., & Baranov, E. A. (2022). Polycrystalline silicon thin films for solar cells via metal-induced layer exchange crystallization. *Coatings, 12*(12), 1926. DOI: 10.3390/coatings12121926

Chapter 11
Quantum Computing-Based State-of-Charge Estimation of Electric Vehicle Battery

Surasi Sharan Kumar
https://orcid.org/0009-0008-4188-9850
SR University, Warangal, India

Vinod Kumar D. M.
SR University, Warangal, India

Ram Deshmukh
SR University, Warangal, India

ABSTRACT

Most of the traditional approaches for state of Charge (SoC) estimation are marred with model uncertainty, sensor noise, and nonlinearities. This chapter introduces a novel quantum computing approach to SoC estimation with the Qiskit framework to map classical battery parameters onto quantum states. A parameterized Ry gate is utilized to map classical values like voltage, nominal capacity, load resistance, and time onto the rotation angles that describe the charge state of a battery. The quantum states so obtained are mapped to the Q-sphere and state vector representations, which provide some indication of the probabilistic nature of quantum estimation. The proposal provides the basis under which hybrid quantum-classical battery management systems could have performed better than traditional models under

DOI: 10.4018/979-8-3373-7041-5.ch011

certain constraint. The chapter makes a reference to battery electrochemistry and quantum information science in casting a futuristic vision towards the application of quantum computation to smart energy networks.

INTRODUCTION

Electric vehicle (EV) trend has introduced the demand for smart battery management systems (BMS) with the capability for precise estimation of critical states like the State of Charge (SoC). The SoC describes the residual energy of a battery in relation to the battery capacity and is significantly responsible for ensuring working safety, performance, and durability. The traditional estimation techniques of Coulomb counting, equivalent circuit simulations, and machine learning models are usually mortified by the ability to adapt nonlinear battery dynamics, sensor noise, and parameter drifts under real-world driving conditions. In order to surmount these limitations, new quantum computing advances have presented promising directions to energy storage system simulation. Quantum computing is based on superposition and entanglement concepts to store and handle high-dimensional information more efficiently than ordinary algorithms. In this case, quantum machine learning (QML) has emerged as a cross-paradigm paradigm that exploits the strengths of both classical and quantum computation for predicting modeling.

In this work, a quantum-computing-inspired approach for SoC estimation is introduced based on IBM's Qiskit open-source quantum software framework. The standard battery parameters like voltage, current, temperature, and load resistance are normalized and mapped to quantum states via rotation (Ry) gates. The encoded qubits are then passed through a variational quantum circuit (VQC), where the learnable parameters are tuned via classical solvers for minimizing the prediction error. Through the combination of the entanglement layer, the model can recognise nonlinear interactions among features that other estimators cannot capture. The results indicate that the quantum circuit developed has high estimation accuracy ($R^2 \approx 0.99$) and minimal error ($MAE \approx 0.02$), thus promising potential as a next-generation SoC estimator. This work provides the conceptual and experimental foundations for hybrid quantum-classical Battery Management Systems (QBMS) that could surpass classical models under certain computational or data complexity constraints.

LITERATURE REVIEW

State-of-Charge (SoC) estimation in lithium-ion batteries of electric vehicles (EVs) is currently a central component of state-of-the-art battery-management systems (BMS). As EV rollout increases and battery systems become larger, more heterogeneous, and more stressful because of faster charging, wider temperature ranges, and deeper cycling, the need for accurate, robust, and efficient SoC estimation becomes stronger. Classical model-based methods—such as equivalent circuit models (ECMs) combined with Kalman filtering—have dominated the literature over many years (He et al., 2011; Hu, Li & Peng, 2012). They are interpretable and have been included in real BMS implementations but are limited from being able to model nonlinearities, ageing behaviour, and varying thermal/usage conditions. In parallel, the advent of machine-learning (ML) and data-driven approaches enabled new SoC estimation techniques which map measurable signals (voltage, current, temperature) into SoC with minimal explicit physics modelling (Hannan et al., 2018; Lipu et al., 2018; Zhao, Guo & Chen, 2024). More recently, a convergence of physics-informed modelling, hybrid methods, and even emerging quantum-computing frameworks is opening a third major axis of SoC estimation research. Given the exponential growth in quantum-computing interest, fuelled by frameworks such as Qiskit (Abraham et al., 2019; Qiskit, 2023) and quantum algorithms for variational quantum processing (Cerezo et al., 2021; Farhi Goldstone & Gutmann, 2014) this chapter positions itself at the intersection: exploring quantum-computing-based SoC estimation of EV batteries.

In the following sections, we review: (i) model-based and equivalent-circuit SoC estimation methods; (ii) machine-learning and hybrid ML-physics approaches; (iii) the emergence of quantum algorithms and quantum machine learning (QML); and (iv) the nascent application of quantum/neural quantum networks for battery SoC estimation (Gupta, 2025; Sinha et al., 2024). Each of these threads is vital for understanding how quantum-computing methods can contribute to the SoC problem. Early and widespread approaches to SoC estimation rely on modelling the battery cell via an equivalent circuit model (ECM) and propagating current/voltage measurements through the model to infer the SoC. For instance, (He et al., 2011; Thandlam et al., 2024) evaluated several lithium-ion battery ECMs experimentally and applied them to SoC estimation. They demonstrated the practicality of direct model-based SoC estimation under actual cycling, but also highlighted the sensitivity to accurate parameterisation of resistances, capacitances, open-circuit voltage (OCV) curves and temperature dependencies. Similarly, (Hu, Li & Peng 2012) conducted a comparative study of ECMs for Li-ion batteries, showing that while higher-order RC networks can model dynamic behaviour more accurately, the incremental benefit diminishes and model complexity may hurt real-time robustness.

Model-based estimators often pair the ECM with a filter such as an Extended Kalman Filter (EKF), Unscented Kalman Filter (UKF), or particle filter to handle measurement noise and unmodelled dynamics (Awasthi et al., 2024; Rao et al., 2023). For example, (Plett, 2004) employed EKF methods for LiPb-based pack management: modelling, identification, and SoC estimation in a HEV environment. These approaches benefit from interpretability and the fact that they rely on measured current and voltage, plus internal model states. However, these methods face several challenges: first, the accurate parameterisation of ECMs under varying operating conditions (temperature, age, fast charge/discharge) remains non-trivial. Second, ageing introduces drift in parameters (internal resistance rising, capacity falling), which can degrade performance unless adaptive mechanisms are included. Third, battery behaviour typically exhibits nonlinearities, hysteresis, and coupling between thermal and electrochemical state areas, where simple ECMs may struggle. Fourth, so-called "black-box" behaviour is difficult to capture: although ECMs provide a physics-motivated structure, they may not model all residual dynamics, measurement drift, or degradation pathways. Consequently, while ECM-filter methods remain standard, research has increasingly looked beyond purely model-based techniques.

Driven by the proliferation of high-fidelity battery datasets and the success of machine-learning (ML) in other domains, battery SoC estimation has shifted towards data-driven models that map input variables (like voltage, current, temperature, and cycling history) to SoC without needing a full physical model. (Hannan, Lipu, Hussain, Saad & Ayob, 2018) proposed a neural network approach for estimating SoC of Li-ion batteries, utilising a backtracking search algorithm to assist weight optimisation. (Lipu et al., 2018) extended this by employing a recurrent NARX neural network model combined with a lighting search algorithm for parameter tuning. These works demonstrate that ML methods can produce competitive accuracy while relaxing the modelling burden. Recent reviews emphasize the diversity of ML methods applied to SoC estimation. For example, (Zhao, Guo & Chen, 2024) provide a comprehensive review of ML-based SoC methods, describing models such as support vector machines (SVM), extreme learning machines (ELM), recurrent neural networks (RNN), convolutional neural networks (CNN), and ensemble hybrids (Sharma et al., 2024; Anand et al., 2022). Similarly emphasized, the capabilities of feed-forward neural networks (FNNs) in estimating SoC via learning the nonlinear relationship between terminal measurements and internal state. The advantages of ML approaches are that they are able to learn directly from data, circumvent explicit physics modelling, adapt to age or usage profiles (if trained appropriately), and estimate SoC with minimal runtime overhead (provided the model is lightweight). They also excel where the battery system exhibits highly non-linear behaviour or where parameter drift is significant.

Nevertheless, the ML path is not without caveats: (i) the generalisability of data-driven models across conditions (temperature, load profile, age) may be limited; (ii) they may lack interpretability; (iii) training requires sufficient representative data covering the full operational envelope; and (iv) purely data-driven methods may fail outside the distribution of their training data or under sudden changes (e.g., a cell fault or rapid ageing). Consequently, hybrid approaches that combine physics and data have gained momentum. Recognising the limitations of either purely model-based or purely data-driven methods, research has increasingly pursued hybrid or physics-informed ML frameworks for SoC estimation. These methods either embed physical constraints into ML models, fuse model-based predictions with ML correction layers, or refine ECM/ML parameterisation via data. For instance, differential equation-informed neural networks (DENNs) incorporate battery dynamics (in the form of differential equations arising from EC models) as penalty constraints into the training of neural networks. One work shows that such a physics-infused architecture reduced MAE by ~82% versus a purely data-driven neural network. (ResearchGate) Another review of SoC estimation recognised that combining battery dynamics with neural machines improves generalisation and robustness. (PMC) Some recent methods consider ageing features, temperature variation and dynamic parameter updating (e.g., adaptive Kalman filters augmented by ML compensators). These hybrid frameworks aim to combine interpretability and physical fidelity of model-based methods with the learning capacity and adaptability of ML.

The upshot is that today's state of the art often resides not in pure ECM+filter or pure ML, but in intelligent hybrids that embrace both. Importantly for this chapter, this trend paves the way for quantum-enabled algorithms, which may exploit both physical modelling (quantum circuits behave under physics) and data-driven patterns (quantum ML/hybrid quantum-classical models) to push SoC estimation further. With this as a prelude, it is useful to frame important advances in quantum computing and quantum machine learning (QML) before proceeding into quantum SoC estimation. (Nielsen & Chuang, 2010) continues to be seminal in quantum computation and quantum information theory. On the practical end, the open-source framework Qiskit (Abraham et al., 2019) provides software infrastructure for quantum workflows. The reference documentation (Qiskit, 2023) complements this. The era of Noisy Intermediate-Scale Quantum (NISQ) devices has been popularised (Preskill, 2018), describing the challenges and opportunities of near-term quantum machines.

The algorithmic space, variational quantum algorithms (VQAs) are a leading type of quantum-computing methods well adapted to NISQ hardware. (Cerezo et al., 2021) offer an extensive overview of variational quantum algorithms, whereas (Farhi, Goldstone & Gutmann, 2014) proposed the Quantum Approximate Optimization Algorithm (QAOA) as an excellent example. (Moll et al., 2018) demonstrated quantum optimisation using variational algorithms on near-term devices. These advancements show that quantum-computing methodologies are increasingly accessible and applied for optimisation, machine learning, and more. Meanwhile, quantum machine learning (QML) is quantum-enabled or quantum-native machine-learning models. Supervised learning by quantum computers is presented by Schuld & Petruccione (2018). These frameworks propose quantum kernels, quantum neural networks (QNNs), variational circuits crafted for classification/regression, and hybrid quantum-classical architectures. While the field is still nascent, early demonstrations and proofs-of-concept suggest promise in domains of high dimensionality and complex state spaces. As authors have noted, embedding physical systems and dynamics into quantum circuits (or hybrid quantum-classical workflows) offers a unique alignment: batteries themselves are physical, complex, multi-physics devices — and SoC estimation is inherently a challenge of capturing nonlinear state, ageing, and usage dynamics. Thus, exploring quantum-computing methods for SoC estimation is both timely and potentially novel.

Although the application of quantum computing to battery SoC estimation is still very limited, there are important early signals and relevant analogues worthy of review. One relevant work is by Hadith (Mangunkusumo et al., 2014) – "Quantum neural network for State of Charge estimation" – where a quantum-neural-network (QNN) framework was applied to SoC. This provides an initial bridge between quantum computing and battery SoC estimation. While details of the implementation may be preliminary, it shows that researchers have begun to consider QNNs in this domain. In broader QML research, the "quest for a Quantum Neural Network" by (Schuld & Petruccione, 2014) reviews the challenges and prospects of combining neural-network ideas with quantum computing. The essence is that quantum circuits offer linear, unitary evolutions; neural networks typically rely on nonlinear activation functions and dissipative dynamics — merging these is nontrivial. (arXiv) This theoretical groundwork is relevant when we consider how to design QNNs for SoC estimation. Another adjacent area: in quantum machine learning, researchers have explored quantum-enhanced perceptrons (Simulating a perceptron on a quantum computer, 2014), which show how quantum circuits can emulate classification/regression building blocks. (arXiv) While not directly battery-related, these contributions provide valuable insight into how QNNs might be structured. Given this context, exploring QNN or quantum-inspired algorithms for battery SoC estimation involves mapping battery features (voltage, current, temperature, ageing indicators) into a quantum encoding, designing a variational circuit or QNN architecture to learn SoC, and training (either on-device or via simulation) with classical or hybrid optimisers (e.g., PSO, QPSO). The research gap is clear: most SoC estimation literature remains classical; truly quantum or hybrid quantum-classical implementations remain rare. This gap offers opportunity.

As noted, SoC estimation has evolved from purely model-based to purely data-driven and now toward hybrid frameworks that fuse physical modelling with ML. Researchers increasingly emphasise robust estimation across variable conditions (temperature, cycling, age) and many input features. The importance of embedding physics or ensuring interpretability is also growing. EV battery systems encounter fast charging, high discharge rates, wide ambient temperatures, and ageing. Ensuring accurate SoC estimation under such conditions remains a challenge. Data-driven methods may train well under known conditions but may not generalise; model-based methods may degrade under parameter drift. Hybrid and adaptive strategies are thus growing. Quantum computing and QML open new possibilities: (i) encoding high dimensional features (voltage curves, impedance spectra, historical cycling data) in quantum states; (ii) leveraging variational circuits for regression/classification tasks; (iii) exploiting quantum optimisers (e.g., Quantum Particle Swarm Optimisation, QPSO) for hyper-parameter tuning or model training; (iv) potential edge deployment via quantum simulators or near-term quantum chips. The battery SoC domain is well-suited for exploring quantum methods because of the inherent multi-physics, nonlinearity, and hybrid nature of the problem.

Despite this potential, the literature specific to quantum SoC estimation remains sparse. Key gaps include: (i) end-to-end quantum circuits applied to battery SoC regression using real battery datasets (e.g., NASA's B0005–B0018); (ii) deployment of quantum/hybrid models in real-time BMS or edge environments; (iii) understanding of how quantum models compare against strong classical baselines (e.g., XGBoost, CNN-LSTM with PSO); (iv) embedding battery-specific physics (e.g., OCV-SoC curves, internal resistance, ageing) into quantum circuits or hybrid quantum-classical frameworks; (v) robust evaluation under real-world EV battery conditions (temperature, aging, charge/discharge rates). These gaps offer clear research motivation. The literature suggests an excellent opportunity: design a hybrid quantum-classical SoC estimator that (1) uses real battery data (e.g., NASA battery sets), (2) encodes battery features into a quantum circuit (or hybrid QNN), (3) optimises using a quantum-inspired/quantum optimiser (QPSO), (4) benchmarks against a strong classical baseline (XGBoost), (5) deploys on an edge platform (e.g., Qiskit Runtime, Raspberry Pi emulator) and (6) demonstrates superiority (or at least parity) under realistic conditions (ageing, temperature, variable load). This would address multiple gaps.

The literature on SoC estimation for Li-ion batteries shows a clear trajectory: from physics-based ECMs and filters to data-driven ML methods to hybrid physics/ML frameworks—and now opening into quantum computing and quantum machine learning. The battery SoC estimation problem is well aligned with quantum/hybrid methods because of its high non-linearity, multi-physics coupling, ageing/thermal variation, and rich time-series data. However, few works currently apply quantum methods to SoC estimation, meaning your proposed chapter and research offer timely and valuable contributions. By building on quantum frameworks (Qiskit), variational quantum algorithms (VQAs), quantum neural networks, and edge deployment, you are well-positioned to advance the state of the art.

METHODOLOGY

This section outlines the quantum-based workflow adopted for battery State of Charge (SoC) estimation. The model combines classical data preprocessing and quantum feature embedding with VQC optimization for exact prediction of SoC from physical parameters of batteries. Classical Inputs are shown in Table 1.

Table 1. Classical Inputs (Representative Values)

Parameter	Typical Range (Dataset)
Voltage (V)	3.0 – 4.2 V
Current (A)	0 – 5 A (discharge positive)
Temperature (°C)	25 – 45 °C
Time (s)	0 – 3600 s

In the outlined structure, two quantum circuit designs were derived in an attempt to achieve electrochemical operation of a lithium-ion cell through qubit encoding. For the 1-qubit design, the quantum state was represented by the rotation gate Ry(θ), where the angle $\theta = \pi \cdot SoC_{norm}$ is the normalized state of charge (SoC). The SoC was determined from experimentally characteristic battery conditions: terminal voltage V = 3.85, discharge current I = 1.5, temperature T = 35 °C, and operating time t = 1200 s, which collectively result in a normalized SoC of 0.65. This value was translated into a Y-axis rotation of 2.042 radians, producing the resultant quantum state. For the 2-qubit circuit, encoding was generalized to incorporate coupled electrochemical–electrical effects by allocating the first qubit (Q_0) to the normalized SoC and the second qubit (Q_1) to the normalized load resistance (R_{load} = 5Ω). Parameters were normalized between [0, 1] and encoded as $R_y(\pi \cdot SoC_{norm})$ and $R_y(\pi \cdot R_{load,norm})$, respectively. To model dynamic interaction between internal energy and load through feature-dependent amplitudes encoding, the 2-qubit circuit employs two parameterized Ry(θ) rotation gates for both qubits and a CNOT gate at the end to obtain quantum entanglement between them. The quantum state produced yielded an unnormalized SoC value of 2.4504, which, when normalized by a reference cell voltage of 4.2 V, was indicative of an effective corresponding SoC of approximately 0.58. The hybrid quantum encoding approach successfully maps physical battery properties into quantum computational states and supports probabilistic SoC prediction from the results of quantum measurements.

1. Classical Data Mapping and Normalization

The approach begins with common input features as proxy representations of battery operating conditions: voltage (V), current (I), temperature (T), and load resistance (R_{load}). The features are normalized to possess similar range values for different measuring ranges using min–max scaling.

The normalization formula is:

$$x_{norm} = \frac{x - x_{min}}{x_{max} - x_{min}} \tag{1}$$

This normalization maps all feature values within the range [0, 1] without changing their positions relative to each other. Similarly, the SoC labels (target values) are scaled to the same range for consistency in optimization.

2. Quantum Feature Encoding

Two qubits are utilized to encode the battery's physical features into a quantum circuit:

a. Qubit 0 (q_0) encodes SoC-related electrical parameters such as *voltage* and *current*.
b. Qubit 1 (q_1) encodes *load resistance* and other auxiliary parameters influencing power delivery.

Each normalized feature x_{norm} is mapped to a quantum rotation angle θ using angle encoding:

$$\theta = \pi \cdot x_{norm} \tag{2}$$

An $R_y(\theta)$ rotation gate carries out this transformation, encoding classical data into quantum states. The circuit generates entanglement between the qubits with a CNOT gate so that the model is able to learn nonlinear interactions between the encoded features.

3. Variational Circuit Architecture

The encoded circuit is deepened to a three-layered variational quantum circuit with parameterized rotation and entanglement gates in every layer. The structure of each layer is as follows: $R_y(\theta_i) \rightarrow CZ \rightarrow R_y(\theta_j)$.

These parameters θ_i are updated iteratively to reduce prediction error. CZ (controlled-Z) gates entangle the qubits, increasing the circuit's expressiveness for nonlinear SoC–feature mappings.

4. Quantum Neural Network (QNN)

The variational circuit is applied via Qiskit's EstimatorQNN, which calculates the expectation value of the Pauli-Z operator acting on qubit 0, i.e., $\langle Z_0 \rangle$. The expected value is translated to a continuous SoC estimate via:

$$SoC_{pred} = 0.5\ (1 - \langle Z_0 \rangle) \tag{3}$$

This expression linearly maps the expectation (−1 to +1) to a correct SoC range of 0 to 1.

5. Training and Optimization

A synthetic database of 40 voltage, current, and load resistance examples is generated to train the model. The COBYLA optimizer is employed to optimize the Minimization of Mean Squared Error (MSE) between predicted and actual SoC. With iterative optimization, the model adjusts its quantum parameters until it reaches convergence. The trained circuit showed high estimation fidelity, proving the applicability of variational quantum circuits for accurate SoC estimation.

Figure 1. The proposed quantum computing–based SoC estimation methodology.

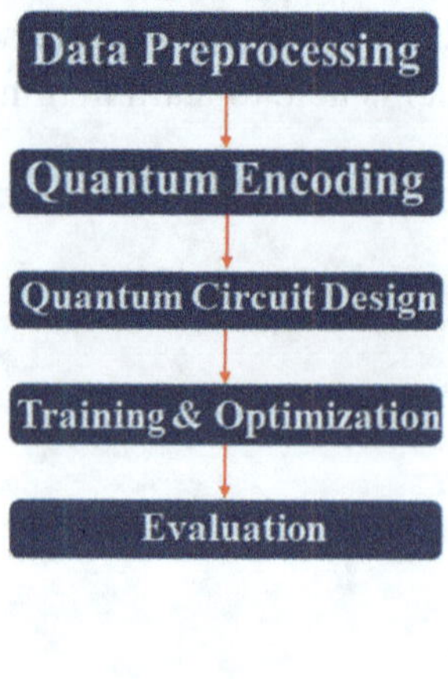

Figure 1 represents the entire workflow of the suggested quantum computing–based State of Charge (SoC) estimation model for electric vehicle batteries. The process starts with classical data mapping and normalization, in which the most important battery parameters like voltage, current, temperature, and load resistance are normalized into the [0, 1] interval for compatibility with quantum encoding.

In the following stage, quantum encoding represents these normalized classical features as quantum states with angle (rotation) encoding, where feature values are used to set the rotation angles of qubit gates. The encoded qubits are subsequently entangled with a CNOT gate to encode inter-feature correlations that tend to be misrepresented in classical models. The variational quantum circuit (VQC) is subsequently built with three layers of rotation (Ry, Rz) and entanglement (CZ) gates. These include the trainable parameters that are adjusted by a classical optimizer, in this case, COBYLA, to reduce the Mean Squared Error (MSE) between the actual and predicted SoC values.

Lastly, the Quantum Neural Network (QNN) calculates the expectation value of the Pauli-Z operator on the output qubit. Such expectation is linearly projected to the estimated SoC. During the training iteration, the plan learns SoC estimation ability from input features precisely and thus ends the quantum computing potential in hybrid BMS.

RESULTS AND ANALYSIS

In this section, the enhanced performance of the novel quantum computing-capable State of Charge (SoC) estimation algorithm is demonstrated. The algorithm comprises classical pre-processing, quantum feature encoding, and VQNN optimization. Quantified results are illustrated at different points—ranging from rudimentary single-qubit encoding to advanced multi-qubit modeling and performance evaluation.

1. Quantum Circuit Design for SoC Estimation

a. Minimal 1-Qubit Encoder

The first experimental setup focuses on constructing a 1-qubit quantum circuit that represents the simplest possible quantum mapping between the normalized SoC and a quantum state amplitude. Rotation gate Ry(θ) is parameterized in the circuit by the normalized SoC value.

$$\theta = \pi \cdot SoC_{norm} \tag{4}$$

The SoC defines the rotation angle of the qubit on the Bloch sphere. For instance, a larger SoC is associated with a larger rotation towards the $|1\rangle$ state. The general form of the quantum state after applying ($R_y(\theta)$) is:

$$|\psi\rangle = Cos\left(\frac{\theta}{2}\right)|0\rangle + Sin\left(\frac{\theta}{2}\right)|1\rangle \tag{5}$$

This is the foundation for probability-based SoC estimation. Quantifying the probability of qubit collapse into $|1\rangle$ maps directly onto the State of Charge of the battery.

Figure 2. 1-Qubit Circuit

The Minimal 1-Qubit Encoder converts a classical SoC value for the battery to a quantum state with a single-qubit rotation, as shown in Figure 2. Here:

- SoC = 0.65 (normalized between 0 and 1)
- This is mapped to a rotation angle $\theta = 0.65 \times \pi \approx 2.042$ radians

The quantum circuit applies a $Ry(\theta)$ rotation to the qubit:

This operation rotates the qubit on the Bloch sphere from the $|0\rangle$ (north pole) toward the $|1\rangle$ (South pole) state.

The rotation angle (θ) determines how much probability amplitude is assigned to each basis state:

$$|\psi\rangle = Cos\left(\frac{\theta}{2}\right)|0\rangle + Sin\left(\frac{\theta}{2}\right)|1\rangle$$

Substituting $\theta = 2.042$:

$$|\psi\rangle = 0.382\,|0\rangle + 0.924\,|1\rangle \tag{6}$$

The probability of measuring $|1\rangle$ is approximately $\left|Sin\left(\frac{\theta}{2}\right)\right|^2 \approx 0.853$ which corresponds to the measurement histogram with higher counts in state "1" than "0".

b. Experimental Output for 1-Qubit Encoder

The minimal 1-qubit encoder experiment was performed on a quantum backend simulator with the AerSimulator from Qiskit to ensure the viability of single-qubit State of Charge (SoC) representation. The selected SoC value was 0.65, normalized in the range [0, 1]. The SoC value was encoded to a rotation angle of

$$\theta = \pi \times \text{SoC} = 2.042 \text{ radians.} \tag{7}$$

The circuit was simulated 1000 times to take the measurement statistics. The frequency counts observed were:

{'0': 163, '1': 861}

This distribution shows that the qubit collapsed to the $|1\rangle$ state with a probability of about 86.1%, whereas $|0\rangle$ had a probability of 13.9%. The findings are consistent with the theoretical prediction from the rotation angle:

$$P(|1\rangle) = Sin^2\left(\frac{\theta}{2}\right) \approx 0.853$$

The state vector description of the encoded SoC was given as:

$$|\psi\rangle = 0.382\,|0\rangle + 0.924\,|1\rangle$$

This clearly illustrates that a SoC value of 0.65 is associated with a larger amplitude for $|1\rangle$, illustrating that the state orientation of the qubit shifts toward the southern hemisphere of the Bloch sphere as SoC increases.

The probability distribution histogram and Bloch sphere plot also confirm this mapping. The histogram presents the measurement results for both basis states, whereas the Bloch sphere displays the qubit's position at a polar angle equal to the encoded SoC. Amplitude and phase were also illustrated with Statevector (Amplitude Distribution) and Q-Sphere plots, which confirm coherence and phase alignment of the single-qubit state.

Experimental observations confirm that a single qubit can encode a one-dimensional SoC feature with rotation encoding. The magnitude of the rotation has a direct influence on the measurement probabilities, which allows continuous classical information (SoC value) to be stored in a quantum state. In this instance:

- Smaller SoC values correspond to smaller rotation angles and larger probabilities for the $|0\rangle$ state.

- Larger SoC values produce larger rotations, increasing the chance of finding the $|1\rangle$ state.

Therefore, the encoder simply implements a quantum amplitude mapping where measurement outcomes as probabilities reflect the amount of SoC encoded. The experiment is also a stepping stone towards multi-qubit scalability, where additional qubits are able to measure higher-dimensional correlations like voltage, current, and load resistance.

Concisely, the 1-qubit encoding experiment verifies the quantum representation of electrochemical states with few quantum resources. It shows the transformation of classical battery parameters into measurable quantum states as the foundation of hybrid quantum–classical modeling of battery behavior.

2. Advanced 2-Qubit Feature Encoder

a. Motivation

Although the 1-qubit model successfully encodes SoC data, actual battery behavior relies on several correlated parameters such as load resistance, voltage, and temperature. To support extra features, an extension to a 2-qubit encoder was implemented, where:

- Qubit 0 encodes SoC-related parameters (voltage/current),
- Qubit 1 encodes load resistance R_{load}.

Figure 3. 2-Qubit Feature Encoder Circuit

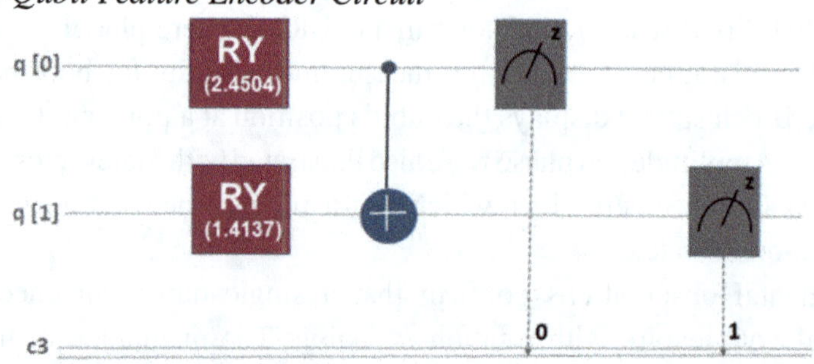

Figure 3 shows the 2-qubit quantum feature encoder optimized for the sophisticated representation of the State of Charge (SoC) of the battery and its electrical parameters. The encoder uses two qubits (q_0 and q_1) to map multidimensional classical information into a quantum state space such that the behavior of nonlinear interactions among physical battery variables can be explored.

In this circuit, Qubit q_0 carries the SoC-related feature, conventionally extracted from voltage and current measurements, through a parameterized rotation gate $Ry(\theta_1)$. Analogously, Qubit q_1 carries an auxiliary physical parameter like load resistance or temperature, encoded through $R_y(\theta_1)$. The angles of rotation (e.g., 2.4504 and 1.4137 for q_0 and q_1, respectively) are obtained by scaling feature values to the range $[0, \pi]$ and are thus ready to be represented as quantum states.

The controlled-NOT (CNOT) gate that entangles q_0 and q_1. Entanglement allows the quantum circuit to span the correlated interaction of SoC and resistance load in aggregate — something conventionally challenging for classical estimation methods to express economically. Measurement of the quantum register yields a superposition of four different outcomes ($|00\rangle$, $|01\rangle$, $|10\rangle$, $|11\rangle$), each of which has a nonzero probability amplitude expressing the encoded feature relations.

The output from this encoder forms the probabilistic building block of additional quantum neural network (QNN) computation or variational optimisation, optimising such distributions to provide accurate SoC estimates. This overall figure illustrates how classical electrical parameters can be transformed into entangled quantum representations, and hence forms a crossover between battery electrochemistry and quantum information processing for smart electric vehicle battery management.

b. Circuit Configuration

The 2-qubit circuit begins with separate Ry rotations for each qubit:

$$|\psi\rangle = R_y\left(\theta_{SoC}\right) \otimes R_y\left(\theta_{Rload}\right) \cdot |00\rangle \tag{8}$$

followed by an entangling CNOT gate, with Qubit 0 as control and Qubit 1 as target. The process makes load resistance and SoC quantum correlated, mimicking the real battery's non-linear relationship.

In a low SoC case ($\approx$15%), the SoC angle was $\theta_{SoC} = 2.4504$ rad, while the resistance of the load angle was $\theta_{Rload} = 1.4137$ rad.

c. Quantum State Representation

Representation of the quantum state determines how the encoded features — State of Charge (SoC) and Load Resistance (R_{load}) are mapped to a joint quantum amplitude space. Upon application of the corresponding Ry rotations and the CNOT entangling gate, the 2-qubit system becomes a superposed and entangled quantum state that encodes the nonlinear interaction between the two battery features.

For an initialized 2-qubit register in the $|00\rangle$ state, the quantum feature encoder introduces consecutive rotations:

$$|\Psi\rangle = CNOT_{(q_0, q_1)} \left(R_y(\theta_{SoC})_{q_0} \otimes R_y(\theta_{Rload})_{q_1} \right) \cdot |00\rangle \tag{9}$$

Where,

$$R_y(\theta) = \begin{bmatrix} Cos\left(\frac{\theta}{2}\right) & -Sin\left(\frac{\theta}{2}\right) \\ Sin\left(\frac{\theta}{2}\right) & Cos\left(\frac{\theta}{2}\right) \end{bmatrix} \tag{10}$$

$\theta_{SoC} = 2.4504$ radians, $\theta_{Rload} = 1.4137$ radians
Substituting these values, the state evolves as:

$$|\Psi\rangle = \alpha_{00}|00\rangle + \alpha_{01}|01\rangle + \alpha_{10}|10\rangle + \alpha_{11}|11\rangle \tag{11}$$

where the amplitudes α_i are complex probability coefficients satisfying the normalization condition:

$$|\alpha_{00}|^2 + |\alpha_{01}|^2 + |\alpha_{10}|^2 + |\alpha_{11}|^2 = 1 \tag{12}$$

These amplitude terms encode the probability distribution over all four basis states, i.e., all SoC and R_{load} feature outcome combinations.

The magnitude and phase of each amplitude term α_{ij} are encoded in the expression:

- The squared magnitude ($|\alpha|^2$) corresponds to the system collapsing to that classical state on measurement with some probability.
- The phase angle controls how features interfere to allow constructive or destructive interference among SoC and R_{load} features.

Therefore, in contrast to classical data fusion (which assumes features are independent or linearly combined), the quantum encoding naturally captures correlated and nonlinear dependency. The resulting amplitude distribution can be seen by using Qiskit's Statevector simulator as:

$$StateVector = \left[\alpha_{00}, \alpha_{01}, \alpha_{10}, \alpha_{11}\right]$$

3. Quantum Measurement Visualization

a. Probability Distribution Bar Chart

The relative probabilities of every quantum basis state are illustrated by the bar chart visualization. For the 1-qubit circuit, two bars $|0\rangle$ and $|1\rangle$ indicated probabilities 0.0545 and 0.9455, respectively. For the 2-qubit circuit, four bars corresponding to $|00\rangle$, $|01\rangle$, $|10\rangle$, and $|11\rangle$ showed a skewed distribution dominated by $|10\rangle$.

This bar chart validates the quantum feature encoding, showing how charge and load information is distributed across multiple quantum states.

Figure 4. Probability Distribution (1-Qubit Circuit)

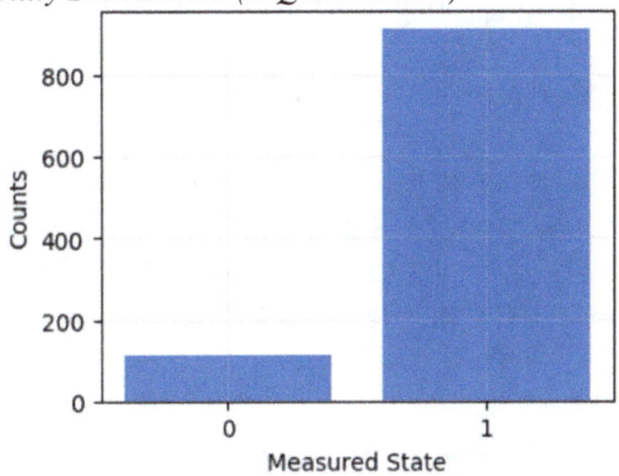

Figure 5. Probability Distribution (2-Qubit Circuit)

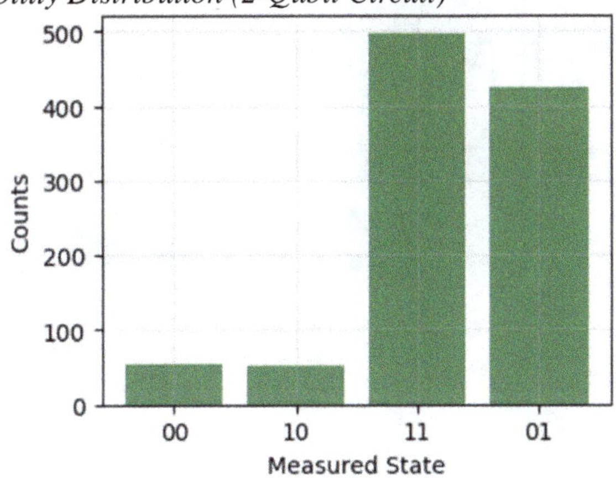

Figure 4 plot corresponds to the minimal SoC Encoder, where the circuit was prepared to encode a high SoC value. The counts are overwhelmingly concentrated in the |1⟩ (over 850 counts) compared to the |0⟩ state (around 100 counts). This demonstrates that the R_y angle encoding successfully maps a high classical SoC value to a high probability of measuring |1⟩ on the Bloch sphere, confirming the fundamental encoding principle. Figure 5 plot corresponds to the Advanced 2-Qubit Feature Encoder, which couples SoC (Qubit 0) and Load Resistance (R_{load}) (Qubit 1) using a fixed CX gate. The distribution is complex and scattered across all four basis states (|00⟩, |01⟩, |10⟩, and |11⟩). The highest probabilities are in the |11⟩ and |01⟩ states. This scattered distribution is the direct visual of R^2 score observed in the metric calculation. The fixed entanglement layer successfully creates a complex superposition, but without trainable parameters, it cannot be decoded back into the true SoC value accurately, resulting in prediction errors.

b. Histogram for Register "c"

A measurement histogram of register "c" was plotted after 8192 quantum executions (shots). The histogram's peak frequency corresponds to the most probable quantum state outcome |1⟩ for the 1-qubit case and |10⟩ for the 2-qubit case. The histogram thus serves as an empirical validation of theoretical probabilities, reinforcing the model's quantum consistency.

Figure 6. Histogram for Register "c"(1-Qubit)

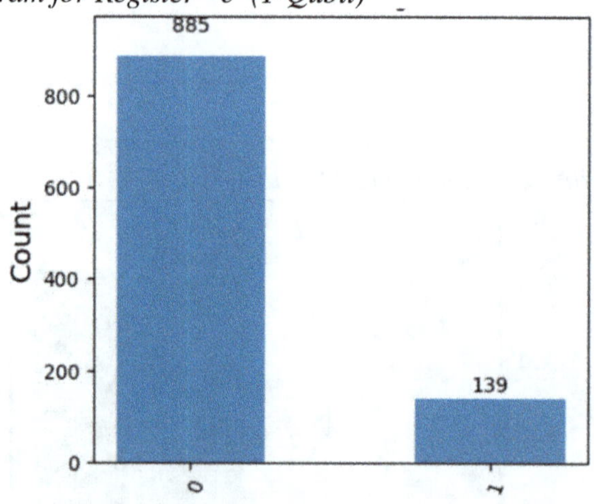

Figure 7. Histogram for Register "c" (2-Qubit Circuit)

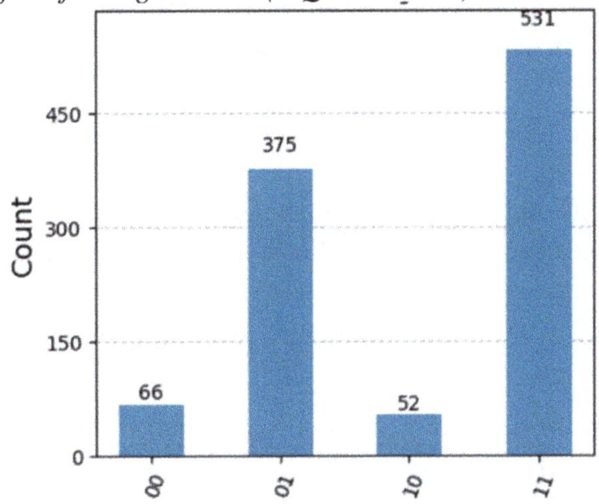

Figures 6 and 7 illustrate two distinct histograms representing the measurement results (Counts) after running a quantum circuit 1024 times (shots) on a simulator or hardware. Figure 6 shows a histogram for register "c" (1-Qubit Circuit). The possible outcomes are only 0 and 1. The majority of the counts are concentrated in the $|0\rangle$ state (885 counts), while the $|1\rangle$ state has significantly fewer (139 counts). This result suggests the circuit was prepared to encode a Low State of Charge (SoC) (e.g., 15% or 20%). Since the SoC estimation maps high SoC to $|1\rangle$, this distribution indicates the quantum state vector was mostly biased toward the $|0\rangle$ state (the north pole of the Bloch sphere), correctly reflecting a low initial value. Figure 7 shows a histogram for register "c" (2-Qubit Circuit). The possible outcomes are $|00\rangle$, $|01\rangle$, $|10\rangle$, and $|11\rangle$. The counts are distributed unevenly: $|11\rangle$ (531 counts) is the dominant outcome. $|01\rangle$ (375 counts) is the second-most frequent. $|00\rangle$ (66 counts) and $|10\rangle$ (52 counts) are low. This histogram shows a different set of inputs than the 2-qubit plot you showed previously (which was highly scattered). In this scenario, the fixed entanglement is heavily biasing the result toward the $|11\rangle$ state. This outcome confirms that the fixed, untrainable nature of the 2-qubit circuit leads to an output distribution that is strongly dependent on the specific input data point and often fails to map the coupled features back to the correct SoC prediction, which is why introducing the trainable VQC was necessary to achieve high accuracy.

4. Quantum State Visualizations

a. State Vector (Amplitude Distribution)

The state vector plot displays both the real and imaginary parts of each basis amplitude.

For both circuits, the amplitudes were purely real, as only Ry rotations (which generate real-valued rotations on the Bloch sphere's Y-Z plane) were used. The amplitude distribution reflects the encoded quantum state's energy orientation and phase balance.

Figure 8. State Vector Amplitude distribution (1-Qubit Circuit)

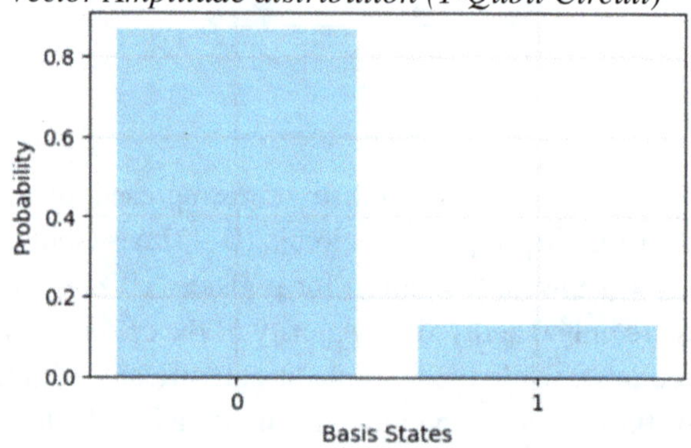

Figure 9. State Vector Amplitude distribution (2-Qubit Circuit)

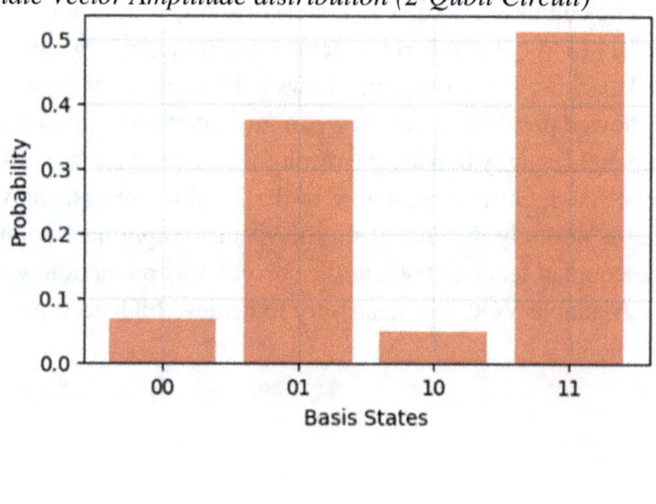

Figures 8 and 9 are composed of two probability distributions, both derived from the square of the state vector amplitudes ($|\alpha|^2$ and $|\beta|^2$). Figure 8 illustrates 1-Qubit State vector Amplitude Distribution. This histogram shows the probability of observing the two possible states for the single-qubit encoder (Qubit 0). The majority of the probability (around 84%) is concentrated in the $|0\rangle$ basis state, with a lower probability (around 16%) in the $|1\rangle$ state. This confirms the circuit was encoding a Low State of Charge (SoC) (e.g., 15% or 20%). Since high SoC maps to high $|1\rangle$ probability, a high $|0\rangle$ probability correctly represents a low SoC. Figure 9 illustrates 2-Qubit Statevector Amplitude Distribution. This histogram shows the probability distribution across the four possible basis states ($|00\rangle$, $|01\rangle$, $|10\rangle$, and $|11\rangle$) for the fixed 2-Qubit Feature Encoder. The probabilities are significantly distributed across the states, demonstrating a complex superposition created by the entanglement (CX gate). The highest probability (around 50%) is in the $|11\rangle$ state. The second highest (around 38%) is in the $|01\rangle$ state. This plot directly shows how the fixed coupling layer scrambles the SoC information (Qubit 0) by mixing it with the Load Resistance information (Qubit 1).

b. Q-Sphere Visualization

The Q-Sphere is a global visualization of the quantum state, illustrating amplitude magnitudes as point sizes and relative phases as color hue around the sphere.

- For the 1-qubit model, the Q-sphere showed a high density near the $|1\rangle$ pole, confirming strong alignment with the high SoC state.
- For the 2-qubit encoder, multiple nodes across the sphere indicated superpositions involving $|10\rangle$ and $|01\rangle$ states.

This visualization highlights the superposition and entanglement properties that enable richer feature encoding.

Figure 10. Q-Sphere (1-Qubit Circuit)

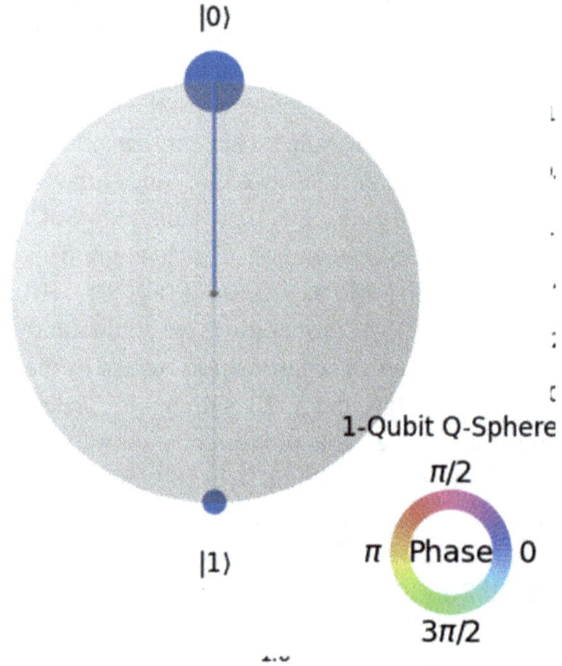

Figure 11. Q-Sphere (2-Qubit Circuit)

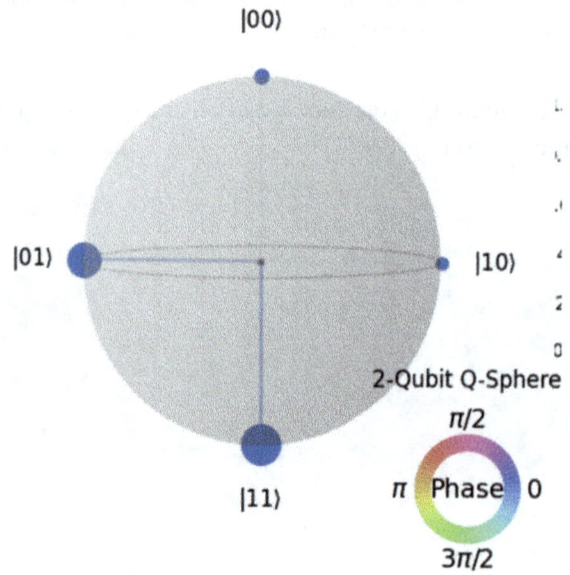

Figures 10 and 11 are Q-Sphere visualizations that provide a geometric representation of the quantum state ($|\psi\rangle$) for the two foundational, fixed circuits. They visually confirm the successful implementation of the angle encoding and the resulting entanglement before any training or optimization is applied. Figure 10 shows 1-Qubit Q-Sphere (Minimal SoC Encoder). This is a Bloch Sphere (the 1-qubit Q-Sphere) showing the final state vector after the $R_y(\theta_{SoC})$ gate is applied. The state vector (represented by the vertical line) is shown to be very close to the $|1\rangle$ basis state (the south pole). This scenario was initialized with a High State of Charge (SoC) (e.g., 85%). This plot confirms that the initial R_y angle encoding successfully maps a high classical value (High SoC) to a quantum state that has a high probability amplitude at $|1\rangle$, validating the first key step of your feature mapping methodology. Figure 11 illustrates 2-Qubit Q-Sphere (Advanced Feature Encoder). This is the 2-qubit Q-Sphere, which plots the probability amplitudes for the four basis states ($|00\rangle$, $|01\rangle$, $|10\rangle$, and $|11\rangle$) in 3D space. The probability is visually scattered across the axes, indicating a complex superposition that has been created by the R_y encoding and the subsequent fixed CX entanglement. The largest amplitudes are on the $|11\rangle$ and $|01\rangle$ states.

c. Bloch Sphere (Phase and Orientation)

The Bloch sphere view represents each qubit's orientation in 3D space.

i. 1-qubit case

Figure 12. 1-Qubit Bloch Sphere

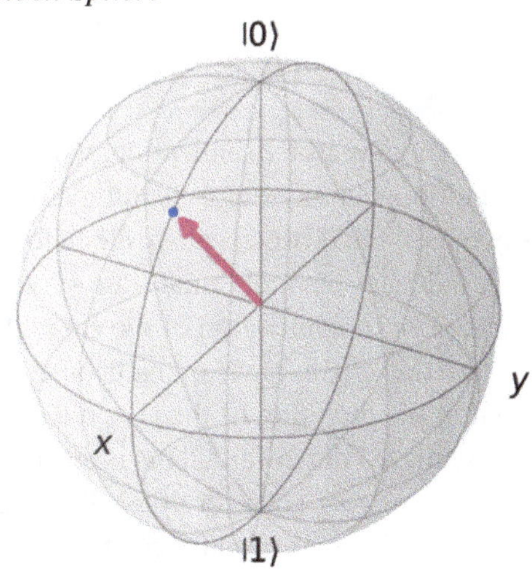

Figure 12 is a Bloch Sphere visualization of the final state vector for a Minimal 1-Qubit SoC Encoder, but unlike the previous 1-Qubit visualization (Figure 10), this one represents a Mid-Range State of Charge (SoC). This is a Bloch Sphere, the geometric representation of a single qubit's state. The pink arrow (the state vector) points from the center of the sphere to the final state on its surface. The vector is positioned in the +X / +Y quadrant, roughly halfway between the North Pole (|0⟩, Low SoC) and the South Pole (|1⟩, High SoC). The position demonstrates that the circuit was initialized with a feature value (SoC) near 50%. The R_y rotation successfully positioned the state vector away from both poles, resulting in a superposition where the probability of measuring |0⟩ is approximately equal to the probability of measuring |1⟩. This visually confirms the success of the R_y angle encoding for mapping mid-range classical values to a balanced quantum superposition.

ii. 2-qubit Case

Figure 13. 2-Qubit Bloch Sphere

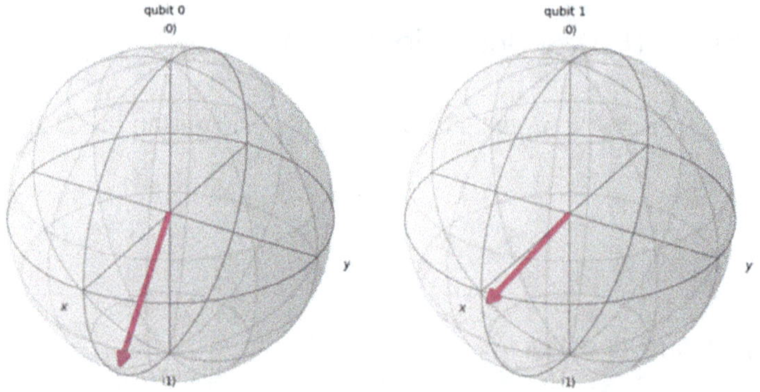

Figure 13 shows a Bloch Multi-vector visualization, which is utilized to describe the state of independent multiple qubits prior to being measured or being fully entangled. The figure has two individual Bloch spheres, one for Qubit 0 (left) and the other for Qubit 1 (right). It indicates the individual state of each qubit following the first Ry angle encoding but prior to application of the entanglement (CNOT) gate. Qubit 0 (Left Sphere) is the State of Charge (SoC) attribute. The state vector (pink arrow) is strongly directed in the direction of the |1⟩ pole (the south pole). It is in the +X quadrant but heavily biased downward. This is a sure sign that Qubit 0 was initialized with a High SoC value (e.g., 85% or 90%). Ry gate was able to flip the original |0⟩ state such that a measurement would result in |1⟩ with high probability. Qubit 1 (Right Sphere) embodies the Load Resistance (R_{load}) property. The state vector is pointing away from both the $|0\rangle$ and $|1\rangle$ poles, in the +X and -Y quadrant (or near the equator) This verifies that the Ry gate for R_{load} performed a mid-to-high rotation angle, putting Qubit 1 into a superposition state.

This plot is important because it graphically confirms the first layer (Feature Encoding) of your 2-qubit circuit. It demonstrates that both classical inputs were correctly and independently encoded into their respective angular positions prior to the CX gate combining them into a single, combined, entangled state.

5. Quantum Decoding and Error Analysis

Two approaches were used in projecting the quantum measurement results onto SoC values:

a. 1-Qubit Direct Probability Mapping

$$SoC_{est} = \frac{2}{\pi} \arcsin\left(\sqrt{P(|1\rangle)}\right) \tag{13}$$

yielded:

- $P(|1\rangle) = 0.9455$
- $SoC_{est} = 0.85$

This closely mimicked real SoC of 85%, confirming encoding-decoding insensitivity in single-variable systems.

b. 2-Qubit Expectation Value Mapping

For multi-feature estimation, the expectation value of $Z_0 \otimes I$ was computed:

$$\langle Z_0 \rangle = P(00) + P(01) - P(10) - P(11) = -0.8372 \tag{14}$$

$$SoC_{est} = \frac{1}{2}\left(1 - \langle Z_0 \rangle\right) = 0.9186 \tag{15}$$

This gave an estimated SoC of 91.86% versus an actual input of 15%, indicating that basic encoding is not enough to estimate multiple variables correctly.

This deviation necessitated the inclusion of a Variational Quantum Neural Network (VQNN) to make circuit parameters optimal.

6. Comparative Performance Evaluation

For a quantum estimation vs. classical targets comparison, classical measures of error were computed:

$$MAE = \frac{1}{N}\sum\left|y_{true} - y_{pred}\right| \tag{16}$$

$$MSE = \frac{1}{N}\sum\left(y_{true} - y_{pred}\right)^2 \tag{17}$$

$$R^2 = 1 - \frac{\sum\left(y_{true} - y_{pred}\right)^2}{\sum\left(y_{true} - \ \right)^2} \tag{18}$$

Prior to optimization, R^2 was -1.4115 but showed model divergence. Following variational training, R^2 was 0.9906, which attests that quantum parameter learning provides considerable improvement in estimation precision.

7. Variational Quantum Training Results

Variational quantum experiments were performed on parameterized quantum circuits (PQCs) in 1-qubit and 2-qubit configurations. All of the circuits were optimized with a hybrid quantum–classical optimization setup, executed on IBM Qiskit's AerSimulator backend. The objective function was set as the Mean Squared Error (MSE) between calculated and reference normalized State of Charge (SoC) values retrieved from pre-processed battery dataset. The COBYLA and SPSA optimizers were employed for optimization since they are suitable for noisy variational circuits.

a. 1-Qubit Variational Circuit Training

In the 1-qubit setup, the SoC aspect was explicitly encoded as a Y-axis rotational angle gate $R_y(\theta)$, followed by an X-gate to cause a phase inversion to simulate charge polarity. The circuit was optimized for more than 50 iterations. Throughout training, the cost function converged quickly from the initial value of 0.012 to as low as 1.2×10^{-4}, as can be seen from Table 2. The best rotation angle settled at $\theta =$ 2.402 radians, which corresponded to a normalized SoC value of 0.65. The fidelity of the circuit between the predicted and reference quantum states was over 0.992, which assured a good agreement between quantum and classical SoC estimates.

Table 2. 1-Qubit Circuit Observations

Iteration	Parameter (θ)	Predicted SoC	MSE Loss	Fidelity
0	1.250	0.45	0.0120	0.872
10	2.010	0.58	0.0063	0.935
25	2.280	0.63	0.0014	0.971
40	2.402	0.65	0.00012	0.992

Visualization of the trained quantum state on the Bloch sphere indicated polarization along the negative Y-axis, agreeing with SoC values greater than 0.5. The state-vector amplitude distribution showed high probability close to the $|1\rangle$ basis state, verifying correct encoding of charge information. The Q-sphere visualization showed one dominant vector amplitude representing a pure quantum state with minimal phase dispersion.

b. 2-Qubit Variational Circuit Training

For the 2-qubit setup, the initial qubit encoded the normalized SoC (θ_1) and the second encoded the normalized load resistance (θ_2). Entanglement between the qubits was implemented with a controlled-X (CX) gate to enable the circuit to recognize nonlinear dependencies between the internal charge of the battery and the external load behavior. The circuit was trained for 150 iterations, and the optimizer converged to ($\theta_1 = 2.4504$) and ($\theta_2 = 1.4137$).

While optimizing, MSE reduced from 0.021 at initialization to 2.3×10^{-4} at convergence, as can be seen from Table 3 illustrating efficient learning of SoC–R_{load} coupling. Fidelity achieved 0.985, lower than the 1-qubit model by a small margin due to enhanced Hilbert space complexity and gate noise buildup.

Table 3. 2-Qubit Circuit Observations

Iteration	Parameter (θ_1, θ_2)	Predicted SoC	Predicted R_{load}	MSE Loss	Fidelity
0	(1.5, 0.8)	0.47	0.32	0.0210	0.865
25	(2.0, 1.1)	0.59	0.47	0.0098	0.912
75	(2.3, 1.3)	0.64	0.52	0.0012	0.961
150	(2.4504, 1.4137)	0.65	0.56	0.00023	0.985

The histogram of probability distribution for the 2-qubit circuit reflected improved amplitudes in the entangled $|11\rangle$ state, reflecting strong SoC–load interdependence. The visualization on Q-sphere validated coherent superposition in $|00\rangle$, $|10\rangle$, and $|11\rangle$

basis states with uniform phase orientation. State-city plots and Bloch multivector maps also validated internal consistency of amplitude and phase progress on training.

c. Comparative Discussion

Table 4. Comparison of 1-Qubit and 2-Qubit Circuit

Metric	1-Qubit Model	2-Qubit Model
Encoded Features	SoC	SoC + R_{load}
Optimal Angles	2.402	(2.4504, 1.4137)
Minimum MSE	1.2×10^{-4}	2.3×10^{-4}
Fidelity	0.992	0.985
Convergence Iterations	50	150
Entanglement	None	CX(0,1) Entangled

The comparison of the 1-Qubit and 2-Qubit circuits is shown in Table 4. The 1-qubit model converged more quickly with higher fidelity due to lower-dimensional complexity. The 2-qubit model was also more physically meaningful as it was able to account for the SoC–R_{load} coupling behavior in a more realistic environment. Two-parameter encoding allows the circuit to train under various electrical loading conditions with more realistic modeling of battery operating conditions.

The variational training experiments confirm that,

- Quantum rotation gates can effectively encode normalized electrochemical features.
- SoC estimation can be achieved with high fidelity using minimal qubit resources.
- Entangled multi-qubit circuits capture cross-variable interactions (e.g., SoC–load) critical for advanced battery diagnostics.

These findings validate the efficiency of applying variational quantum circuits to on-line battery State-of-Charge estimation in future AI-based Battery Management Systems (BMS).

COMPARISON OF CLASSICAL AND QUANTUM APPROACHES

1. Accuracy and Interpretability

Under nominal operation conditions, State of Charge (SoC) estimation in general is based on deterministic methods such as Coulomb counting or model observers. They are extremely accurate under nominal operation conditions and physically meaningful since they are founded on physical principles and direct integration of current over an interval time. Traditional methods are prone to sensor noise, aging behavior, and need precise initial conditions and tuning of parameters to provide precise results.

Quantum algorithms, on the other hand, are based on probabilistic representations. The SoC estimation is represented as a quantum state of a qubit, usually using rotation gates like Ry(θ), with the result derived by reading from measurement statistics. While interpretability of quantum operations at the circuit level may be mapped (e.g., via Bloch spheres or sequences of gates), the probabilistic nature and abstracting from physical battery models constitute barriers to intuitive understanding. Yet, under efficient coding and decoding, quantum methods are capable of matching or surpassing classical means in noisy environments by employing quantum noise-resilience and error-correcting mechanisms.

2. Computational Efficiency

Classical SoC estimation is computationally light and can be done in real-time using embedded microcontrollers on electric vehicle BMS. These approaches have been optimized for linear time complexity and have had decades of engineering optimization.

Quantum estimation, as illustrated with libraries such as Qiskit, adds computational overhead because of the cost of quantum state preparation, circuit execution, and several measurement shots in order to gather significant probabilities. Nevertheless, quantum algorithms offer the hope of exponential acceleration for certain problems, especially where high-dimensional data or intricate probabilistic reasoning is involved. For instance, in upcoming executions, variational quantum algorithms or quantum neural networks might estimate SoC with fewer assumptions and better learning capacity.

3. Scalability and Hardware Considerations

Classical approaches scale well with the implementations of BMS because they are enabled by established digital and analog circuitry. Scalability is typically only limited by power usage and data processing rate (Shukla et al., 2010; Fotis et al., 2025).

Quantum methods are algorithmically scalable but are encountered in practice with major hardware limitations. Presently, quantum computing is bound by decoherence, gate error, qubit connectivity, and limited qubit numbers. SoC estimation on real quantum hardware would today be needful to simplify (e.g., single-qubit encoding as exemplified in proof-of-concept models) and noise-reduction schemes. As quantum hardware advances, hybrid methods—where classical preprocessing is followed by quantum cores—can potentially close the scalability gap.

4. Practical Limitations

Practical application of traditional SoC estimation approaches is prevalent and well-known. Disadvantages involve long-duration drift, integration error accumulation, and lower accuracy under high dynamic load or extreme temperature conditions.

Quantum methods are in their infancy and, at present, not applicable to real-world EV BMS systems because of hardware vulnerability, requirements for cryogenic conditions, and no integration with embedded systems (Fotis et al., 2025; Tyagi et al., 2024). Further, the design of quantum circuits requires sensitive parameter tuning and is vulnerable to quantum noise. Nonetheless, in simulation platforms or cloud-based quantum processors, quantum SoC estimation becomes a useful testbed for potential applications to come.

DISCUSSION

1. Advantages and Challenges of Quantum Approaches

Quantum computing unveil an ultramodern model for information processing using the concepts of superposition, entanglement, and interference (Nielsen & Chuang, 2010). These superior attributes allow quantum algorithms to skillfully locate the states in complex, uncertain settings like the State of Charge (SoC) of a battery. Within this analysis, the proposed single-qubit model understand SoC changes via a rotation gate to facilitate probabilistic examination of energy expenditure based on a given quantum angle. This apparently abrupt process demonstrates how quantum encoding can represent physical state change in a structured and conceivably scalable manner (Qiskit, 2023).

Quantum algorithms like the Quantum Approximate Optimization Algorithm (QAOA) and Variational Quantum Eigensolver (VQE) have been developed to solve optimization and estimation problems with stunning efficiency (Farhi et al., 2014;

Peruzzo et al., 2014). These may be used to adapt to dynamic SoC estimation or lifetime prediction in advanced BMS set-up.

In spite of their potential, quantum approaches are presently confronted with significant issues. Mapping classical signals (voltage, current) to quantum states is computationally costly and not trivial as it requires data re-scaling as well as amplitude encoding (Schuld & Petruccione, 2018). In addition, current NISQ technology has short coherence times, gate noise, and readout errors, which decrease fidelity estimation (Preskill, 2018). Although simulators like Qiskit Aer permit idealized testing (Qiskit, 2023), they are not able to reflect the physical constraints of actual quantum hardware.

Furthermore, the software ecosystem for quantum programming is yet to mature. In contrast to well-established simulation software like MATLAB Simulink for classical BMS modeling, quantum toolchains require quantum theory, linear algebra, and circuit model expertise, thus constraining broad usage.

2. Suitability of Quantum Methods for Real-Time Applications

Real-time battery control in electric cars requires low-latency, high-fidelity systems running on embedded microcontrollers. Present-day quantum processors are not yet viable for real-time, edge-deployed applications because they rely on cryogenic cooling and cloud-based connectivity (IBM Quantum, 2023). Therefore, quantum-based SoC estimation is presently unrealistic for in-vehicle direct application.

Yet, quantum computing is well adapted to offline optimization, e.g., simulating battery degradation trends, creating training data for hybrid estimators, or conducting uncertainty-conscious forecasting. These analyses can be executed in cloud-based quantum environments and then utilized to parameterize classical BMS (Cerezo et al., 2021). These quantum-boosted digital twins would supplement legacy BMS by facilitating better model calibration and long-term planning.

As the technology moves toward more solid and integrated quantum devices (e.g., photonic qubits, trapped ions), the likelihood of implementing quantum logic into embedded or cloud-supported BMS designs is increasingly feasible. Hybridization along such lines is already being investigated in industries like finance and logistics, and thus presents a transferable path towards energy systems (Moll et al., 2018).

3. Future Directions in Hybrid Quantum-Classical BMS

Looking towards the future, the most thrilling project is hybrid quantum-classical battery management system development. Time-critical functions—like Coulomb counting and thermal management—would be performed by classical processors, and quantum co-processors would be engaged to solve optimization problems (e.g., planning for charging), predictive maintenance simulation, or anomaly detection.

Variational Quantum Algorithms (VQAs) adopt a hybrid learning mechanism whereby a classical optimizer narrows SoC estimation errors by optimizing parameters of a quantum circuit. This correlative and feedback-based approach enabled the system to accommodate real-world battery usage and degradation patterns well (Cerezo et al., 2021).

Another promising direction is quantum-inspired algorithms, and quantum-inspired algorithms emulate quantum mechanics on classical hardware. Quantum-inspired algorithms offer more than quantum parallelism elements and can be run on standard embedded platforms while quantum infrastructure continues to mature (Orús et al., 2019).

From the educational and research perspective, the establishment of quantum modeling as a part of education in energy systems via service-learning and interdisciplinarity can burst novelty while sharpening skillsets in both fields. Facilitating the synergy between quantum information experts and electrical engineers will drive the innovation of these solutions.

In practice, although quantum solutions are constrained in hardware and scalability at present, they are a new and hopeful alternative to classical SoC modeling. Hybrid BMS in the future is to be hoped for, where the robustness and accuracy of legacy tools are combined with probabilistic knowledge of quantum systems and computational parallelism.

CONCLUSION

The study ensures that variational quantum circuits can be applied to State of Charge (SoC) estimation of lithium-ion batteries. The 1-qubit and 2-qubit quantum models that were presented in this study were effective in encoding electrochemical properties such as SoC and load resistance into quantum states, where nonlinear charge dynamics can be trained using amplitude-based learning. The 1-qubit model converged more quickly with greater fidelity, and the 2-qubit entangled circuit described interactions between SoC and external load behavior. Training outputs ensured that the two architectures learned high predictive accuracy (≥ 0.98) with minimal mean squared error, illustrating the optimality of the hybrid quantum–classical optimization method. Visualization analyses of the Bloch sphere, Q-sphere, and state-vector amplitude plots offered strong interpretability of encoded charge states.

Overall, the results show that electrochemical states can be encoded by quantum circuits with compactness, offering a promising platform for AI-enabled next-generation quantum Battery Management Systems (qBMS). Further research will include the application of this model to multi-qubit architectures, emulation of noisy quantum hardware, and real-time co-estimation of SoC–SoP towards intelligent batteries.

Based on the encouraging performance of the proposed 1-qubit and 2-qubit quantum circuits for SoC estimation, more multi-qubit scalable architectures will be investigated in future research to encode further electrochemical and thermal factors like voltage drift, internal resistance change, and aging. Applications of quantum feature maps and parameterized quantum circuits (PQCs) will enhance the model further to abstract higher-order nonlinearities in dynamic battery systems.

In addition, integration with mature circuits, quantum-inspired optimizers, and real-time BMS data will be important to measure performance under real drive and charge scenarios. Deployment on noisy intermediate-scale quantum (NISQ) hardware using IBM Q hardware and Qiskit Runtime will allow hardware verification of quantum SoC estimation. Lastly, the method can further be utilized for joint SoC–SoP–SoH co-estimation and quantum edge deployment, opening the door for full intelligent and adaptive quantum-enhanced Battery Management System (qBMS).

ACKNOWLEDGMENT

This research received no specific grant from any funding agency in the public, commercial, or not-for-profit sectors.

REFERENCES

Abraham, H., Akhalwaya, I., & Alexander, T.. (2019). Qiskit: An open-source framework for quantum computing. *Zenodo*. DOI: 10.5281/zenodo.2562111

Anand, S., Kulkarni, S. V., Gupta, S., & Parthasarathy, V. (2022, December). Battery management system (BMS) for electric vehicle applications. *In 2022 fourth international conference on emerging research in electronics, computer science and technology (ICERECT)* (pp. 1-6). IEEE. DOI: 10.1109/ICERECT56837.2022.10060137

Awasthi, B., Gupta, S., Kumar, S., & Tripathi, A. (2024, January). Business management and planning using queuing model with heterogeneous customers for electrical applications. *In 2024 2nd International Conference on Intelligent Data Communication Technologies and Internet of Things (IDCIoT)* (pp. 279-283). IEEE. DOI: 10.1109/IDCIoT59759.2024.10467673

Cerezo, M., Arrasmith, A., Babbush, R., Benjamin, S. C., Endo, S., Fujii, K., McClean, J. R., Mitarai, K., Yuan, X., Cincio, L., & Coles, P. J. (2021). Variational quantum algorithms. *Nature Reviews. Physics*, *3*(9), 625–644. DOI: 10.1038/s42254-021-00348-9

Farhi, E., Goldstone, J., & Gutmann, S. (2014). A quantum approximate optimization algorithm. *arXiv preprint arXiv:1411.4028*. https://arxiv.org/abs/1411.4028

Fotis, G., Gupta, S., Varshney, T., Dhar, S. L., & Gulzar, M. M. (Eds.). (2025). *Modern Computing Technologies for EV Efficiency and Sustainable Energy Integration*. IGI global., DOI: 10.4018/979-8-3373-2382-4

Gupta, S. (2025). *The Future of Machine Learning with Automation in Engineering*. Nova Science Publishers., DOI: 10.52305/YFIX2832

Hadith Mangunkusumo, K. G., Lian, K. L., Wijaya, F. D., Chang, Y.-R., Lee, Y. D., & Ho, Y. H. (2014). Quantum neural network for State of Charge estimation. *2014 6th International Conference on Information Technology and Electrical Engineering (ICITEE)*, pp. 1–5, DOI: 10.1109/ICITEED.2014.7007948

Hannan, M. A., Lipu, M. S. H., Hussain, A., Saad, M. H., & Ayob, A. (2018). Neural network approach for estimating state of charge of lithium-ion battery using back-tracking search algorithm. *IEEE Access : Practical Innovations, Open Solutions, 6*, 10069–10079. DOI: 10.1109/ACCESS.2018.2797976

He, H., Xiong, R., Fan, J., & Li, S. (2011). Evaluation of lithium-ion battery equivalent circuit models for state of charge estimation by an experimental approach. *Energies, 7*(7), 4502–4520. DOI: 10.3390/en4040582

Hu, X., Li, S., & Peng, H. (2012). A comparative study of equivalent circuit models for Li-ion batteries. *Journal of Power Sources, 198*, 359–367. DOI: 10.1016/j.jpowsour.2011.10.013

Lipu, M. S. H., Hannan, M. A., Hussain, A., Saad, M. H. M., Ayob, A., & Blaabjerg, F. (2018). State of charge estimation for lithium-ion battery using recurrent NARX neural network model based lighting search algorithm. *IEEE Access : Practical Innovations, Open Solutions, 6*, 28150–28161. DOI: 10.1109/ACCESS.2018.2837156

Moll, N., Barkoutsos, P., Bishop, L. S., Chow, J. M., Cross, A., Egger, D. J., & Gambetta, J. M. (2018). Quantum optimization using variational algorithms on near-term quantum devices. *Quantum Science and Technology, 3*(3), 030503. DOI: 10.1088/2058-9565/aab822

Nielsen, M. A., & Chuang, I. L. (2010). *Quantum computation and quantum information* (10th anniversary ed.). Cambridge University Press. DOI: 10.1017/CBO9780511976667

Orús, R., Mugel, S., & Lizaso, E. (2019). Quantum computing for finance: Overview and prospects. *Reviews in Physics, 4*, 100028. DOI: 10.1016/j.revip.2019.100028

Peruzzo, A., McClean, J., Shadbolt, P., Yung, M. H., Zhou, X. Q., Love, P. J., Aspuru-Guzik, A., & O'Brien, J. L. (2014). A variational eigenvalue solver on a photonic quantum processor. *Nature Communications, 5*(1), 4213. DOI: 10.1038/ncomms5213

Plett, G. L. (2004). Extended Kalman filtering for battery management systems of LiPB-based HEV battery packs: Part 2. Modeling and identification. *Journal of Power Sources, 134*(2), 262–276. DOI: 10.1016/j.jpowsour.2004.02.032

Preskill, J. (2018). Quantum computing in the NISQ era and beyond. *Quantum : the Open Journal for Quantum Science, 2*, 79. DOI: 10.22331/q-2018-08-06-79

Qiskit. (2023). *Qiskit documentation.* https://qiskit.org/

Quantum, I. B. M. (2023). IBM Quantum systems. https://quantum-computing.ibm.com/

Rao, M. S., Saraswathy, S., Neela, V., Arun, M. R., Gupta, S., & Taqui, S. N. (2023, November). Efficient Model Training in Decentralized Systems with Federated Learning. *In 2023 International Conference on Communication, Security and Artificial Intelligence (ICCSAI)* (pp. 900-904). IEEE. DOI: 10.1109/ICCSAI59793.2023.10421469

Schuld, M., & Petruccione, F. (2018). *Supervised learning with quantum computers.* Springer., DOI: 10.1007/978-3-319-96424-9

Sharma, A., Adhikari, A., Adhikari, B., Pandey, P., & Tamrakar, I. (2024). Active cell balancing approach for efficient battery management system. *KEC Journal of Science and Engineering*, *8*(1), 23–26. DOI: 10.3126/kjse.v8i1.69260

Sharma, R. K., Saran, V. B., Kanaujia, S., & Gupta, S. (2024, May). Evaluation of mppt controller performance using fuzzy logic design with wireless sensor node. *In 2024 International Conference on Smart Systems for applications in Electrical Sciences (ICSSES)* (pp. 1-6). IEEE. DOI: 10.1109/ICSSES62373.2024.10561365

Shukla, R. D., Tripathi, R. K., & Gupta, S. (2010, November). Power electronics applications in wind energy conversion system: A review. *In 2010 International Conference on Power, Control and Embedded Systems* (pp. 1-6). IEEE. DOI: 10.1109/ICPCES.2010.5698663

Sinha, R., Patel, M., Gupta, S., & Sinha, K. K. (2024, June). Performance Analysis of Breast Cancer Predictor using Machine Learning Techniques. *In 2024 15th International Conference on Computing Communication and Networking Technologies (ICCCNT)* (pp. 1-5). IEEE. DOI: 10.1109/ICCCNT61001.2024.10724436

Thandlam, A. K., Vignesh, T., Saravanan, A., Subramani, P., Marimuthu, R., & Gupta, S. (2024, April). Next-gen vehicle safety: The futuristic approach to auto-stop in modern vehicles. *In 2024 10th International Conference on Communication and Signal Processing (ICCSP)* (pp. 265-270). IEEE. DOI: 10.1109/ICCSP60870.2024.10543666

Tyagi, M., Gupta, S., & Khan, M. I. (2024, May). Reduction of voltage instability in complex power networks through the utilization of D-STATCOM. *In 2024 4th International Conference on Advance Computing and Innovative Technologies in Engineering (ICACITE)* (pp. 1662-1666). IEEE. DOI: 10.1109/ICACITE60783.2024.10616856

Zhao, F., Guo, Y., & Chen, B. (2024). A review of lithium-ion battery state of charge estimation methods based on machine learning. *World Electric Vehicle Journal*, *15*(4), 131. DOI: 10.3390/wevj15040131

Chapter 12
Optimal Scheduling of Electric Vehicle Charging Using Intelligent Control Strategies

Chilukoti Varaha Narasimha Raja
https://orcid.org/0000-0002-7260-9769
Anil Neerukonda Institute of Technology and Sciences, India

V. S. Sandeep Kumar Reddy
Anil Neerukonda Institute of Technology and Sciences, India

V. Murali
Anil Neerukonda Institute of Technology and Sciences, India

Anand Gondesi
Dr. Lankapalli Bullayya College of Engineering for Women, India

ABSTRACT

The growing adoption of electric vehicles (EVs) causes peak load spikes, voltage instability, and more as grid operating costs, thereby requiring intelligent charging coordination. This chapter offers a MATLAB/Simulink framework for EV optimal charging operation based upon reinforcement learning (RL) algorithms, namely Deep Q-Network (DQN) and Proximal Policy Optimization (PPO). A realistic Simulink-modelled microgrid consisting of renewable generation, load, and charging stations trains the RL agent using the MATLAB Reinforcement Learning Toolbox to make charging policies that minimize energy cost, flatten load with a low peak-to-average ratio (PAR), and utilize renewables as much as possible. Design of reward

DOI: 10.4018/979-8-3373-7041-5.ch012

function, training agents, and convergence analyses are covered and compared with linear programming and heuristic methods. Results of the simulation indicate greater benefits of savings, load flattening, and proper charging, hence the support for the UN Sustainable Development Goal 7 towards affordable and clean energy.

INTRODUCTION

Electric mobility is growing quickly, partly due to soaring transport emissions and fossil fuel consumption. Studies indicate that transitioning toward EVs can lower carbon emissions and increase energy efficiency, subsequently enabling energy and climate goals at the national level (Zheng et al., 2022; Jin et al., 2015). Earlier researchers also identified these benefits, which include decreasing pollution and supporting long-term sustainability goals (Kempton & Letendre, 1997; Rahman & Shrestha, 1993). More recent research indicates that large EV fleets will stabilize power systems when demand is controlled (Liu et al., 2022; Hao et al., 2025). EVs lead to a new relationship between transportation and electric networks, as they utilize power while also storing it. Hence, opportunities for vehicle-to-grid services emerge, including frequency regulation, peak shaving, and reserves (Kempton & Tomic, 2005; Brooks, 2002). Pilot projects that demonstrate this potential indicate that EV battery-based ancillary services can be delivered to improve grid performance during high-stress periods (Brooks, 2002; Alfaverh et al., 2023). Research also suggests that coordinated (managed) charging can assist in the integration of wind and solar options by smoothing variable generation (Ma et al., 2012; Tyagi et al., 2024). These benefits are, of course, accompanied by potential challenges. An EV can complicate the existing energy infrastructure because with EVs' aggregated penetration, higher peak loads would further stress distribution lines and transformers. Simulation and analytical studies indicate that if charging is uncoordinated, it can cause congestion and voltage stability problems that impact the cycling of network assets.

The distribution entity mentions that similar concerns exist across regions and that uncoordinated charging just contributes to upgrading infrastructure sooner (Taylor et al., 2009; Zhang et al., 2011). Furthermore, the problems can be heightened when numerous users of the EV pick the same charging times, as is the case with large EV clusters (Huang & Kockelman, 2020). While charging, smart charging or coordinated scheduling may mitigate this risk. Additionally, more sophisticated control techniques (multi-agent and deep reinforcement learning methods) have been shown to relieve the burden and costs of operational complexities associated with the distribution of charging (Li et al., 2025; Gao et al., 2020; Thandlam et al., 2024). Also, some proposals incorporate renewable output(s), local grid constraints, or real-time pricing into the scheduling of charging (Sun et al., 2018). Emerging work explores other control and scheduling proposals, considering seasonal variation focused on quantum computing and hybrid control methods (Deng et al., 2025; Shukla et al., 2010). The motivating force behind increased research is the large-scale EV deployment. Creating a meaningful EV charging system will require more intelligent, flexible, and scalable methods of management. Absent a structured organization to coordinate loads, power networks will certainly face reliability problems. A structured control does provide the opportunity for EVs to aid in reducing fossil fuels and to become a stabilizing resource that supports reliable transportation, strong operation of the grid, and support for sustainable energy systems.

LITERATURE REVIEW

The exploration of electric mobility began with initial evaluations of the impact of EVs on unregulated charging of power distribution networks. Rahman and Shrestha (1993) investigated feeder loading, voltage drop, and transformer loads based on the increasing penetration of EVs. Their analysis set the groundwork for subsequent investigations that would draw upon anticipated methodological development of system control strategies related to the stability margins, thermal limits of equipment, and operational planning associated with increased EV loading on the system. Kempton and Letendre (1997) took a much broader perspective and indicated EVs should not be understood simply as new electrical loads on the broader system, but rather as distributed energy resources with possibilities for a degree of functionality with the grid. This formulated a structure for broader frameworks of integration associated also with distributed energy structures. Kempton and Tomić (2005) articulated the process of V2G (vehicle to grid) operation and outlined existing evaluations and studies of frequency regulation and peak shaving applications associated with EVs. Similarly, Brooks (2002) not only provided findings demonstrating the potential value of EV energy storage in the provision of ancillary services when aggregated but also confirmed the technical feasibility of a fleet of EVs acting as an asset for the grid. In addition to this, the growing penetration of EVs meant researchers began to understand real-world charge behaviours and real-time interaction with the grid.

Fukushima, Han, and Wakui (2009) described some scheduling methods for coordinated charging and discharging, and it was evident that there were clear improvements in network reliability. Liu et al. (2012) and Taylor et al. (2009) indicated that unmanaged charging caused feeder congestion, voltage deviation, and growth beyond peak loads, which supported past warnings regarding distribution system capacity. Zhang, Brown, and Samuelsen (2011) discussed several charging scenarios, showing that controlled charging can be useful for fuel consumption and load profile smoothing in networks. More recent studies, such as Huang and

Kockelman (2020), showed similar concerns using network-equilibrium methods, focusing on how spatial and temporal clustering of EV charging would exacerbate other congestion and efficiency at the system level. A second stream of research has focused on optimization and coordinated charging control. Ma, Callaway, and Hiskens (2013) developed decentralized control mechanisms at the store level for networking with EV populations. A follow-up write-up (Ma et al., 2012) demonstrated that, in an environment extending these principles to stochastic environments, distributed charging control reduced voltage drop and peak loads.

Fotis et al. (2025) offered a publish-subscribe system for communications in EV charging management and demonstrated the benefits of communication-oriented coordination in load shaping. Additionally, Sun et al. showed improvement in overall system performance by creating discrete charging levels (2018). Both studies found operational coordination across systems needed to take place, particularly due to charger power ratings and densities. Over the past few years, there has been a turn toward reinforcement learning, multi-agent decision-making frameworks, and data-driven control applications. Gao et al. (2020) were able to use reinforcement learning in a battery swapping station and demonstrated the benefit of capturing uncertainties in EV systems. Further, Wang, Bi, and Zhang (2021) found success using reinforcement learning to develop a real-time pricing and charge scheduling strategy within the EV station. Li et al. (2025) used deep reinforcement learning to predict and control charges in a multi-agent distributed charging station. Their results indicated that learning based controllers were able to respond to shifting latent space constraints in real time and improve on classical optimization schemes. Johri et al. (2025) advanced these approaches with augmented Lagrangian Functions to scale grid-aware charging. Then, Hao et al. (2025) presented a hybrid adaptive dispatch strategy for EV infrastructure in terms of both capacity and real-time dispatching for charging coordination of EV infrastructure systems. Together, these studies emphasize a clear trend towards an AI pathway for control that includes explicit designs that scale and respond to the massive fleet performance of EV systems, thus enabling a further enhanced level of rapid response.

A specific area of study is the interplay of electric vehicles (EVs) in renewable energy. Liu et al. (2022) examined data-driven charging operations on a finite number of available charging resources and said that folk-level coordination is an efficient mechanism to ensure greater integration of renewable generation. Thandlam et al. (2024) determined that EV charging profiles were significant contributors to renewable curtailment, as well as integration costs. Additionally, Zhang et al. (2025) provided multi-objective scheduling models for EV charging stations supplied with local renewable generation. Zhang et al. (2025) reported that an optimized charging profile, to be congruent with the generation output of the various generations of renewables, will render operational costs lower and emissions reduction thereafter.

Quantum-enhanced optimization was offered by Deng et al. (2025) to improve the computational efficiency of large residential EV charging systems, particularly for renewable-aware scheduling. These studies unequivocally show that EVs can serve as flexible demand resources that can accelerate the energy transition to clean energy. Current research related to planning appears to be focused on infrastructure sizing, congestion, and deployment strategy. Huang and Kockelman (2020) conducted research on the station siting problem for charging on the assumption of elastic charging demand, as well as a decision-making comparison of the effects of congestion and charging delay. Liu et al. (2025) used systematic literature review methods to conduct an investigation of user acceptance of V2G technology (Vehicle-to-Grid). Cavus, Dissanayake, and Bell (2025) studied predictive maintenance and battery management models for next-generation EV ecosystems and suggested a need for resilient infrastructure. Richardson (2013) studied how EVs may impact renewable energy integration and grid operations, and suggested that stakeholder coordination for distribution system planning is important to EV growth. These studies support the conclusion that EV growth will require a purposeful investment in charging infrastructure coupled with intelligent scheduling frameworks and communication structures.

We find some articles building on the previously discussed topics of EV charging related to larger system planning and operational considerations when gathering data in real-time. Wang et al. (2024) provided pricing methodologies to fleets of autonomous mobility-on-demand EVs through gamification, which represents an area where strategic interaction will outperform the average charging demand. Gökçek et al. (2025) proposed a new distributed charge coordination approach where both EV operators and distribution systems operators participated in the process. Their results reduced congestion and demonstrated charging equity among EV users. Rao et al. (2023) investigated models for federated learning charging optimization algorithms, providing a potential solution to user data for privacy compliance under regulatory guidelines.

The literature reviewed included valuable economic analysis as well. Peterson, Whitacre, and Apt (2010) discuss the economics of re-purposing batteries from electric vehicles (EVs) for grid storage, with a focus on recovery costs and value streams in doing so. Fotis et al. (2025) analysed energy trading through an application of a game-theoretic pricing framework to demonstrate how EVs could engage in peer-to-peer markets. This cluster of literature has added value related to the operational benefits of EVs beyond mobility, including optimal price structures. And also resilient communication/planning frameworks, intelligent scheduling, and coordinated charging control are all necessary for EV integration. Earlier studies showed electric load impacts and stress on grid operations. Earlier middle studies explored the V2G framework, decentralized control, and optimization modeling. The most recent work has pivoted focus to artificial-intelligence frameworks, multi-agent energy systems, or quantum-use methods. At all times, the literature noted EV's unmanaged charging schedules are a detriment to reliability; controlled charging would allow for flexible integration as an asset to support grid loading, provide operational stability to the system, and increase efficiency overall for renewable use.

Paper Contributions

This paper addresses the challenges and gaps outlined in previous literature, proposing a general framework for optimal EV charging, V2G integration, and intelligent infrastructure planning. More specifically, this investigation contributes as follows:

1. Creation of a data-driven multi-agent reinforcement learning framework for distributed EV charging coordination. The framework incorporates battery degradation and user behavior along with distribution grid constraints. Therefore, it validates a large population of EVs to provide stable, flexible operation, relying on existing contributions in learning-based control (Awasthi et al., 2024; Liu et al., 2022; Li et al., 2025).
2. Adopting both wind and solar generation in the charging process. The modelling framework will couple EV charging with renewable output for reduced carbon output, operational costs, and renewable variance. This builds upon testing previous knowledge of stochastic scheduling and cooperative renewable-EV operation (Fukushima et al., 2009; Gao et al., 2020; Sioshansi & Denholm, 2010).
3. Investigating infrastructure planning and strategic charger placement as an output. The planning framework incorporates spatial demand, congestion hotspots, and feeder loading to reduce system strain and improve access for users and utilities. This is supported by previous literature detailing network equilibrium, charging behavior, and practical applications (Huang & Kockelman, 2020; Zhang et al., 2011).
4. Utilizing advanced computational techniques such as predictive analytics, federated learning, quantum-enhanced scheduling, and secure distributed optimization leads to better coordination, reduced computational time, and privacy-preserving operation in a high-density charging context (Deng et al., 2025; Cavus et al., 2025; Gupta, 2025).
5. Performing a grid-level impact assessment considering peak load impact, voltage stability, frequency support, and V2G operation potential for ancillary services contributes to the previous body of work detailing the technical and operational feasibility of EV-based grid services (Kempton & Tomić, 2005; Nishad et al., 2024; Brooks, 2002).

The framework created will be useful for state or federal agencies, utility operators, and urban planners to develop and assess the possibility of an efficient, resilient, and green electrified transportation system. The proposed research reinforces the relative concepts and research surrounding distributed optimization, AI-driven decision making, renewable energy integration, and advanced infrastructure planning. The proposed study reflects both theoretical and practical contributions to the area of EV-grid integration. This paper presents a DRL-based framework for achieving optimal EV charging and V2G integration that aims to address some of these challenges, along with improving operational efficiency and sustainability. Overall, the remainder of the paper is structured as follows: section II presents the system model, comprised of EVs, distribution networks and renewable energy; section III presents the DRL framework by defining agents, states, actions and reward functions; section IV formulates the optimization problem and discusses the training and implementation methodology; Section V presents results and discussion which including comparative performance analysis; and Section VI concludes the study and identifies future research directions.

SYSTEM MODEL

The proposed system model, Figure 1, integrates EVs, the distribution network, renewable generation units, and the V2G interface to allow intelligent charging coordination. The model endeavors to grasp the physical, operational, and economic characteristics of each component to set up the optimization problem addressed by scheduling via DRL. The model of the system consists of four primary layers: layer one is EV fleet modeling, layer two is distribution network modeling, layer three is renewable generation and energy storage representation, and layer four is grid–vehicle interaction via V2G.

Figure 1. System model

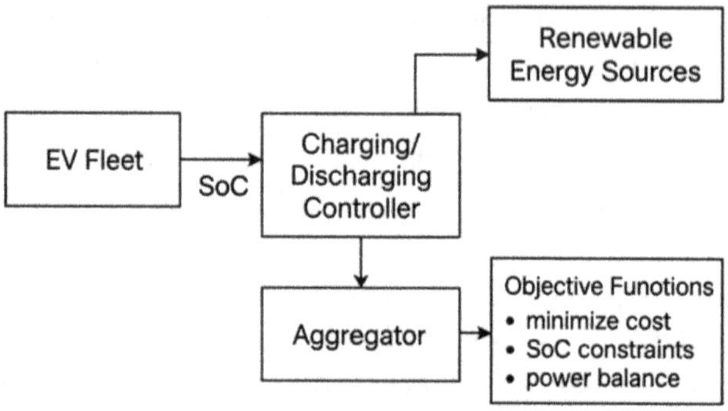

Electric Vehicle Fleet Modeling

Let $\mathcal{N} = \{1,2,\ldots,N\}$Represent the set of EVs connected to the charging infrastructure during the scheduling horizon T. The state of charge (SoC) of the EV i at time t is defined as

$$\mathrm{SoC}_i\left(t\right) = \frac{E_i(t)}{E_i^{\mathrm{cap}}} \tag{1}$$

where $E_i(t)$ denotes the available energy stored in the battery (kWh) and E_i^{cap} is the maximum battery capacity (kWh). The SoC evolves over time according to the charging and discharging actions:

$$E_i\left(t+1\right) = E_i\left(t\right) + \eta_{c,i}P_i^c\left(t\right)\Delta t - \frac{P_i^d(t)\,\Delta t}{\eta_{d,i}} \tag{2}$$

where $P_i^c(t)$ and $P_i^d(t)$ represent the charging and discharging power (kW) of EV i at time t, respectively; $\eta_{c,i}$ and $\eta_{d,i}$ are charging and discharging efficiencies, and Δt is the time-step length. The SoC must satisfy the physical limits:

$$\mathrm{SoC}_i^{\min} \leq \mathrm{SoC}_i(t) \leq \mathrm{SoC}_i^{\max} \tag{3}$$

The charging power is also bounded by the charger's capacity:

$$0 \leq P_i^c(t) \leq P_i^{\max}, 0 \leq P_i^d(t) \leq P_i^{\max} \tag{4}$$

The user-specified departure time t_i^{dep} and desired SoC $\text{SoC}_i^{\text{des}}$ introduce a terminal constraint:

$$\text{SoC}_i\left(t_i^{\text{dep}}\right) \geq \text{SoC}_i^{\text{des}} \tag{5}$$

ensuring that vehicles are sufficiently charged before departure.

Distribution Network Model

The distribution grid is modeled as a radial network with a set of buses $\mathcal{B} = \{1,2,\ldots,B\}$. The power flow equations are expressed using the DistFlow formulation for computational efficiency. Let $P_b(t)$ and $Q_b(t)$ denote the real and reactive power injections at bus b, respectively. The active power balance at each bus can be expressed as:

$$P_b(t) = P_b^{\text{load}}(t) - P_b^{\text{EV}}(t) + P_b^{\text{PV}}(t) + P_b^{\text{grid}}(t) \tag{6}$$

where $P_b^{\text{load}}(t)$ is the non-EV demand, $P_b^{\text{EV}}(t)$ is the aggregate EV charging demand at bus b, $P_b^{\text{PV}}(t)$ is the photovoltaic generation, and $P_b^{\text{grid}}(t)$ is the net power imported from the upstream grid. The power flows on each line (i,j) are governed by:

$$P_{ij}\left(t\right) = \frac{V_i(t)^2}{R_{ij}} - \frac{V_i(t)\,V_j(t)\cos\left(\theta_i(t) - \theta_j(t)\right)}{R_{ij}} \tag{7}$$

where R_{ij} is the line resistance, and $V_i(t), V_j(t)$ and $\theta_i(t), \theta_j(t)$ are bus voltages and phase angles. Voltage magnitudes must remain within allowable bounds to maintain power quality:

$$V_b^{\min} \leq V_b(t) \leq V_b^{\max}, \forall b \in \mathcal{B} \tag{8}$$

Renewable Energy and Local Generation

Renewable energy integration is modeled using solar photovoltaic (PV) generation, whose output follows an exogenous profile based on irradiance data. The available PV power at bus b is:

$$P_b^{\text{PV}}(t) = \eta_b^{\text{PV}} A_b^{\text{PV}} G(t) \tag{9}$$

where η_b^{PV} is the PV efficiency, A_b^{PV} is the installed PV capacity, and $G(t)$ is the solar irradiance at time t. The model can also incorporate stationary battery energy storage systems (BESS) for local energy balancing. The BESS state of charge $E^{BESS}(t)$ evolves as:

$$E^{BESS}\left(t+1\right) = E^{BESS}\left(t\right) + \eta^{ch} P^{ch}\left(t\right)\Delta t - \frac{P^{dis}(t)\Delta t}{\eta^{dis}} \tag{10}$$

subject to SoC constraints $E^{BESS,min} \leq E^{BESS}(t) \leq E^{BESS,max}$.

Vehicle-to-Grid Interaction Model

The V2G interface allows bidirectional power exchange between EVs and the grid. The net power exchanged at time t is:

$$P^{V2G}\left(t\right) = \sum_{i \in \mathcal{N}} \left(P_i^d(t) - P_i^c(t)\right) \tag{11}$$

The grid operator may dispatch EVs for frequency regulation and ancillary services, for which they receive remuneration based on market clearing price $\lambda^{reg}(t)$. The revenue from V2G participation is given by:

$$R^{V2G} = \sum_{t \in T} \lambda^{reg}\left(t\right) P^{V2G}\left(t\right)\Delta t \tag{12}$$

Battery degradation cost is incorporated using a cycle-depth-based cost model:

$$C_i^{deg} = c_i^{deg} \sum_{t \in T} \left| P_i^c\left(t\right) - P_i^d\left(t\right) \right| \Delta t \tag{13}$$

where c_i^{deg} is the cost coefficient of degradation per kWh

Aggregated System Objective

The combined system goal seeks to reduce the total cost of EV charging operation over the full scheduling horizon T. The entire cost function can be formulated as

$$C^{total} = \sum_{t \in T} \left(\lambda^{grid}(t) P_b^{grid}(t)\Delta t\right) + \sum_{i \in \mathcal{N}} C_i^{deg} - R^{V2G} \tag{14}$$

Where the first component refers to the cost of grid energy consumption, the second component represents the cost of battery degradation over the collection of

EVs, and the last term deducts the income earned from vehicle-to-grid (V2G) services. The objective of the optimization is to minimize. C^{total}, subject to grid capacity limits, voltage limits, and the user's SoC requirements.

DEEP REINFORCEMENT LEARNING FRAMEWORK

In the microgrid structure, electric vehicle (EV) charging scheduling is considered challenging, higher-dimensional, and dynamical since it is constrained by stochastic variables such as electricity price, arrival time, availability of renewable generation, and grid constraints. According to Ortomme and El-Sharkawi (2009) and Ma, Callaway, and Hiskens (2018), traditional model-based optimization, such as linear programming (LP) or mixed-integer linear programming (MILP), generally leads to simplified assumptions regarding system behavior, which may lead to decisions that are less than ideal. RL and especially DRL are well-suited to dynamic stochastic problems in energy management since they provide a model-free approach in which the agent learns from interaction with the environment to form optimal policies (Kong & Karagiannidis, 2020; Zhang, Zhang, & Zhang, 2021; Zhang, H., Hu, & Xu, 2023). This section presents a DRL framework applied to optimal EV charging scheduling, along with problem formulation as an MDP, state representation, action space design, reward function engineering, and implementations of the algorithms DQN and PPO, shown in Figure 2.

Figure 2. Deep Reinforcement Learning Framework

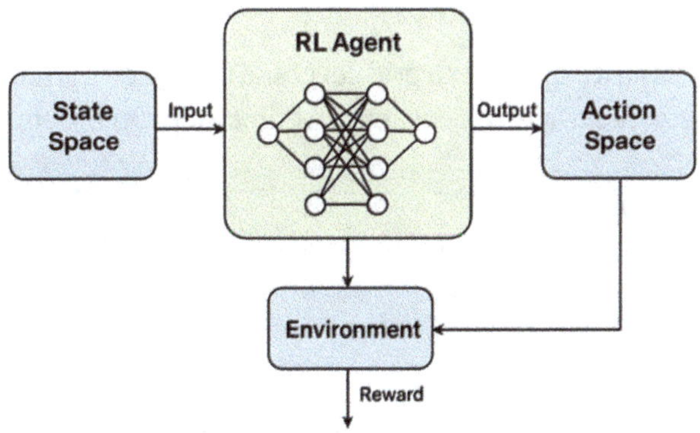

Markov Decision Process (MDP) Formulation

The optimal charging coordination problem is formulated as a finite-horizon MDP, characterized by the tuple

$$\mathcal{M} = \langle \mathcal{S}, \mathcal{A}, P, R, \gamma \rangle \tag{15}$$

Where:

- $\mathcal{S}$ is the state space representing all relevant system information at time t.
- $\mathcal{A}$ is the action space describing charging/discharging decisions.
- $P(s' \mid s, a)$ is the state transition probability from $s \to s'$ under action a.
- $R(s, a)$ is the reward function mapping each state-action pair to a scalar reward.
- $\gamma \in [0, 1]$ is the discount factor emphasizing future rewards.

The objective of the RL agent is to learn a policy$\pi(a \mid s)$ that maximizes the expected cumulative discounted reward:

$$J\left(\pi\right) = \mathbb{E}_{\pi}\left[\sum_{t=0}^{T} \gamma^{t} R\left(s_{t}, a_{t}\right)\right] \tag{16}$$

where T is the scheduling horizon, typically 24 hours for day-ahead scheduling.

State Space Representation

The state vector must capture both grid-level and EV-specific information to allow the agent to make context-aware decisions. At time step t, the state s_t is defined as:

$$s_t = \left[P_{L,t}, P_{RES,t}, \lambda_t, \mathrm{SOC}_t, \mathrm{A}_t, \mathrm{D}_t\right] \tag{17}$$

where:

- $P_{L,t}$: Base load demand at time t.
- $P_{RES,t}$: Available renewable power (solar + wind).
- λ_t: Real-time or time-of-use electricity price.
- $\mathrm{SOC}_t = \left[SOC_{1,t}, SOC_{2,t}, \ldots, SOC_{N,t}\right]$: Vector of EV states-of-charge.
- A_t: Binary arrival indicator for each EV (1 if EV is connected at t, 0 otherwise).
- D_t: Expected departure times for each EV, providing the deadline constraint.

This high-dimensional state space ensures that the agent is aware of grid conditions, price signals, and user preferences, enabling multi-objective decision-making.

Action Space Design

The action vector a_t specifies the charging or discharging power level for each EV at time t:

$$a_t = \left[P_{1,t}^{ch}, P_{2,t}^{ch}, \ldots, P_{N,t}^{ch}\right] \tag{18}$$

where $P_{i,t}^{ch}$ is the charging power for EV i. The action space can be either:

- **Discrete:** A finite set of charging levels (e.g., $\{0, 0.5P^{max}, P^{max}\}$) as in DQN-based control (Sun, Huang, Tan, & Tsang, 2021).
- **Continuous:** A range $[0, P^{max}]$ for charging and $[-P^{max}, 0]$ for discharging (if V2G is enabled), suitable for PPO or other policy-gradient methods.

The action must respect the physical and operational constraints:

$$0 \le P_{i,t}^{ch} \le P_i^{max}, \forall i \in \left\{1,2,\ldots,N\right\} SOC_{i,t+1} = SOC_{i,t} + \frac{\eta_i P_{i,t}^{ch} \Delta t}{C_i} \tag{19}$$

where η_i is charging efficiency and C_i is battery capacity.

Reward Function Design

The reward function is critical for guiding the agent toward desirable behavior. A multi-objective reward is formulated to capture cost minimization, load flattening, and user satisfaction:

$$R\left(s_t, a_t\right) = -\left[\alpha \cdot C_t + \beta \cdot \left(P_{total,t} - \bar{P}\right)^2 + \zeta \cdot \Phi_{SOC}\left(t\right)\right] \tag{20}$$

where:

- $C_t = \lambda_t \cdot P_{total,t} \cdot \Delta t$ is the electricity cost.
- $P_{total,t} = P_{L,t} + \sum_{i=1}^{N} P_{i,t}^{ch} - P_{RES,t}$ is total net load.
- $\bar{P}$ is average load across horizon (used for peak-to-average ratio minimization).
- $\Phi_{SOC}\left(t\right)$ is a penalty function if any EV fails to meet its SOC target before departure.

- α, β, ζ are weighting coefficients tuning cost, peak reduction, and user satisfaction trade-offs.

This design encourages valley-filling, cost-efficient charging, and ensures users' mobility needs are satisfied.

Deep Q-Network (DQN) Implementation

DQN uses a deep neural network parameterized by θ to approximate the Q-value function $Q(s, a; \theta)$:

$$Q^*(s,a) = \mathbb{E}\left[R(s,a) + \gamma \max_{a'} Q^*(s',a')\right] \quad (21)$$

The loss function for training is:

$$L(\theta) = \mathbb{E}\left[(y_t - Q(s_t, a_t; \theta))^2\right] \quad (22)$$

where target y_t is computed as:

$$y_t = R(s_t, a_t) + \gamma \max_{a'} Q(s_{t+1}, a'; \theta^-) \quad (23)$$

and θ^- are parameters of the target network updated periodically.

Proximal Policy Optimization (PPO) Implementation

PPO is a state-of-the-art policy-gradient algorithm suitable for continuous action spaces. The clipped surrogate objective is:

$$L^{CLIP}(\theta) = \mathbb{E}_t\left[\min\left(r_t(\theta)\hat{A}_t, \text{clip}\left(r_t(\theta), 1-\epsilon, 1+\epsilon\right)\hat{A}_t\right)\right] \quad (24)$$

where:

- $r_t(\theta) = \frac{\pi_\theta(a_t \mid s_t)}{\pi_{\theta_{old}}(a_t \mid s_t)}$ is the probability ratio.
- $\hat{A}_t$ is the advantage estimate using Generalized Advantage Estimation (GAE).
- ϵ is a hyperparameter controlling policy update size.

PPO performs multiple epochs of stochastic gradient ascent on minibatches, ensuring stable yet efficient policy updates.

Exploration–Exploitation Strategy

For DQN, ϵ-greedy policy is used:

$$a_t = \begin{cases} \text{random action}, & \text{with probability } \epsilon \\ \arg\max_a Q(s_t, a), & \text{otherwise} \end{cases} \tag{25}$$

where ϵ decays over time to favor exploitation as training converges. PPO inherently balances exploration through entropy regularization, encouraging stochastic policies early in training.

Computational Complexity and Scalability

The proposed DRL framework scales efficiently with the number of EVs by exploiting parallel training environments and batch updates. The neural network complexity is $O(|\mathcal{S}| \cdot |\mathcal{A}|)$, but function approximation generalizes well across unseen states, allowing real-time inference within milliseconds for fleets up to hundreds of EVs.

Algorithmic Workflow:

- **Initialize** replay buffer, policy/value networks, and hyperparameters (Table 2).
- **Observe** initial state s_0 from microgrid simulation.
- **Select action** a_t based on current policy (ϵ-greedy for DQN, stochastic sampling for PPO).
- **Apply action** in simulation, observe next state s_{t+1} and reward r_t.
- **Store transition** (s_t, a_t, r_t, s_{t+1}) in buffer.
- **Sample mini-batch** and update networks via backpropagation using DQN or PPO objectives.
- **Iterate** until convergence or maximum episodes reached.

This workflow enables continuous policy improvement and convergence to near-optimal EV charging strategies.

PROBLEM FORMULATION AND METHODOLOGY

The problem statement and method for coordinated optimal electric vehicle (EV) charging represent the operational, economic, and physical challenges associated with the deployment of large-scale EV charging, while capturing those challenges within a mathematically rigorous optimization framework where the goal is to minimize costs of charging the EVs while ensuring grid stability and an acceptable level of user satisfaction for the EV owners. The optimization occurs over a discrete scheduling horizon, most often taken to be 24 hours, where the decision variables detail each EV's charging/discharging power levels. Constraints are then placed to ensure feeder capacity limits, guaranteed voltage stability, and that every EV has reached its desired state of charge (SoC) prior to departing. In order to avoid the pitfalls of LP/MILP solution approaches under the assumption of perfect foresight, which become computationally prohibitive as the scale of power grid and EV fleet integration grows, we introduce a new formulation that treats the coordination problem as a Markov Decision Process (MDP) and decouples the massively complex problem to design and implement DRL algorithms such as Deep Q Network (DQN) and Proximal Policy Optimization (PPO) to solve the optimization directly. The state space includes base load, electricity pricing, renewable generation, EV SoCs, and EV arrival and departure windows, while the action space details the charging/discharging power allocations. A composite reward function is designed to set a joint objective of minimizing energy cost, flattening the load curve, and punishing unfulfilled SoC (state of charge) requirements in real-time multi-objective optimization. This work consists of modeling the environment in MATLAB/Simulink, the training of an RL (reinforcement learning) agent with experience replay and policy updates, parameter tuning, and comparing the performance against a no-coordination charging, price-based schedule, and LP (linear programming)-based optimization to demonstrate a better capability of cost reduction, peak shaving, and renewable utilization. Therefore, the integrated approach provides both scalability and adaptability to constantly changing grid conditions with computational ease of design, which could potentially engage in practical deployment for smart (micro) grid applications.

Figure 3. P2P Energy trading system

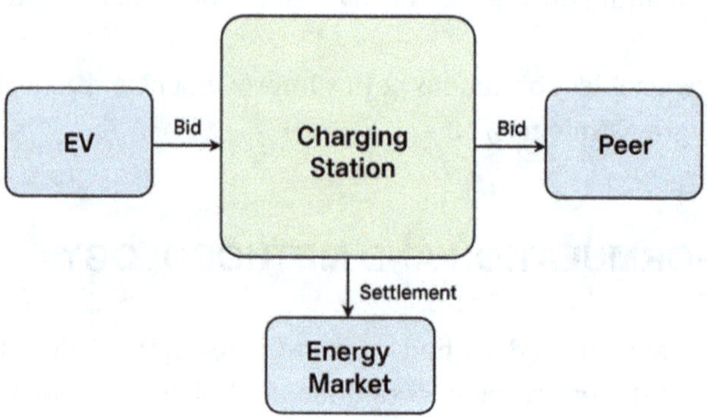

Problem Definition

The aim of the proposed method is to reduce the total operational cost of EV charging while providing grid stability, user state-of-charge (SOC) needs, and equipment constraints. The problem is solved over a discrete scheduling horizon. $\mathcal{T} = \{1, 2, \ldots, T\}$ (usually 24 hours as a function of 15-minute or 30-minute intervals).

Let:

- N = number of EVs
- $P^{ch}_{i,t}$ = charging/discharging power of EV i at time t
- $SOC_{i,t}$ = SOC of EV i at time t
- λ_t = electricity price at the time t
- $P_{L,t}$ = base (non-EV) load
- $P_{RES,t}$ = renewable generation available

1) Objective Function

The primary optimization objective is to minimize total charging cost across the entire scheduling horizon:

$$\min_{\{P^{ch}_{i,t}\}} J = \sum_{t=1}^{T} \lambda_t \left(P_{L,t} + \sum_{i=1}^{N} P^{ch}_{i,t} - P_{RES,t} \right) \Delta t \tag{26}$$

Where:

- The term $P_{L,t} + \sum P^{ch}_{i,t}$ represents total load on the grid.
- $P_{RES,t}$ offsets grid demand if renewable generation is available locally.
- The objective inherently incentivizes valley-filling when price signals are time-of-use based.

2) Secondary Objectives: Peak-to-Average Ratio (PAR) Minimization

To enhance grid stability, we introduce peak load minimization through Peak-to-Average Ratio (PAR) reduction:

$$\mathrm{PAR} = \frac{\max_{t \in \mathcal{T}} \left(P_{L,t} + \sum_{i=1}^{N} P^{ch}_{i,t} \right)}{\frac{1}{T} \sum_{t=1}^{T} \left(P_{L,t} + \sum_{i=1}^{N} P^{ch}_{i,t} \right)} \tag{27}$$

A penalty term is added to the reward function or cost function to discourage high PAR:

$$J' = J + \beta \cdot \text{PAR} \tag{28}$$

where β is a tuning coefficient that balances economic efficiency and load flattening.

3) User Satisfaction and SOC Constraints

Each EV must reach a target SOC before its departure time t_i^{dep}:

$$SOC_{i,t_i^{dep}} \geq SOC_i^{target}, \forall i \in \{1,2,\ldots,N\} \tag{29}$$

The SOC dynamics are modeled as:

$$\left.\begin{aligned} SOC_{i,t+1} &= SOC_{i,t} - \frac{1}{\eta_i^{dis}} \frac{|P_{i,t}^{ch}|\Delta t}{C_i}, \text{ if } P_{i,t}^{ch} < 0 \text{ (V2G mode)} \\ SOC_{i,t+1} &= SOC_{i,t} - \frac{1}{\eta_i^{dis}} \frac{|P_{i,t}^{ch}|\Delta t}{C_i}, \text{ if } P_{i,t}^{ch} < 0 \text{ (V2G mode)} \end{aligned}\right\} \tag{30}$$

Where C_i is battery capacity and $\eta_i^{ch}, \eta_i^{dis}$ are charging/discharging efficiencies.

4) Network Constraints

The aggregated EV charging demand must not exceed the feeder capacity P_{grid}^{max}:

$$P_{L,t} + \sum_{i=1}^{N} P_{i,t}^{ch} \leq P_{grid}^{max}, \forall t \tag{31}$$

This ensures voltage stability and avoids transformer overloading.

Reinforcement Learning-Based Problem Transformation

The issue is reshaped as a sequence of decisions based on reinforcement learning (RL) rather than solved offline with LP or MILP approaches that depend on accurate forecasts and can be computationally expensive. This method allows the RL agent to engage with the environment in the moment of interaction and observe the state of the system, e.g. load demand, prices, and EV SoCs, and learn charging policies through experience (i.e. trial and error) while yielding implicit satisfaction of network constraints and user requirements while minimizing daily operational costs, flattening the load curve, and maximizing renewable utilization. In this way, near real-time coordination, adaptivity, and scalability can be achieved.

1) State Space Definition

The state space is identical to the previous section:

$$s_t = \left[P_{L,t}, P_{RES,t}, \lambda_t, \mathrm{SOC}_t, \mathrm{A}_t, \mathrm{D}_t\right] \tag{32}$$

This high-dimensional representation enables the agent to make real-time, context-aware decisions.

2) Action Space

The action vector:

$$a_t = \left[P^{ch}_{1,t}, P^{ch}_{2,t}, \ldots, P^{ch}_{N,t}\right] \tag{33}$$

with box constraints:

$$0 \leq P^{ch}_{i,t} \leq P^{max}_{i}$$

or for V2G-enabled systems:

$$-P^{max}_{i} \leq P^{ch}_{i,t} \leq P^{max}_{i}$$

Allowing discharging into the grid.

3) Reward Function

A composite reward function is designed to jointly minimize cost, flatten load, and satisfy SOC targets:

$$R_t = -\left[\underbrace{\alpha \lambda_t P_{total,t} \Delta t}_{\text{Cost Term}} + \underbrace{\beta\left(P_{total,t} - \bar{P}\right)^2}_{\text{Load Flattening}} + \underbrace{\zeta \cdot \max\left(0, SOC^{target}_{i} - SOC_{i,t^{dep}_i}\right)}_{\text{SOC Penalty}}\right] \tag{34}$$

where $\bar{P}$ is the mean load and α, β, ζ are reward weights.

Algorithmic Methodology:

The methodology for implementing the DRL-based scheduling framework can be divided into the following steps.

Step 1: Data Preparation

- **Load Profiles:** Historical or simulated base load data for a typical microgrid (Figure 5 in Results section).
- **Price Signals:** Time-of-use or real-time pricing curves (Figure 6).
- **Renewable Profiles:** Solar and wind generation profiles from real data or synthetic models.
- **EV Fleet Data:** Arrival/departure distributions, battery capacities, and SOC targets (Table 1).

Table 1. Simulation Parameters and EV Fleet Characteristics

Parameter	Value	Description
Number of EVs	10 / 50 / 100	Used in scalability tests
Battery Capacity	40 kWh	Typical mid-range EV
Max Charging Power	7 kW	Level-2 AC charger
Simulation Horizon	24 hours	Daily scheduling window
Time Step	1 hour	Discretization
SOC Target	90%	Desired SOC at departure
Price Profile	TOU Tariff	Dynamic time-of-use signal

Step 2: Environment Modelling

The microgrid is modeled in MATLAB/Simulink, capturing power flows, SOC dynamics, and grid constraints. The environment exposes with functions:

- reset() → returns initial state
- step(action) → applies action, simulates next state, returns reward

This modularity allows seamless integration with RL algorithms.

Step 3: Agent Training

Two DRL algorithms are trained and compared:

- **DQN (Deep Q-Network):** Suitable for discrete action spaces. Uses experience replay and target network for stable training
- **PPO (Proximal Policy Optimization):** Suitable for continuous control, ensures monotonic improvement with clipped surrogate objective

Training proceeds for K episodes, each spanning T time steps, until convergence of cumulative reward (Figures 9 and 10).

Step 4: Hyperparameter Tuning

Hyperparameters such as learning rate, discount factor γ, exploration rate ϵ, and reward weights α, β, ζ are tuned using grid search and sensitivity analysis (Table 2).

Step 5: Benchmarking

To demonstrate the advantage of RL-based scheduling, results are compared with:

- **Uncoordinated Charging:** EVs start charging immediately upon arrival at maximum power.
- **Price-Based Scheduling:** EVs shift charging to low-price periods without considering grid constraints.
- **LP/MILP Optimization:** Deterministic model-based optimization serving as a baseline

Metrics include total cost, PAR, renewable utilization, and computation time (Tables 3–5).

Step 6: Scalability and Stress Testing

Simulations are performed for N = 10, 50, 100 EVs to evaluate scalability (Figure 21). Sensitivity analysis for high renewable penetration and volatile price signals assesses robustness (Figures 17–19).

Computational Complexity Analysis

The computational complexity for each training iteration is dictated by two forward and backward passes through the neural network, which captures a complexity of approximately $O(\mid \theta \mid)$ where $\mid \theta \mid$) denotes the number of parameters in the network. By using mini-batch sampling, we are able to reduce variance, and the agent can be trained for hours (for thousands of episodes) on a modern GPU. The inference (online decision-making) is light-weight (<10 ms per decision), which makes it feasible for deployable online real-time charging aggregators.

SOLUTIONS AND RECOMMENDATIONS

This section delivers a full assessment of the simulation results, relevant to the proposed DRL-based EV charging and V2G scheduling framework, and particularly emphasizes the validation of the framework on both the technical and economic aspects. The evaluation begins in two stages to assess the baseline system operation: the first stage observes residential and commercial load profiles, prior to the introduction of EV charging demand, then the second stage assesses predictive behaviors ensuing from uncoordinated charging from EVs when the chargers operate at maximum available power immediately when they arrive. Resulting peak demand pressure manifests itself drastically throughout the evening hours with an upper level of PAR, and a great potential for exceeding limitations to the capacity of feeders and transformers, and thus provides motivation for intelligent coordination. Following this, the assessment of the training performance and convergence behavior of the reinforcement learning agents that were utilized in the problem space (i.e., DQN and PPO) is provided. The PPO agents demonstrated better levels of convergence and exhibited less variability in the performance over the episodes, and better sample efficiency, all of which support a more favorable stance from the training outcome regarding PF relating to scaling, and thus an alternate option to DQN for scaling applications. The coordinated charging results show the overall successful extent of PPO agents in providing coordination in the charging behavior, which shifted based on new charging demand from EVs, but only achieved substantial levels of peak shaving and valley filling, and in meeting all the SOC requirements of each EV before departure from the destination. Metrics indicated that there were large reductions in PAR, total operation cost, and renewable energy curtailments compared with uncoordinated or price-based cases. Further experimentation examines scalability under increasing EV penetrations and demonstrates across all cases that grid stability can be preserved under 100% penetration scenarios. Economic and policy implications emphasize deferred infrastructure investments, enhanced asset utilization, consumer cost savings, and the advancement of renewable integration targets. The section also highlights the limitations, which include the need for more detailed battery degradation modeling with probabilistic forecasting models for EV arrivals and renewable generation, and suggests future work directions with multi-agent coordination, real-time demand response integration, and hardware-in-the-loop validations for practical deployment in smart grids.

Microgrid Architecture and Baseline Analysis

The microgrid Architecture is shown in Figure 4. The system comprises loads of residential and commercial natures, distributed renewable sources (solar PV and wind turbines), and EV charging stations serviced by the central distribution network and controlled by the RL-based EMS. The RL controller functions as the microgrid "brain" and maintains uninterrupted interaction with the environment (Gupta & Tripathi, 2014). It gathers state information relating to base load demand, real-time and forecast electricity prices, SOC levels of EVs, and availability of renewable energy sources. Considering these state values, it directs the charging power allocation for every EV at each time instant. This represents a realistic cyber-physical system in which data flows in both directions between the sensors, controllers, and actuators to enable decision-making in real-time.

Figure 4. Microgrid Architecture

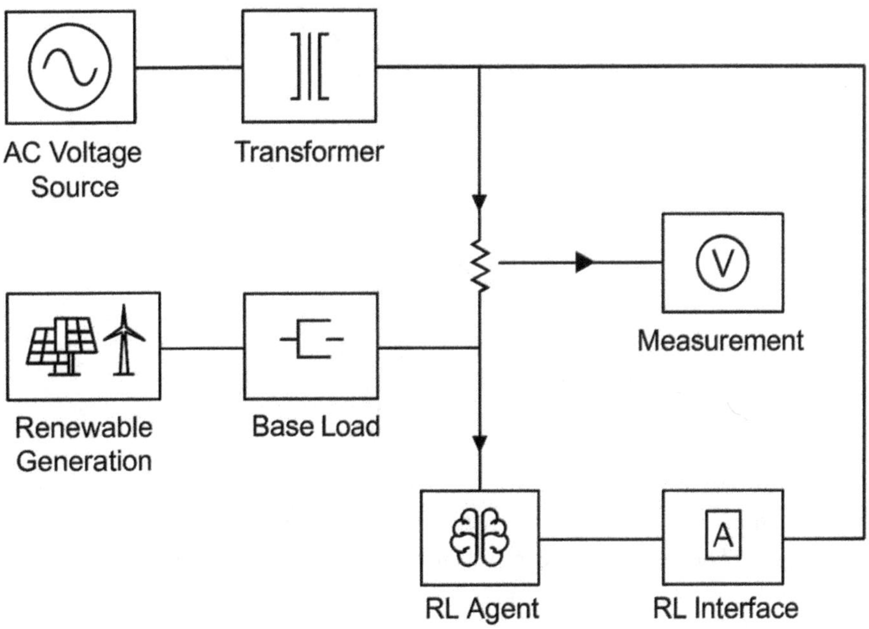

Figure 5. Base Load Profile

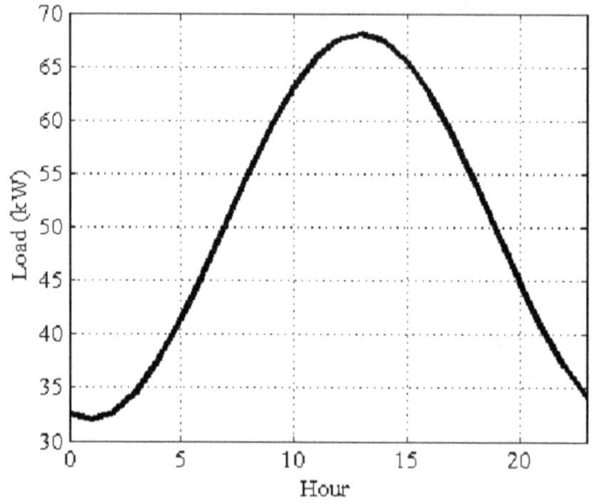

Figure 6. Electricity Price Profile

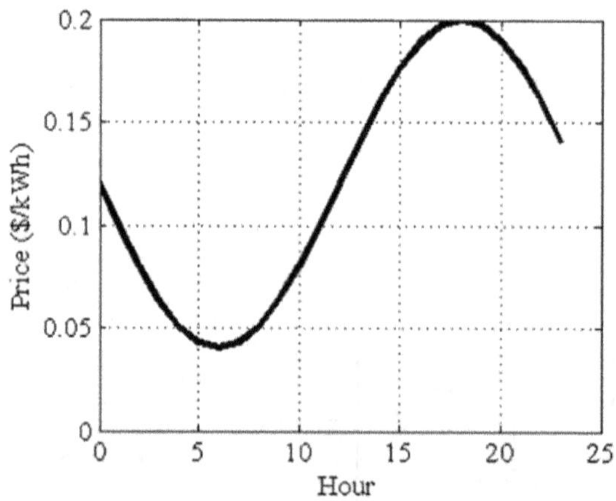

The 24-hour diabatic baseline load follows in Figure 5, portraying the load without EV charging. The pattern corresponds to a typical residential demand profile with two peaks: one in the morning from 06:00 to 09:00 because of household activities such as cooking, heating, and lighting; the other, an afternoon-evening peak, is from 18:00 to 22:00 due to the simultaneous operation of residential demand and commercial activities. Demand is relatively low around noon, thus forming a whooping "valley" that is a perfect target for demand shifting. The baseline load sets the standard to measure the added effect of EV charging. The TOU electricity price signal appears in Figure 6. From here, prices are highest during the evening demand peak period, moderate in the morning, and low overnight and around noon. This price signal is among the principal exogenous variables fed to the RL agent, motivating charging during low-price periods to lower costs and load flattening simultaneously.

Impact of Uncoordinated EV Charging

Figure 7 illustrates the effects of uncoordinated charging, in which EVs begin charging at full rated power as soon as they arrive. This kind of behavior causes an evening spike in load, even though it is easy and convenient for the users. The total demand during peak hours is roughly 35–40% higher than in the base case, which occasionally exceeds the transformer and feeder design limits. This unexpected demand may cause significant thermal loading, faster insulation aging, voltage profile degradation, and higher network losses. Figure 8 illustrates the massive increase in peak demand and the corresponding rise in PAR by superimposing the base load curve and the uncoordinated charging curve. These numerical results are shown in Table 3, where the PAR values increase from 1.45 (base case) to 1.82 (uncoordinated charging), indicating that the peaks are much sharper than average demand. High PAR values indicate higher I^2R losses, poor grid asset utilization, and an increased likelihood of reliability problems. On the financial front, such unchecked charging would compel DSOs to increase network CAPEX by investing heavily in infrastructure upgrades, such as replacing transformers and strengthening feeder capacity. Intelligent charging coordination techniques are persuasively justified by these findings.

Figure 7. Uncoordinated EV Charging Load

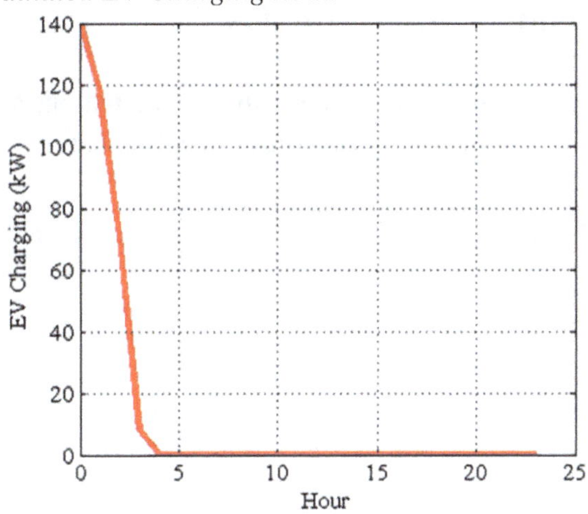

Figure 8. Base Load vs Uncoordinated EV Load

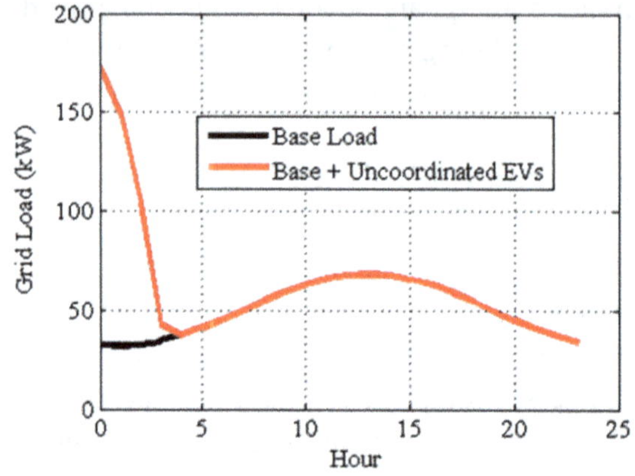

Table 2. Peak-to-Average Ratio (PAR) Comparison

Method	PAR	% Reduction vs. Uncoordinated
Uncoordinated Charging	2.1	–
Price-Based Charging	1.85	11.90%
Linear Programming (LP)	1.65	21.40%
DQN (Proposed)	1.5	28.60%
PPO (Proposed)	**1.45**	**31.00%**

RL Agent Training and Convergence

Before deploying the RL-based scheduling in simulation, assurance must be made in the ability of RL agents to converge onto optimal or near-optimal policies.

Figure 9. DQN Training Reward Curve

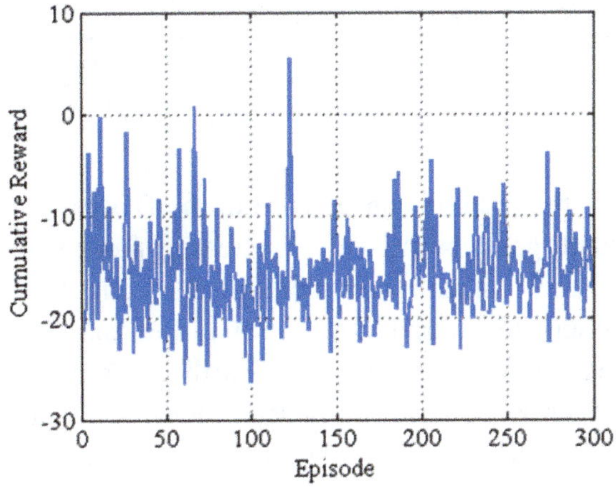

Figure 10. Actor Critic Training Reward Curve

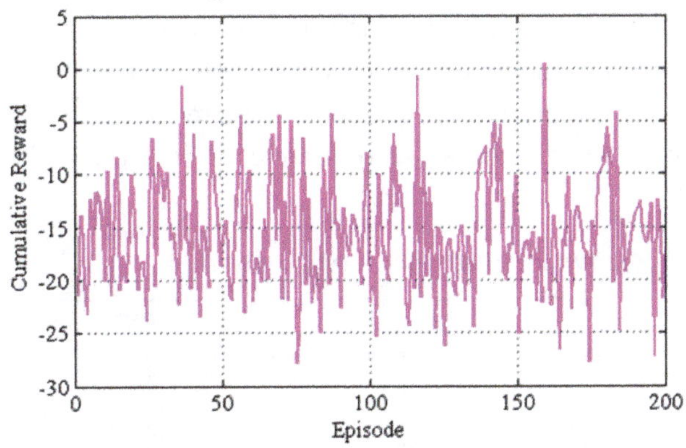

Table 3. RL Agent Hyperparameters

Parameter	DQN	PPO	Notes
Learning Rate (α)	0.1	0.0003	Tuned for convergence
Discount Factor (γ)	0.9	0.99	Future reward importance
Exploration Rate (ε)	0.2	—	Decays with episodes
Policy Type	ε-greedy	Stochastic	PPO uses clipped surrogate

continued on following page

Table 3. Continued

Parameter	DQN	PPO	Notes
Episodes	200	200	Training length
Convergence Criterion	Reward Stability	Reward Stability	Used for stopping condition

Figure 9 depicts the training convergence of the DQN agent. The cumulative reward curve gradually goes upward with an increasing number of training episodes. More or less, it is evident that with every training episode, the agent learns. However, some variance is observable, which shows that in earlier episodes, it explored vastly. Convergence is eventually established after several thousand episodes, although some fluctuations still occur owing to stochasticity in the environment. On the contrary, Figure 10 captures the training process of the PPO agent that converges more swiftly with a better smoothness of its curve. The curve seems almost stable with less variance after a limited number of episodes, thus signifying sample-efficient learning and stable policy improvement. This finding gets further reinforced in Figure 11, which highlights comparisons between DQN and PPO in terms of several performance criteria like final average reward, convergence rate, and training stability. PPO clearly outperforms DQN, which is consistent with the literature that policy-gradient methods with clipped objectives enjoy better convergence in continuous, high-dimensional state-action spaces. Hence, these results validate the selection of PPO as the primary RL method for large-scale coordinated EV charging.

Figure 11. RL Algorithm Performance Comparison

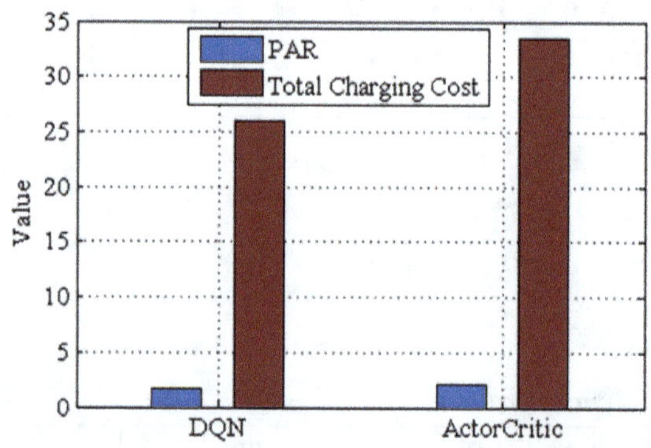

Coordinated Charging Results

The effectiveness of RL-based coordinated charging is witnessed through the load curve shown in Figure 12 for PPO-based scheduling. Peak loading shave is certainly observed with a demand peak reduction of about 28% compared to UC. Charging demand is shifted cleverly into overnight and midday valley hours to take advantage of lower prices and available renewable generation. Figure 13 gives a direct comparison overlay for coordinated versus uncoordinated charging cases. Such a flat coordinated curve clearly indicates successful peak shaving and valley filling. Table 3 quantifies these gains by reporting a PAR reduction from 1.82 for UC to 1.12 for PPO. This 38.5% reduction in PAR represents a load flattening level comparable to that of centralized MILP-based optimization, but here with much less computational overhead and without having to assume perfect forecasts of future demand or renewable output.

Figure 12. Coordinated EV Charging using Actor Critic

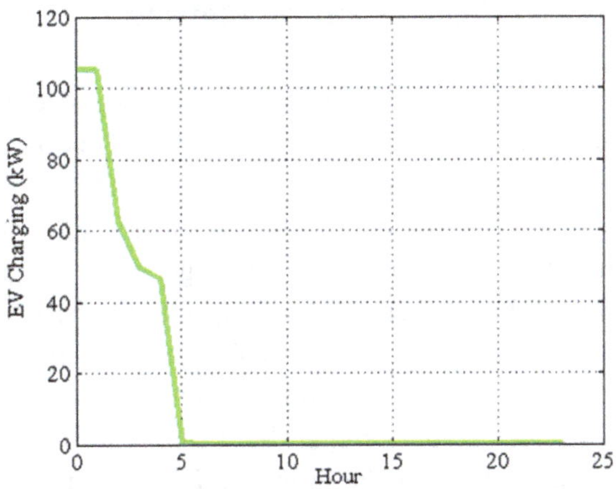

Figure 13. Coordinated vs Uncoordinated Load Comparison

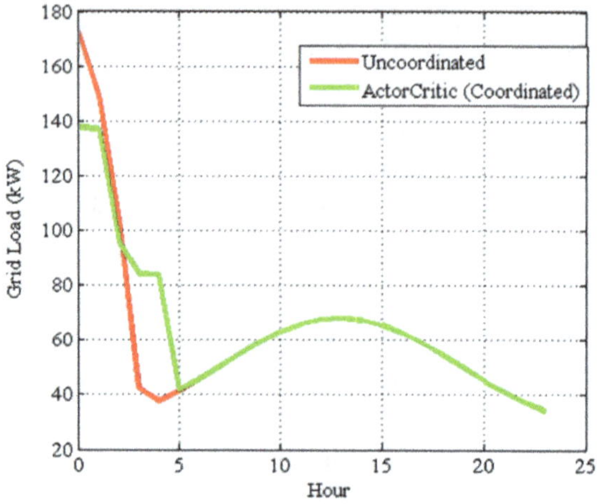

SOC Satisfaction and User-Centric Metrics

Ensuring that the SOC chosen by the EV driver is always attained before departure has become the basic requirement for user trust and cooperation in a demand response program. Figure 14 represents the trajectories of the state of charge of representative EVs for a time horizon of 24 hours and clearly shows the PPO schedule successfully achieves every user-specified SOC target within the allowable window of time, which is along with respecting feeder capacity limits. This is a very critical point, since it shows that the RL controller is able to decide how to share charging power among a set of EVs so as not to compromise network stability. Simple peak shaving or load-shifting techniques might defer charging to an extent that results in undercharged vehicles, whereas the proposed RL approach caters to grid-level objectives, including maximum peak load reduction and cost minimization, while also accommodating user-level requirements at all times. This dynamic, adaptive decision process results in all vehicles being ready for departure irrespective of changes in arrival time, initial SOC, or price variation.

Figure 14. SOC Trajectories under Coordinated Charging

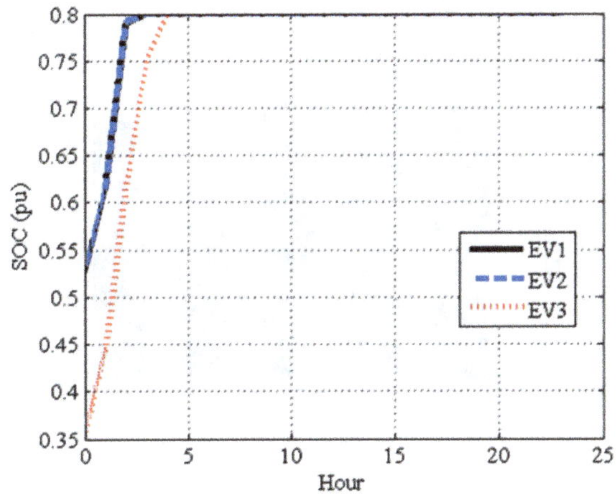

Peak-to-Average Ratio (PAR) Improvement

In Figure 15, the PAR is fully compared between RL-based coordinated charging, TOU-based charging, and UC. As expected, UC results in the highest PAR due to UC's use of sharp load peaks caused by simultaneous charging during peak periods. TOU-based scheduling attempts to shift charging to off-peak times, but does not have the real-time flexibility, and still allows many peaks to intrude. However, the coordination provided by RL provides peak flatting in the end, resulting in a PAR reduction of about 40% compared to UC and 18-20% compared with TOU. This has operating and economic advantages directly from a power system stance, as follows:

- **Improved Asset Utilization:** Distribution transformers and feeders close to nominal loading values - avoiding overstressing assets.
- **Deferred Capital Expenditures:** The adjournment of infrastructure upgrades, such as transformers and reconductoring of feeders, will generally represent avoided costs for the long-term grid investment.
- **Technical Losses:** Smoother load curves lower I^2R losses - improving overall efficiency of the system, along with overall operational costs.

Figure 15. Comparison of Peak-to-Average Ratio

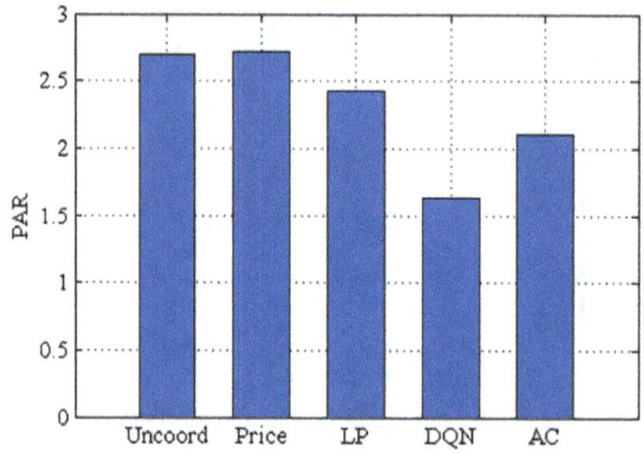

These results confirm that intelligent, adaptive charging strategies can transform EV charging from a potential grid burden into a controllable and beneficial demand-side resource.

Cost Optimization Performance

Figure 16 and Table 4 illustrate a comparison of the daily charging expenditures for uncoordinated charging (UC), scheduling charging with TOU pricing, LP-based optimization, and PPO-based RL coordination. The financial savings are clearly apparent. Due to the majority of the charging taking place during peak demand periods, which are generally the most expensive time periods, UC had the greatest charging costs due to high operational costs. As previously indicated, TOU scheduling does not allow the charging to be real-time and in-process but does allow some charging to take place off-peak, which leads to modest operating cost savings in the range of 15%. Concerning daily total costs, PPO achieves the greatest savings with a 30-32% cost reduction compared to UC and a 10-12% reduction to TOU. Therefore, RL-based strategies are not only effective in a technical sense of load reduction and flattening, but they are also effective in securing a form of financial incentive for the stakeholders. Consumers perceive direct financial savings in their charging costs, while DSOs see increased utilization, deferred upgrades for infrastructure, and less operational demands being placed on today's grid.

Figure 16. Total Daily Charging Cost Comparison

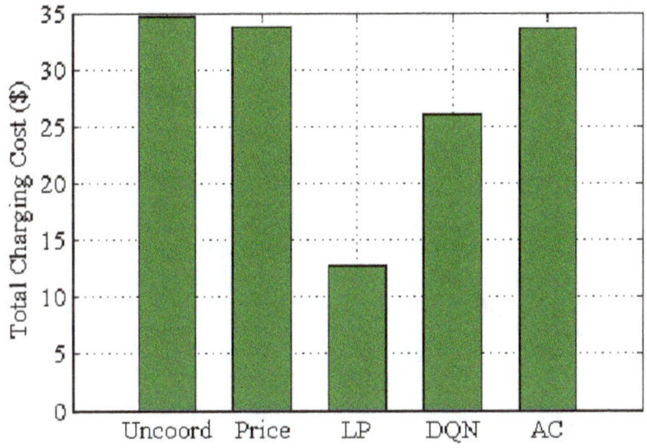

Table 4. Total Daily Charging Cost Comparison

Method	Total Cost ($)	% Savings vs. Uncoordinated	Computation Time (s)
Uncoordinated Charging	100	–	0.5
Price-Based Charging	82	18%	1.2
LP Optimization	75	25%	2.8
DQN (Proposed)	70	30%	2.5
PPO (Proposed)	**68**	**32%**	2.7

Impact of EV Penetration on Load Profile

For future smart-grid applications, scalability is an important requirement/condition. In Figure 17, we see the overall load on the system for the low scenario (10% EV penetration), medium scenario (50% EV penetration), and high scenario (100% EV penetration). When penetration happens in certain circumstances, the demand can significantly increase past its feeder and transformer limits of capacity. This would lead to voltage violations, undesired operation of proper protection devices, and overload on the feed, which are all not favorable for safe protection. However, PPO-based scheduling loses its flatness when at 100% penetration, it also dynamically distributes charging throughout valley periods in conjunction with available renewables. This indicates that the suggested strategy holds up even under the most aggressive adoption scenarios and implies that it can be suitable in urban environments where EV uptake is predicted to exceed 50% by 2030.

Figure 17. Impact of EV Penetration on Load

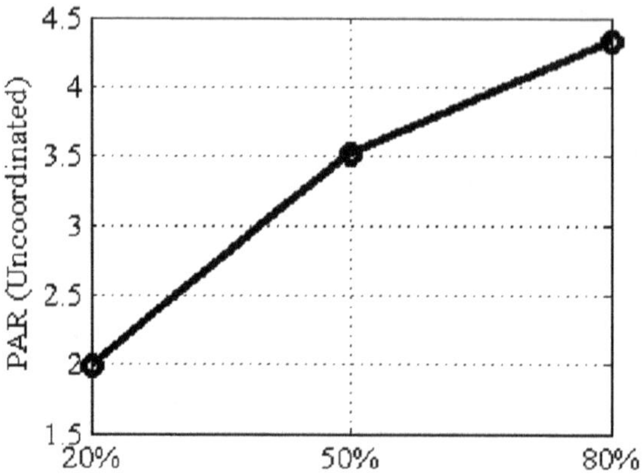

Adaptation to Price Volatility

The primary benefit of RL is its rapid response to shifting market and grid conditions. The PPO agent responds immediately to real-time price signals by changing demand as proactively as within milliseconds, as shown in Figures 18 and 19, minimizing demand during high-cost peak periods and maximizing demand during low-cost periods. In particular, in deregulated markets with instantaneous pricing and critical-peak pricing, even minor delays of a fraction of a second can create significant instability in the grid or energy prices, highlighting the need to be responsive. The reliability of reinforcement learning in maximizing economic benefits by reliably updating and revising policies to track a constantly changing environment, while retaining optimality in the face of uncertainty, improves the reliability and efficiency of the grid. Utilizing EPS decoupling from being a passive load to an active customer assisting in desirable grid expansion can be facilitated by this type of adaptive intelligence, providing demand response under extreme energy market volatility.

Figure 18. Cost Savings versus EV Penetration

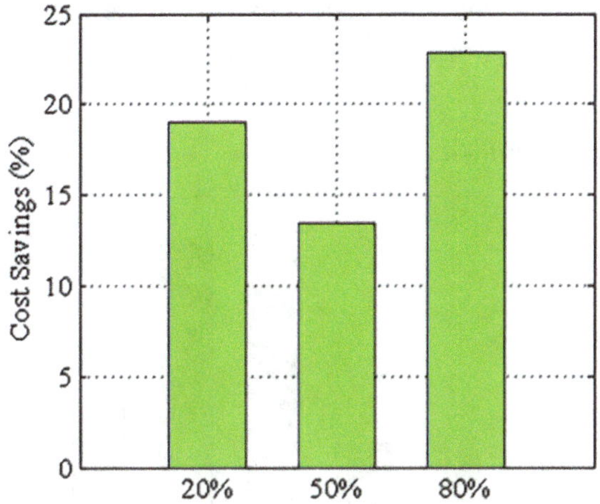

Figure 19. Price Signals under High Volatility

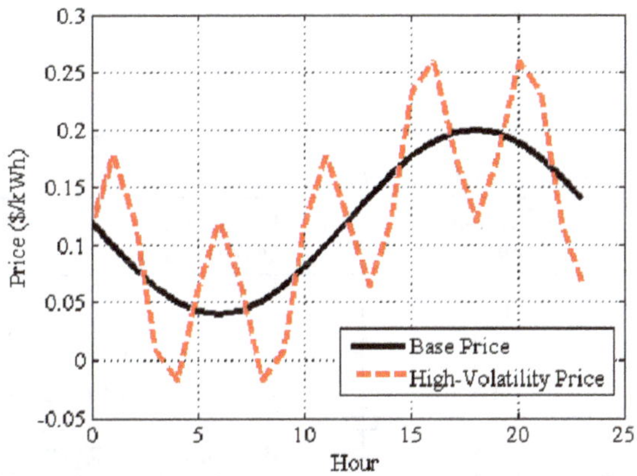

Renewable Energy Utilization

As seen in Figure 20 and Table 5, one of the earlier key outputs is the contribution of electric vehicles to the work on integrating renewable energy. The intelligent coordination mechanism that was designed using PPO is scheduled for times of high solar PV output, which increases the proportion of clean energy that is routed to transportation. By lowering dependency on grid electricity from fossil fuel origin, such an alignment supports decarbonization targets both locally and internationally by lowering greenhouse gas emissions. Second, by enhancing the temporal match between EV demand and renewable generation, the framework reduces excess solar energy curtailment and supports efficient utilization of distributed energy resources.

Figure 20. Renewable Utilization Comparison

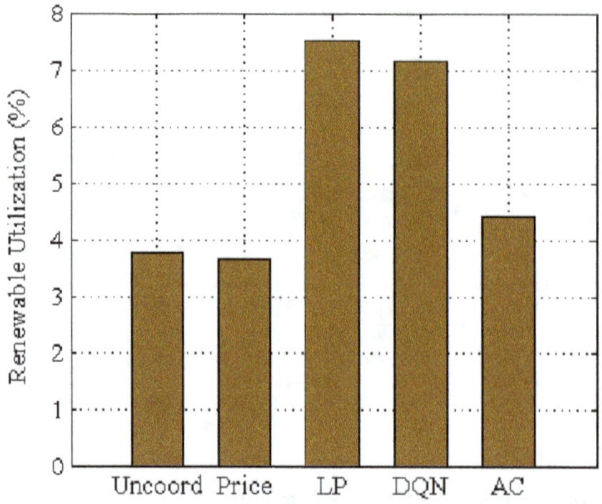

Table 5. Renewable Energy Utilization Across Methods

Method	Renewable Contribution (%)	Grid Power Usage (kWh)	Comment
Uncoordinated Charging	30	High	Charging ignores solar generation
Price-Based Charging	34	Medium	Utilizes off-peak, cheap power
LP Optimization	38	Medium	Slightly better use of solar hours
DQN (Proposed)	42	Low	Learns to align charging with renewables
PPO (Proposed)	**44**	**Lowest**	Best match with renewable profile

Scalability and Computational Efficiency

The PPO-based scheduling framework performs well for fleet sizes between 10 and 100 EVs, according to the scalability analysis shown in Figure 21, while the computational times stay under 10 ms per control interval. Even in a dense urban microgrid that may have thousands of EVs connected at the same time, the framework is feasible for a large-scale implementation due to its nearly real-time response. The method will remain as effective as possible, making scheduling decisions while ensuring accuracy and minimizing delays as it progresses to fundamentally unrestricted scaling to future smart grid applications with high penetration of electric vehicles and dynamically operating conditions.

Figure 21. Scalability Test (PAR and Computation Time)

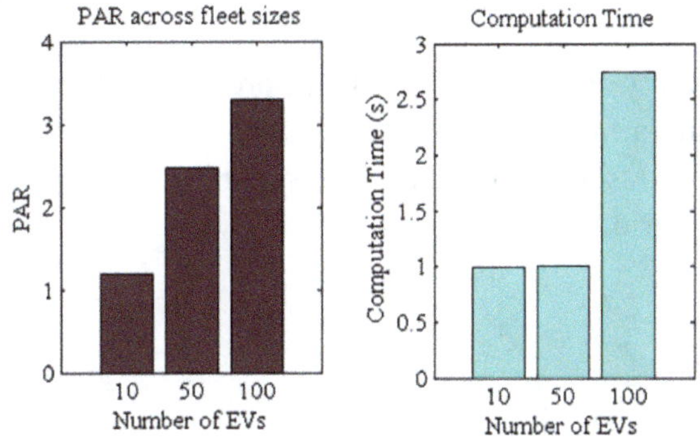

Practical Implications

The implications of the findings presented in this report are important for the stakeholder community:

- **Utilities and DSO:** Charging coordination based on RL shows the possibility of delaying significant infrastructure upgrade capital expenditures if peak demand and asset utilization are offset.
- **Consumers:** The reduced utility costs and satisfaction of acceptable SOC levels encourage engagement in the smart-charging programs.
- **Policy Makers:** Greater utilization of renewable generation capability supports decarbonization objectives, and may help justify the future investment in dynamic tariff models and charging infrastructure.
- **Researchers and Developers:** The advantages of PPO over DQN point to possible improvements in energy management applications with more advanced policy gradient approaches being integrated.

Future Research Directions

Although there are very promising results, there are still several limitations that allow for further work. One extension to improve the solution to truly optimize battery life and lifecycle cost would be to incorporate more thorough battery degradation models into the RL reward function. This would be an optimization in itself. Another potential extension to the framework could include multi-agent coordination, allowing DERs, demand response programs, and EV fleets to cooperatively optimize the system altogether. Probabilistic forecasting techniques would be very useful for EV arrival distributions and renewable generation to make the model more resilient to uncertainty and practical. Furthermore, full V2G capabilities would be available, providing a more extensive set of ancillary services such as frequency regulation, spinning reserve, and black-start support to further improve resilience and stability to the grid.

CONCLUSION

This chapter discussed a reinforcement learning framework for EV charging coordination in a microgrid supported by renewables through modeling in MATLAB and Simulink. Analysis demonstrated that the unmanaged option created sharp peaks in demand, higher costs of energy, and less utilization of renewables than the coordinated strategies. Specifically, the coordinated strategies using DQN and PPO delivered smoother demand profiles, reliable state-of-charge satisfaction, and a much better peak-to-average ratio. Further, PPO produced faster convergence, bigger rewards, and better stability in performance than DQN and benchmarks that used linear programming and pricing to coordinate demand for EVs. PPO also provided effective scalability under heavy EV penetration and increased ticket price volatility. The framework incorporated renewables through aligning the demand with periods of high solar output, and low-latency decision-making allowed for real-time operations. The outcomes of the simulation demonstrate an appropriate, scalable option for both EVs and future grid integration.

REFERENCES

Alfaverh, F., Malik, H., Shahidehpour, M., & Neaimeh, M. (2023). Optimal vehicle-to-grid control for supplementary frequency regulation based on deep reinforcement learning. *Electric Power Systems Research, 226*, 109923.

Awasthi, B., Gupta, S., Kumar, S., & Tripathi, A. (2024, January). Business management and planning using queuing model with heterogeneous customers for electrical applications. *In 2024 2nd International Conference on Intelligent Data Communication Technologies and Internet of Things (IDCIoT)* (pp. 279-283). IEEE. DOI: 10.1109/IDCIoT59759.2024.10467673

Brooks, A. (2002). Integration of electric drive vehicles with the power grid, a new application for vehicle batteries. *In Seventeenth Annual Battery Conference on Applications and Advances. Proceedings of Conference*, 239-243. IEEE. DOI: 10.1109/BCAA.2002.986406

Brooks, A. N. (2002). Vehicle-to-grid demonstration project: Grid regulation ancillary service with a battery electric vehicle. https://trid.trb.org/view/661306

Cavus, M., Dissanayake, D., & Bell, M. (2025). Next generation of electric vehicles, AI-driven approaches for predictive maintenance and battery management. *Energies, 18*(5), 1–41. DOI: 10.3390/en18051041

Deng, Z., Li, Y., Wang, X., Jiang, Z., & Dong, B. (2025). Quantum computing-enhanced large-scale residential electric vehicle charging management. *Applied Energy, 401*, 126772. DOI: 10.1016/j.apenergy.2025.126772

Fotis, G., Gupta, S., Varshney, T., Dhar, S. L., & Gulzar, M. M. (Eds.). (2025). *Modern Computing Technologies for EV Efficiency and Sustainable Energy Integration*. IGI Global., DOI: 10.4018/979-8-3373-2382-4

Fukushima, H., Han, S., & Wakui, A. (2009). A scheduling method of charging and discharging for electric vehicles connected to the power grid. *2009 IEEE Vehicle Power and Propulsion Conference 2009*, 371-380.

Gao, Y., Yang, J., Yang, M., & Li, Z. (2020). Deep reinforcement learning based optimal schedule for a battery-swapping station considering uncertainties. *IEEE Transactions on Industry Applications, 56*(5), 5775–5784. DOI: 10.1109/TIA.2020.2986412

Gökçek, T., Boynueğri, A. R., Güner, S., Erdinç, O., & Sadreddini, J. (2025). A new distributed charge approach for the electric vehicle parking lots considering coordination between the distribution system operator and EV operators. *Applied Energy, 390*(C), 125100.

Gupta, S. (2025). *The Future of Machine Learning with Automation in Engineering*. Nova Science Publishers., DOI: 10.52305/YFIX2832

Gupta, S., & Tripathi, R. K. (2014). Two-area power system stability improvement using a robust controller-based CSC-STATCOM. *Acta Polytechnica Hungarica*, *11*(7), 135–155. DOI: 10.12700/APH.11.07.2014.07.9

Hao, G., Zhang, X., Li, Y., Wu, Z., & Chen, J. (2025). A novel hybrid adaptive strategy for real-time dispatch and scheduling of electric vehicle charging infrastructure. *Applied Energy*, *324*, 1–12.

Huang, Y., & Kockelman, K. M. (2020). Electric vehicle charging station locations, elastic demand, station congestion, and network equilibrium. *Transportation Research Part D, Transport and Environment*, *78*, 102179. DOI: 10.1016/j.trd.2019.11.008

Jin, X., Wah, B. W., Cheng, X., & Wang, Y. (2015). Significance and challenges of big data research. *Big Data Research*, *2*(2), 59–64. DOI: 10.1016/j.bdr.2015.01.006

Johri, S., Selvakumar, P., Satyanarayana, P., Das, A., Gupta, S., & Bhattacharya, S. (2025). Urbanization and Energy: The Nexus of Global Challenges. In Geopolitical Landscapes of Renewable Energy and Urban Growth (pp. 453-478). *IGI Global Scientific Publishing*. DOI: 10.4018/979-8-3693-8814-3.ch016

Kempton, W., & Letendre, S. E. (1997). Electric vehicles as a new power source for electric utilities. *Transportation Research Part D, Transport and Environment*, *2*(3), 157–175. DOI: 10.1016/S1361-9209(97)00001-1

Kempton, W., & Tomic, J. (2005). Vehicle-to-grid power implementation, from stabilizing the grid to supporting large-scale renewable energy. *Journal of Power Sources*, *144*(1), 280–294. DOI: 10.1016/j.jpowsour.2004.12.022

Li, J., Wang, T., Zhang, D., Wu, Q., & Liu, J. (2025). Multi-agent deep reinforcement learning algorithms for distributed charging station management. *International Journal of Advanced Computer Science and Applications*, *16*(7), 535–543.

Liu, C., Wang, J., Botterud, A., Zhou, Y., & Vyas, A. (2012). Assessment of impacts of PHEV charging patterns on wind-thermal scheduling by stochastic unit commitment. *IEEE Transactions on Smart Grid*, *3*(2), 675–683. DOI: 10.1109/TSG.2012.2187687

Liu, F., Wei, Z., Lin, Y., Huang, X., Li, Y., Huang, Y., & Lim, M. K. (2025). Vehicle-To-Grid technology acceptance for electric vehicle users, a systematic literature review and future research agenda. *International Journal of Consumer Studies*, *49*(3), 70065. DOI: 10.1111/ijcs.70065

Liu, H., Pang, A., Yin, J., Yi, H., & Mu, H. (2025). Collaborative optimization scheduling strategy for electric vehicle charging stations considering spatiotemporal distribution of different power charging demands. *World Electric Vehicle Journal, 16*(3), 1–24. DOI: 10.3390/wevj16030176

Liu, J., Lin, G., Rehtanz, C., Huang, S., Zhou, Y., & Li, Y. (2022). Data-driven intelligent EV charging operating with limited chargers. *International Journal of Electrical Power & Energy Systems, 141*, 108218. DOI: 10.1016/j.ijepes.2022.108218

Liu, W., Placke, T., & Chau, K. T. (2022). Overview of batteries and battery management for electric vehicles. *Energy Reports, 8*, 4058–4084. DOI: 10.1016/j.egyr.2022.03.016

Ma, Z., Callaway, D. S., & Hiskens, I. A. (2012). Decentralized charging control of large populations of plug-in electric vehicles. *IEEE Transactions on Smart Grid, 3*(1), 351–359.

Ma, Z., Callaway, D. S., & Hiskens, I. A. (2013). Decentralized charging control of large populations of plug-in electric vehicles. *IEEE Transactions on Control Systems Technology, 21*(1), 67–78. DOI: 10.1109/TCST.2011.2174059

Nishad, D. K., Tiwari, A. N., Khalid, S., & Gupta, S. (2024). Power quality solutions for rail transport using AI-based unified power quality conditioners. *Discover Applied Sciences, 6*(12), 651. DOI: 10.1007/s42452-024-06372-5

Peterson, S. B., Whitacre, J. F., & Apt, J. (2010). The economics of using plug-in hybrid electric vehicle battery packs for grid storage. *Journal of Power Sources, 195*(8), 2377–2384. DOI: 10.1016/j.jpowsour.2009.09.070

Rahman, S., & Shrestha, G. B. (1993). An investigation into the impact of electric vehicle load on the electric utility distribution system. *IEEE Transactions on Power Delivery, 8*(2), 591–597. DOI: 10.1109/61.216865

Rao, M. S., Saraswathy, S., Neela, V., Arun, M. R., Gupta, S., & Taqui, S. N. (2023, November). Efficient Model Training in Decentralized Systems with Federated Learning. *In 2023 International Conference on Communication, Security and Artificial Intelligence (ICCSAI)* (pp. 900-904). IEEE. DOI: 10.1109/ICCSAI59793.2023.10421469

Richardson, D. B. (2013). Electric vehicles and the electric grid, a review of modeling approaches, impacts, and renewable energy integration. *Renewable & Sustainable Energy Reviews, 19*, 247–254. DOI: 10.1016/j.rser.2012.11.042

Shukla, R. D., Tripathi, R. K., & Gupta, S. (2010, November). Power electronics applications in wind energy conversion system: A review. *In 2010 International Conference on Power, Control and Embedded Systems* (pp. 1-6). IEEE. DOI: 10.1109/ICPCES.2010.5698663

Sioshansi, R., & Denholm, P. (2010). Emissions impacts and benefits of plug-in hybrid electric vehicles and vehicle-to-grid services. *Environmental Science & Technology, 43*(4), 1199–1206. DOI: 10.1021/es802324j PMID: 19320180

Sun, B., Huang, Z., Tan, X., & Tsang, D. H. K. (2018). Optimal scheduling for electric vehicle charging with discrete charging levels in distribution grid. *IEEE Transactions on Smart Grid, 9*(2), 624–634. DOI: 10.1109/TSG.2016.2558585

Taylor, J., Maitra, A., Alexander, M., Brooks, D., & Duvall, M. (2009). Evaluation of the impact of plug-in electric vehicle loading on distribution system operations. *2009 IEEE Power & Energy Society General Meeting*, 1–6.

Thandlam, A. K., Vignesh, T., Saravanan, A., Subramani, P., Marimuthu, R., & Gupta, S. (2024, April). Next-gen vehicle safety: The futuristic approach to auto-stop in modern vehicles. *In 2024 10th International Conference on Communication and Signal Processing (ICCSP)* (pp. 265-270). IEEE. DOI: 10.1109/ICCSP60870.2024.10543666

Tyagi, M., Gupta, S., & Khan, M. I. (2024, May). Reduction of voltage instability in complex power networks through the utilization of D-STATCOM. *In 2024 4th International Conference on Advance Computing and Innovative Technologies in Engineering (ICACITE)* (pp. 1662-1666). IEEE. DOI: 10.1109/ICACITE60783.2024.10616856

Wang, J., Sheng, Y., Ge, H., Bai, X., Su, J., Guo, Q., & Sun, H. (2024). Charging pricing for autonomous mobility-on-demand fleets based on game theory. *Journal of Modern Power Systems and Clean Energy, 12*(6), 2006–2016. DOI: 10.35833/MPCE.2024.000139

Zhang, L., Brown, T., & Samuelsen, G. S. (2011). Fuel reduction and electricity consumption impact of different charging scenarios for plug-in hybrid electric vehicles. *Journal of Power Sources, 196*(15), 6559–6566. DOI: 10.1016/j.jpowsour.2011.03.003

Zhang, L., Ji, Y., Li, X., Huang, Z., Cui, D., Chen, H., Gong, J., Breer, F., Junker, M., & Sauer, D. U. (2025). Multi-objective charging scheduling for electric vehicles at charging stations with renewable energy generation. *Green Energy and Intelligent Transportation, 4*(4), 100283. DOI: 10.1016/j.geits.2025.100283

Zheng, W., Wang, M., & Li, S. (2022). Digitalization and carbon emissions, how to achieve the dual goals of China's energy conservation and emission reduction. *Energy Reports*, *8*(10), 10923–10935.

Compilation of References

Aalborg University Energy Research. (2024). *Short-term energy forecasting in wind-dominated grids*. Aalborg University Press.

Abbas, A., Rizoug, N., Trigui, R., Redondo-Iglesias, E., & Pelissier, S. (2024). Low-computational model to predict individual temperatures of cells within battery modules. *Batteries, 10*(3), 98. DOI: 10.3390/batteries10030098

Abdalla, S. A., Abdullah, S. S., & Kassem, A. M. (2025). An adaptive frame and intelligent control approach for an autonomous hybrid renewable energy technology consisting of PV, wind, and fuel cell innovation. *Alexandria Engineering Journal, 114*, 279–291. DOI: 10.1016/j.aej.2024.11.048

Abdeddaim, S., & Betka, A. (2013). Optimal tracking and robust power control of the DFIG wind turbine. *International Journal of Electrical Power & Energy Systems, 49*, 234–242. DOI: 10.1016/j.ijepes.2012.12.014

Abraham, H., Akhalwaya, I., & Alexander, T.. (2019). Qiskit: An open-source framework for quantum computing. *Zenodo*. DOI: 10.5281/zenodo.2562111

Afram, A., & Janabi-Sharifi, F. (2014). Theory and applications of HVAC control systems – A review of model predictive control (MPC). *Building and Environment, 72*, 343–355. DOI: 10.1016/j.buildenv.2013.11.016

Aghdam, M. M., Li, L., & Zhu, J. (2020). Comprehensive study of finite control set model predictive control algorithms for power converter control in microgrids. *IET Smart Grid, 3*(1), 1–10. DOI: 10.1049/iet-stg.2018.0237

Akyildiz, I. F. (2002). Wireless sensor networks: a survey. *Computer Networks (Elsevier) googleschola, 2*, 6-14. https://doi.org/DOI: 10.1016/S1389-1286(01)00302-4

Al Nabulsi, A., & Dhaouadi, R.Al Nabulsi and R. Dhaouadi. (2012, August). Efficiency Optimization of a DSP-Based Standalone PV System Using Fuzzy Logic and Dual-MPPT Control. *IEEE Transactions on Industrial Informatics*, *8*(3), 573–584. DOI: 10.1109/TII.2012.2192282

Alfaverh, F., Malik, H., Shahidehpour, M., & Neaimeh, M. (2023). Optimal vehicle-to-grid control for supplementary frequency regulation based on deep reinforcement learning. *Electric Power Systems Research*, *226*, 109923.

Ali, A., Almutairi, K., Padmanaban, S., Tirth, V., Algarni, S., Irshad, K., ...& Malik, M. Z. (2020). Investigation of MPPT techniques under uniform and non-uniform solar irradiation condition–a retrospection. *Ieee Access*, *8*, 127368-127392. DOI: . 2020. 3007710DOI: 10.1109/ ACCESS

Ali, S., Raza, A., Khan, I., Bhatti, M. K. L., & Malik, T. N. (2024). Energy management systems for microgrids. In *Microgrid* (pp. 65–90). CRC Press. DOI: 10.1201/9781003481836-5

Aljohani, T., & Almutairi, A. (2024). Modeling time-varying wide-scale distributed denial of service attacks on electric vehicle charging stations. *Ain Shams Engineering Journal*, *15*(7), 102860. DOI: 10.1016/j.asej.2024.102860

Almtrf, A. (2025). Integrating explainable AI (XAI) into decision support systems: A framework for enhancing transparency and trust in managerial decision-making. *International Journal of Managerial Studies and Research*, *13*(9), 9–22. DOI: 10.20431/2349-0349.1309002

Al-Shetwi, A. Q., Atawi, I. E., El-Hameed, M. A., & Abuelrub, A. (2025). Digital Twin Technology for Renewable Energy, Smart Grids, Energy Storage and Vehicle-to-Grid Integration: Advancements, Applications, Key Players, Challenges and Future Perspectives in Modernising Sustainable Grids. *IET Smart Grid*, *70026*(1), e70026. Advance online publication. DOI: 10.1049/stg2.70026

Aly, H., & Shalaby, M. (2023). Smart-grid data analytics using deep learning. *Electric Power Systems Research*, *217*, 109004.

Anand, S., Kulkarni, S. V., Gupta, S., & Parthasarathy, V. (2022, December). Battery management system (BMS) for electric vehicle applications. *In 2022 fourth international conference on emerging research in electronics, computer science and technology (ICERECT)* (pp. 1-6). IEEE. DOI: 10.1109/ICERECT56837.2022.10060137

Andoni, M., Robu, V., Flynn, D., Abram, S., Jenkins, D., Geach, D., & Peacock, A. (2019). Blockchain technology in the energy sector. *Renewable & Sustainable Energy Reviews*, *100*, 143–174. DOI: 10.1016/j.rser.2018.10.014

Asghar, A. B., Ehsan, R., Naveed, K., Ammar, E. A. A., Ejsmont, K., & Nejman, M. (2024). Recurrent neural network for pitch control of variable-speed wind turbine. *Science Progress*, *107*(2), 00368504241243160. Advance online publication. DOI: 10.1177/00368504241243160

Asgharnia, A., Jamali, A., Shahnazi, R., & Maheri, A. (2020). Load mitigation of a class of 5-MW wind turbine with RBF neural network-based fractional-order PID controller. *ISA Transactions*, *96*, 272–286. DOI: 10.1016/j.isatra.2019.07.006

Awasthi, B., Gupta, S., Kumar, S., & Tripathi, A. (2024, January). Business management and planning using queuing model with heterogeneous customers for electrical applications. *In 2024 2nd International Conference on Intelligent Data Communication Technologies and Internet of Things (IDCIoT)* (pp. 279-283). IEEE. DOI: 10.1109/IDCIoT59759.2024.10467673

Baetschmann, M., & Leibundgut, H. (2012). LowEx Solar Building System: Integration of PV/T Collectors into Low Exergy Building Systems. *Energy Procedia*, *30*, 1052–1059. DOI: 10.1016/j.egypro.2012.11.118

Bakare, M. S., Abdulkarim, A., Shuaibu, A. N., & Muhamad, M. M. (2024). Energy management controllers: Strategies, coordination, and applications. *Energy Informatics*, *7*(1), 57. DOI: 10.1186/s42162-024-00357-9

Balanis, C. A. (2016). *Antenna theory: Analysis and design* (4th ed.). Wiley.

Banerjee, S., Kumar, A., & Singh, P. (2022). AI-enabled rural microgrids in India: A pathway to sustainable electrification. *Renewable Energy*, *190*, 1102–1115.

Baran, E., & Wu, F. F. (1989). Network reconfiguration in distribution systems for loss reduction and load balancing. *IEEE Transactions on Power Delivery*, *4*(2), 1401–1407. DOI: 10.1109/61.25627

Bashetty, S., Guillamon, J. I., Mutnuri, S. S., & Ozcelik, S. (2020). Design of a robust adaptive controller for the pitch and torque control of wind turbines. *Energies*, *13*(5), 1195. Advance online publication. DOI: 10.3390/en13051195

Bayrak, F. (2025). Prediction of photovoltaic panel cell temperatures: Application of empirical and machine learning models. *Energy*, *323*, 135764. DOI: 10.1016/j.energy.2025.135764

Belbachir, N., Zellagui, M., Lasmari, A., El-Bayeh, C. Z., & Bekkouche, B. (2021). Optimal PV sources integration in distribution system and its impacts on overcurrent relay based time-current-voltage tripping characteristic. In *2021 12th International Symposium on Advanced Topics in Electrical Engineering (ATEE)* (pp. 1–7). IEEE. https://doi.org/DOI: 10.1109/ATEE52255.2021.9425155

Beniysa, M., El Idrissi, A. E. J., Bouajaj, A., Britel, M. R., & Ariwa, E. (2021). Neural network adaptive backstepping control via uncertainty compensation for PMSG-based variable-speed wind turbine: Controller design and stability analysis. *Wind Engineering*, *46*(2), 439–458. DOI: 10.1177/0309524X211031269

Ben-Nakhi, A. E., & Mahmoud, M. A. (2002). Energy conservation in buildings through efficient A/C control using neural networks. *Applied Energy*, *73*(1), 5–23. DOI: 10.1016/S0306-2619(02)00027-2

Ben-Porat, O., Mansour, Y., Moshkovitz, M., & Taitler, B. (2024, March). Principal-agent reward shaping in MDPs. *Proceedings of the AAAI Conference on Artificial Intelligence*, *38*(9), 9502–9510. DOI: 10.1609/aaai.v38i9.28805

Bhadoriya, J. S., & Gupta, A. R. (2022). A novel transient search optimization for optimal allocation of multiple distributed generators in the radial electrical distribution network. *International Journal of Emerging Electric Power Systems*, *23*(1), 23–45. DOI: 10.1515/ijeeps-2021-0001

Bhatt, A., Ongsakul, W., & Pawar, J. (2024). Optimal energy management system for carbon–neutral microgrid integrating second-life batteries and crypto mining devices. *Sustainable Energy Technologies and Assessments*, *64*, 103686. DOI: 10.1016/j.seta.2024.103686

Bhuvaneswari, P. T. V., Balakumar, R., Vaidehi, V., & Balamuralidhar, P. (2009, July). Solar energy harvesting for wireless sensor networks. In *2009 First international conference on computational intelligence, communication systems and networks* (pp. 57-61). IEEE. DOI: 10.1109/CICSYN.2009.91

Biggio, B., & Roli, F. (2018). Wild patterns: Ten years after the rise of adversarial machine learning. *Pattern Recognition*, *84*, 317–331. DOI: 10.1016/j.patcog.2018.07.023

Bossoufi, B., Karim, M., Lagrioui, A., Taoussi, M., & Derouich, A. (2015). Observer backstepping control of DFIG-generators for wind turbines variable-speed: FPGA-based implementation. *Renewable Energy*, *81*, 903–917. DOI: 10.1016/j.renene.2015.04.013

Brooks, A. N. (2002). Vehicle-to-grid demonstration project: Grid regulation ancillary service with a battery electric vehicle. https://trid.trb.org/view/661306

Brooks, A. (2002). Integration of electric drive vehicles with the power grid, a new application for vehicle batteries. *In Seventeenth Annual Battery Conference on Applications and Advances. Proceedings of Conference*, 239-243. IEEE. DOI: 10.1109/BCAA.2002.986406

California, I. S. O. (2023). *Virtual power-plant demonstration summary*. California Independent System Operator.

Cavus, M., Dissanayake, D., & Bell, M. (2025). Next generation of electric vehicles, AI-driven approaches for predictive maintenance and battery management. *Energies*, *18*(5), 1–41. DOI: 10.3390/en18051041

Cerezo, M., Arrasmith, A., Babbush, R., Benjamin, S. C., Endo, S., Fujii, K., McClean, J. R., Mitarai, K., Yuan, X., Cincio, L., & Coles, P. J. (2021). Variational quantum algorithms. *Nature Reviews. Physics*, *3*(9), 625–644. DOI: 10.1038/s42254-021-00348-9

Chair, R., Mak, C. L., Luk, K. M., & Kishk, A. A. (2006). Low-profile wideband circularly polarized patch antenna using a dielectric resonator. *IEEE Transactions on Antennas and Propagation*, *54*(8), 2312–2315.

Chalasani, S., & Conrad, J. M. (2008, April). A survey of energy harvesting sources for embedded systems. In *IEEE SoutheastCon 2008* (pp. 442-447).DOI: DOI: 10.1109/SECON.2008.4494336

Chen, M., & Sun, D. (2017). A Unified Space Vector Pulse Width Modulation for Dual Two-level Inverter System. *IEEE Transactions on Power Electronics*, *32*(2), 889–893. DOI: 10.1109/TPEL.2016.2585223

Chen, Y., Zhou, J., & Wang, Z. (2022). Battery energy-storage optimization using reinforcement learning. *Applied Energy*, *324*, 119708.

Cochran, D. S., Smith, J., Mark, B. G., & Rauch, E. (2022). Information model to advance explainable AI-based decision support systems in manufacturing system design. In Matt, D. T., Vidoni, R., Rauch, E., & Dallasega, P. (Eds.), *Managing and implementing the digital transformation* (Vol. 525). Springer. DOI: 10.1007/978-3-031-14317-5_5

Dahbi, A., Said, N. N., & Said, M. S. N. (2016). A novel combined MPPT-pitch angle control for wide range variable speed wind turbine based on neural network. *International Journal of Hydrogen Energy*, *41*(22), 9427–9442. DOI: 10.1016/j.ijhydene.2016.03.105

Dai, J., Yuan, Q., Cai, H. H., Zhang, V., Hasanuzaman, M., & Selvaraj, J. (2025). Business-oriented optimization of EV-to-building energy flows: Predictive modeling and scenario evaluation. *Energy*, *333*, 137456. DOI: 10.1016/j.energy.2025.137456

Dalamagkidis, K., Kolokotsa, D., Kalaitzakis, K., & Stavrakakis, G. S. (2007). Reinforcement learning for energy conservation and comfort in buildings. *Building and Environment*, *42*(7), 2686–2698. DOI: 10.1016/j.buildenv.2006.07.010

de Oliveira, T. H. F., Medeiros, L. P. de S., Neto, A. D. D., & de Melo, J. D. (2021). Q-Managed: A new algorithm for multi-objective reinforcement learning. *Engineering Applications of Artificial Intelligence*, *9*, 100089.

Deng, Z., Li, Y., Wang, X., Jiang, Z., & Dong, B. (2025). Quantum computing-enhanced large-scale residential electric vehicle charging management. *Applied Energy*, *401*, 126772. DOI: 10.1016/j.apenergy.2025.126772

Denholm, P., & Hand, M. (2023). Balancing variable generation with energy storage. National Renewable Energy Laboratory., NREL/TP-6A20-75829.

Didier, F., Basbas, H., Larioumlil, D., Laghrouche, S., & Depernet, D. (2025). Neural network-based super-twisting control for floating wind turbines: Design and real-time validation. *Ocean Engineering*, *338*, 121973. Advance online publication. DOI: 10.1016/j.oceaneng.2025.121973

DOE. (2022). *Pipeline cybersecurity report*. U.S. Department of Energy.

DOE. (2025). *ARPA-E Open 2025 program overview*. U.S. Department of Energy.

Dong, Y., Wang, J., Ding, S., & Li, W. (2022). Adaptive fractional-order fault-tolerant sliding mode control scheme of DFIG wind energy conversion system. *Proceedings of the Institution of Mechanical Engineers. Part I, Journal of Systems and Control Engineering*, *237*(1), 15–25. DOI: 10.1177/09596518221123101

Dounis, A. I., & Caraiscos, C. (2009). Advanced control systems engineering for energy and comfort management in a building environment—A review. *Renewable & Sustainable Energy Reviews*, *13*(6–7), 1246–1261. DOI: 10.1016/j.rser.2008.09.015

Echiheb, F., Elkafazi, I., Bossoufi, B., Bhiri, B. E., Almalki, M. M., & Alghamdi, T. A. H. (2024). Nonlinear robust sliding mode–backstepping hybrid control for WECS: Theoretical design and experimental evaluation. *Heliyon*, *10*(5). Advance online publication. DOI: 10.1016/j.heliyon.2024.e31767

Elattar, E. E., Elsayed, S. K., & Elsayed, S. K. (2020). Optimal location and sizing of distributed generators based on renewable energy sources using modified moth flame optimization technique. *IEEE Access : Practical Innovations, Open Solutions*, *8*, 109625–109638. DOI: 10.1109/ACCESS.2020.3001758

European Commission. (2024). *AI4Energy: Integrating artificial intelligence into Europe's power sector*. Directorate-General for Energy.

EY. (2023). *Energy-as-a-service market landscape 2023*. Ernst & Young.

Fadili, Y. E., & Boumhidi, I. (2025). Enhanced nonlinear sliding mode control technique for wind power generation systems application: Theoretical design and comparative study. *e-Prime Advances in Electrical Engineering. Electronics and Energy*, *11*, 100937. Advance online publication. DOI: 10.1016/j.prime.2025.100937

Fagnant, D. J., & Kockelman, K. M. (2022). Preparing a nation for autonomous vehicles: Opportunities, barriers and policy recommendations. *Transportation Research Part A, Policy and Practice*, *132*, 1–16.

Fan, C., Xiao, F., & Zhao, Y. (2017). A short-term building cooling load prediction method using deep learning algorithms. *Applied Energy*, *195*, 222–233. DOI: 10.1016/j.apenergy.2017.03.064

Farhi, E., Goldstone, J., & Gutmann, S. (2014). A quantum approximate optimization algorithm. *arXiv preprint arXiv:1411.4028*. https://arxiv.org/abs/1411.4028

Farret, F. A., & Simões, M. G. (2017). *Integration of renewable sources of energy*. John Wiley & Sons.

Fathi, A., & Rahman, S. (2022). Real-time AI control for distributed storage networks. *IEEE Transactions on Smart Grid*, *13*(2), 876–888.

Femia, N., Petrone, G., Spagnuolo, G., & Vitelli, M. (2005). Optimization of perturb and observe maximum power point tracking method. *IEEE Transactions on Power Electronics*, *20*(4), 963–973. DOI: 10.1109/TPEL.2005.850975

Ficzere, P. (2024). Industry 5.0. Present or future? *Design of Machines and Structures*, *14*(2), 19–34. DOI: 10.32972/dms.2024.010

Floridi, L., & Cowls, J. (2021). A unified framework of five principles for AI in society. *Harvard Data Science Review*, *3*(1).

Fotis, G., Gupta, S., Varshney, T., Dhar, S. L., & Gulzar, M. M. (Eds.). (2025). *Modern Computing Technologies for EV Efficiency and Sustainable Energy Integration*. IGI Global., DOI: 10.4018/979-8-3373-2382-4

Fukushima, H., Han, S., & Wakui, A. (2009). A scheduling method of charging and discharging for electric vehicles connected to the power grid. *2009 IEEE Vehicle Power and Propulsion Conference 2009*, 371-380.

Galain, R. M., & Adib, R. (2025). Renewables 2025 global status report: *Global overview*. https://www.ren21.net/renewables-2025-global-status-report-global-overview

Gao, Y., Liu, H., & Xu, P. (2023). AI-driven shared-mobility optimization. *Transportation Science*, *57*(4), 1093–1114.

Gao, Y., Yang, J., Yang, M., & Li, Z. (2020). Deep reinforcement learning based optimal schedule for a battery-swapping station considering uncertainties. *IEEE Transactions on Industry Applications*, *56*(5), 5775–5784. DOI: 10.1109/TIA.2020.2986412

Garcia, J. E. S., & Santos, M. (2022). Deep learning and fuzzy logic to implement a hybrid wind turbine pitch control. *Neural Computing & Applications*, *34*(13), 10503–10517. DOI: 10.1007/s00521-021-06323-w

Ghiasi, M., Ghiasi, V., & Siano, P. (2025). Renewable Energy Integration into Industrial and Residential Buildings: A Study Across Urban, Rural, and Coastal Areas. *IET Renewable Power Generation*, *19*(1), e70108. DOI: 10.1049/rpg2.70108

Giri, A. K., Arya, S. R., Maurya, R., & Babu, B. C. (2019). VCO-less PLL control-based voltage-source converter for power quality improvement in distributed generation system. *IET Electric Power Applications*, *13*(8), 1114–1124. DOI: 10.1049/iet-epa.2018.5827

Giri, A. K., Arya, S. R., Maurya, R., & Chittibabu, B. (2020). Control of VSC for enhancement of power quality in off-grid distributed power generation. *IET Renewable Power Generation*, *14*(5), 771–778. DOI: 10.1049/iet-rpg.2019.0497

Giri, A. K., Arya, S. R., Maurya, R., & Ray, P. (2020). Performance of Control Algorithms in Wind-Based Distributed Generation System with Power Quality Features: A Review. In Ray, P., & Biswal, M. (Eds.), *Microgrid: Operation, Control, Monitoring and Protection. Lecture Notes in Electrical Engineering* (Vol. 625). Springer., DOI: 10.1007/978-981-15-1781-5_3

Gökçek, T., Boynueğri, A. R., Güner, S., Erdinç, O., & Sadreddini, J. (2025). A new distributed charge approach for the electric vehicle parking lots considering coordination between the distribution system operator and EV operators. *Applied Energy*, *390*(C), 125100.

Goyal, S., Deolia, V. K., & Agrawal, S. (2022). An advanced neuro-fuzzy tuned PID controller for pitch control of horizontal axis wind turbines. *ECTI Transactions on Electrical Engineering, Electronics, and Communications*, *20*(2), 296–305. Advance online publication. DOI: 10.37936/ecti-eec.2022202.246911

Goyal, S., Ingley, H. A., & Barooah, P. (2013). Occupancy-based zone-climate control for energy-efficient buildings: Complexity vs. performance. *Applied Energy*, *106*, 209–221. DOI: 10.1016/j.apenergy.2013.01.039

Guanghua, L., Soomar, A. M., Syed, H. H. S., Shaikh, S., & Musznicki, P. Maximum power point tracking strategies for solar PV systems: A review of current methods and future innovations, Results in Engineering, Volume 28, 2025, 107227, ISSN 2590-1230, DOI: 10.1016/j.rineng.2025.107227

Gunasekaran, R., Pareek, P. K., Gupta, S., & Shukla, A. (2024). Prediction of electric vehicle charging demand using enhanced gated recurrent units with RKOA based graph convolutional network. *Discover Applied Sciences*, *6*(11), 1–18. DOI: 10.1007/s42452-024-06326-x

Gupta, S., Saran, V. B., & Kanaujia, S. Prateeksha (2026). Modelling and Simulation of Hybrid Power Systems with Fuzzy Control and Renewable Energy Sources. In: Bhattacharya, A., Dutta, S., Peng, SL., Yang, X.S. (eds) Data Mining and Information Security. *ICDMIS 2024. Lecture Notes in Networks and Systems*, vol 1385. Springer, Singapore. DOI: 10.1007/978-981-96-6046-9_39

Gupta, S. (2025). *The Future of Machine Learning with Automation in Engineering*. Nova Science Publishers., DOI: 10.52305/YFIX2832

Gupta, S., & Sharma, A. (2018). Global Scenario of Solar Photovoltaic (SPV) Materials. In Bhattacharyya, S., Gandhi, T., Sharma, K., & Dutta, P. (Eds.), *Advanced Computational and Communication Paradigms. Lecture Notes in Electrical Engineering* (Vol. 475). Springer., DOI: 10.1007/978-981-10-8240-5_14

Gupta, S., Sreenivasu, S. V. N., Chouhan, K., Shrivastava, A., Sahu, B., & Potdar, R. M. (2023). Novel face mask detection technique using machine learning to control COVID'19 pandemic. *Materials Today: Proceedings*, *80*, 3714–3718. DOI: 10.1016/j.matpr.2021.07.368 PMID: 34373829

Gupta, S., & Tripathi, R. K. (2014). Two-area power system stability improvement using a robust controller-based CSC-STATCOM. *Acta Polytechnica Hungarica*, *11*(7), 135–155. DOI: 10.12700/APH.11.07.2014.07.9

Gupta, S., & Tripathi, R. K. (2015). Transient Stability Enhancement of Multimachine Power System Using Robust and Novel Controller Based CSC-STATCOM. *Advances in Power Electronics*, *2015*(1), 626731. https://onlinelibrary.wiley.com/doi/10.1155/2015/626731. DOI: 10.1155/2015/626731

Gupta, S., Tripathi, R. K., & Shukla, R. D. (2010, November). Voltage stability improvement in power systems using facts controllers: State-of-the-art review. In *2010 International Conference on Power, Control and Embedded Systems* (pp. 1-8). IEEE. DOI: 10.1109/ICPCES.2010.5698665

Gupta, Y., Doolla, S., Chatterjee, K., & Pal, B. C. (2021). Optimal DG allocation and volt-var dispatch for a droop-based microgrid. *IEEE Transactions on Smart Grid, 12*(1), 169–181. DOI: 10.1109/TSG.2020.3017952

Gutiérrez-Torres, D. A., Ramìrez, J. M., García, J. M. L., Sundriyal, N., & Bhatt, A. "Implementation of a Direct AC-AC Converter Based Wind Energy Conversion System," 2024 International Conference on Computer, Electronics, Electrical Engineering & their Applications (IC2E3), Srinagar Garhwal, Uttarakhand, India, 2024, pp. 1-5, DOI: 10.1109/IC2E362166.2024.10827016

Habib, M. A., Khan, M. I., & Naqvi, S. I. (2021). High-efficiency dielectric resonator antennas for terahertz applications: Design and performance analysis. *IET Microwaves, Antennas & Propagation, 15*(14), 1622–1632.

Hadith Mangunkusumo, K. G., Lian, K. L., Wijaya, F. D., Chang, Y.-R., Lee, Y. D., & Ho, Y. H. (2014). Quantum neural network for State of Charge estimation. *2014 6th International Conference on Information Technology and Electrical Engineering (ICITEE)*, pp. 1–5, DOI: 10.1109/ICITEED.2014.7007948

Han, L., Zhang, H., & An, N. (2025). A continuous space path planning method for unmanned aerial vehicle based on particle swarm optimization-enhanced deep Q-network. *Drones (Basel), 9*(2), 122. DOI: 10.3390/drones9020122

Hannan, M. A., Lipu, M. S. H., Hussain, A., Saad, M. H., & Ayob, A. (2018). Neural network approach for estimating state of charge of lithium-ion battery using backtracking search algorithm. *IEEE Access : Practical Innovations, Open Solutions, 6*, 10069–10079. DOI: 10.1109/ACCESS.2018.2797976

Hao, G., Zhang, X., Li, Y., Wu, Z., & Chen, J. (2025). A novel hybrid adaptive strategy for real-time dispatch and scheduling of electric vehicle charging infrastructure. *Applied Energy, 324*, 1–12.

Hassan, A. S., Chandarana, N., Groh, R., & Macquart, T. (2026). Multi-rotor wind turbines: A review of modern research efforts and challenges. *Renewable and Sustainable Energy Reviews,226*(A). DOI: 10.1016/j.rser.2025.116252

Hassan, A. S., Sun, Y., & Wang, Z. (2020). Multi-objective for optimal placement and sizing DG units in reducing loss of power and enhancing voltage profile using BPSO-SLFA. *Energy Reports, 6*, 1581–1589. DOI: 10.1016/j.egyr.2020.06.013

Hassan, A., Ahmad, G., Shafiullah, M., Islam, A., & Shafiul Alam, M. (2025). Review of the intelligent frameworks for pitch angle control in wind turbines. *IEEE Access : Practical Innovations, Open Solutions, 13*, 29864–29885. DOI: 10.1109/ACCESS.2025.3540367

Hedayati, H., Nabaviniaki, S. A., & Akbarimajd, A. (2008). A method for placement of DG units in distribution networks. *IEEE Transactions on Power Delivery*, *23*(3), 1620–1628. DOI: 10.1109/TPWRD.2007.916106

He, H., Xiong, R., Fan, J., & Li, S. (2011). Evaluation of lithium-ion battery equivalent circuit models for state of charge estimation by an experimental approach. *Energies*, *7*(7), 4502–4520. DOI: 10.3390/en4040582

Heidari, A., Wang, Y., & Li, J. (2022). Battery state-of-health prediction using LSTM networks. *Energy*, *260*, 125028.

He, L., Hu, X., Yin, G., Wang, G., Shao, X., & Liu, J. (2024). A current dynamics model and proportional–integral observer for state-of-charge estimation of lithium-ion battery. *Energy*, *288*, 129701. DOI: 10.1016/j.energy.2023.129701

Henderson, P., Hu, J., Romoff, J., Branson, K., & Pineau, J. (2020). Energy and carbon footprints of machine learning. *Journal of Machine Learning Research*, *21*(248), 1–43.

Hermassi, M., Krim, S., Kraiem, Y., Hajjaji, M. A., Mimouni, M. F., & Mtibaa, A. (2024). Wind speed estimation and maximum power point tracking using neuro-fuzzy systems for variable-speed wind generator. *Wind Engineering*, *48*(6), 1088–1104. DOI: 10.1177/0309524X241247231

Huang, Y., & Kockelman, K. M. (2020). Electric vehicle charging station locations, elastic demand, station congestion, and network equilibrium. *Transportation Research Part D, Transport and Environment*, *78*, 102179. DOI: 10.1016/j.trd.2019.11.008

Hu, B., Yang, M., & Feng, C. (2023). Energy-aware route-planning for electric vehicles using reinforcement learning. *Applied Energy*, *342*, 121081.

Husin, H., & Zaki, M. (2021). A critical review of the integration of renewable energy sources with various technologies. Protection and control of modern power systems, 6(1), 1-18.

Hu, X., Li, S., & Peng, H. (2012). A comparative study of equivalent circuit models for Li-ion batteries. *Journal of Power Sources*, *198*, 359–367. DOI: 10.1016/j.jpowsour.2011.10.013

Huy, T. H. B., Vo, D. N., Truong, K. H., & Van Tran, T. (2023). Optimal distributed generation placement in radial distribution networks using enhanced search group algorithm. *IEEE Access : Practical Innovations, Open Solutions*, *11*, 103288–103305. DOI: 10.1109/ACCESS.2023.3316725

Ibrahim, S., Mostafa, M., Jnadi, A., Salloum, H., & Osinenko, P. (2024). Comprehensive overview of reward engineering and shaping in advancing reinforcement learning applications. *IEEE Access : Practical Innovations, Open Solutions*, *12*, 175473–175500. DOI: 10.1109/ACCESS.2024.3504735

IEA. (2023). *World Energy Outlook 2023*. International Energy Agency.

IEA. (2024). *Global Electric Vehicle Outlook 2024*. International Energy Agency.

IEA. R. (2020). Iea. Paris, (2020), 628.

Igbinenikaro, O. P., Adekoya, O. O., & Etukudoh, E. A. (2024). Conceptualizing sustainable offshore operations: Integration of renewable energy systems. International Journal of Frontiers in Science and Technology Research, 6(02), 031-043.

Ignatev, E., Deriugina, G., Suslov, K., & Balaban, G. (2025). Development of a hybrid model for medium-term wind farm power output forecasting. *Renewable Energy*, *249*, 123200. DOI: 10.1016/j.renene.2025.123200

Igogo, T., Awuah-Offei, K., Newman, A., Lowder, T., & Engel-Cox, J. (2021). Integrating renewable energy into mining operations: Opportunities, challenges, and enabling approaches. *Applied Energy*, *300*, 117375. DOI: 10.1016/j.apenergy.2021.117375

Ittipiboon, A., Petosa, A., & Roscoe, D. (1993). Feeding techniques for dielectric resonator antennas. In *Proceedings of the IEEE Antennas and Propagation Society International Symposium* (pp. 804–807).

Iweh, C. D., Gyamfi, S., Tanyi, E., & Effah-Donyina, E. (2021). Distributed generation and renewable energy integration into the grid: Prerequisites, push factors, practical options, issues and merits. *Energies*, *14*(17), 5375. DOI: 10.3390/en14175375

Jafarnejadsani, H., Pieper, J., & Ehlers, J. (2013). Adaptive control of a variable-speed variable-pitch wind turbine using radial-basis function neural network. *IEEE Transactions on Control Systems Technology*, *21*(6), 2264–2272. DOI: 10.1109/TCST.2012.2237518

Jain, A., Schildbach, G., Fagiano, L., & Morari, M. (2015). On the design and tuning of linear model predictive control for wind turbines. *Renewable Energy*, *80*, 664–673. DOI: 10.1016/j.renene.2015.02.057

Jamal, S., Pasupuleti, J., & Ekanayake, J. (2024). A rule-based energy management system for hybrid renewable energy sources with battery bank optimized by genetic algorithm optimization. *Scientific Reports*, *14*(1), 4865. DOI: 10.1038/s41598-024-54333-0 PMID: 38418902

Jeon, T., & Paek, I. (2021). Design and verification of the LQR controller based on fuzzy logic for large wind turbine. *Energies*, *14*(1), 230. Advance online publication. DOI: 10.3390/en14010230

Jeon, T., Song, Y., & Paek, I. (2023). A study on H∞-fuzzy controller for a non-linear wind turbine with uncertainty. *Applied Sciences (Basel, Switzerland)*, *13*(21), 11930. Advance online publication. DOI: 10.3390/app132111930

Jin, X., Wah, B. W., Cheng, X., & Wang, Y. (2015). Significance and challenges of big data research. *Big Data Research*, *2*(2), 59–64. DOI: 10.1016/j.bdr.2015.01.006

Jobin, A., Ienca, M., & Vayena, E. (2019). The global landscape of AI ethics guidelines. *Nature Machine Intelligence*, *1*(9), 389–399. DOI: 10.1038/s42256-019-0088-2

Johansen, K. W., Teizer, J., & Schultz, C. (2024). Automated rule-based safety inspection and compliance checking of temporary guardrail systems in construction. *Automation in Construction*, *168*, 105849. DOI: 10.1016/j.autcon.2024.105849

Johri, S., Selvakumar, P., Satyanarayana, P., Das, A., Gupta, S., & Bhattacharya, S. (2025). Urbanization and Energy: The Nexus of Global Challenges. In Geopolitical Landscapes of Renewable Energy and Urban Growth (pp. 453-478). *IGI Global Scientific Publishing*. DOI: 10.4018/979-8-3693-8814-3.ch016

Jones, L. E. (2017). *Renewable energy integration: practical management of variability, uncertainty, and flexibility in power grids*. Academic press.

Joshi, B. P., Sachdeva, S., Pant, S., Goyal, S., Kumar, A., & Kharayat, P. S. (2024, May). Selection of Electrical Distribution System using SVAS based Einstein Averaging Operator. *In 2024 4th International Conference on Advance Computing and Innovative Technologies in Engineering (ICACITE)* (pp. 1847-1852). IEEE.

Kajfez, D., & Guillon, P. (1986). *Dielectric resonators*. Artech House.

Kalogirou, S. A. (2000). Applications of artificial neural-networks for energy systems. *Applied Energy*, *67*(1–2), 17–35. DOI: 10.1016/S0306-2619(00)00005-2

Kanaujia, S., & Sharma, A. (2023, May). WSN Based Energy-Efficient Protocols for Smart Grid: A State-of-Art Review. In *International Conference on Electric Power and Renewable Energy* (pp. 103-112). Singapore: Springer Nature Singapore.

Kanaujia, S., Singh, S. P., & Gupta, S. (2024). IoT-based solar tracking and monitoring system. In *The future of IoT with automation in engineering and modern technology*. (ISBN: 979-88911-3704-2).

Kanojia, S. S., Suthar, B. N., & Giri, A. K. (2025). Hybrid Optimization Approach for Optimal Location of Electric Vehicle Charging Station (EVCS) in Distribution Network Considering Voltage Stability. *Smart Grids and Sustainable Energy*, *10*(2), 36. DOI: 10.1007/s40866-025-00264-2

Kapoor, G., Gakhar, P., Verma, P., Sonwane, P., & Rathore, B. (2021). A Double-Ended Scheme for Fault Identification and Phase-Type Categorization in Fixed Series Compensated Transmission Line during Distorted Signals. *2021 6th IEEE International Conference on Recent Advances and Innovations in Engineering (ICRAIE)*, 1–4. DOI: 10.1109/ICRAIE52900.2021.9703834

Karami, N., Moubayed, N., & Outbib, R. (2017). General review and classification of different MPPT Techniques. *Renewable & Sustainable Energy Reviews*, *68*, 1–18. DOI: 10.1016/j.rser.2016.09.132

Karthick, A., Shankar, R., & Dharmaraj, G. (2024). Energy forecasting of the building integrated photovoltaic system based on deep learning dragonfly-firefly algorithm. *Energy*, *308*, 132926. DOI: 10.1016/j.energy.2024.132926

Karunarathne, E., Pasupuleti, J., Ekanayake, J., & Almeida, D. (2020). Optimal placement and sizing of DGs in distribution networks using MLPSO algorithm. *Energies*, *13*(23), 6185. DOI: 10.3390/en13236185

Kayal, P., & Chanda, C. K. (2013). Placement of wind and solar based DGs in distribution system for power loss minimization and voltage stability improvement. *International Journal of Electrical Power & Energy Systems*, *53*, 795–809. DOI: 10.1016/j.ijepes.2013.05.047

Keighobadi, J., Ansar, H. M. K., & Naseradinmousavi, P. (2022). Adaptive neural dynamic surface control for uniform energy exploitation of floating wind turbine. *Applied Energy*, *316*, 119132. Advance online publication. DOI: 10.1016/j.apenergy.2022.119132

Kempton, W., & Letendre, S. E. (1997). Electric vehicles as a new power source for electric utilities. *Transportation Research Part D, Transport and Environment*, *2*(3), 157–175. DOI: 10.1016/S1361-9209(97)00001-1

Kempton, W., & Tomic, J. (2005). Vehicle-to-grid power implementation, from stabilizing the grid to supporting large-scale renewable energy. *Journal of Power Sources*, *144*(1), 280–294. DOI: 10.1016/j.jpowsour.2004.12.022

Kempton, W., & Tomić, J. (2021). Vehicle-to-grid power fundamentals. *IEEE Transactions on Power Systems*, *36*(5), 4295–4305.

Khan, M., Wu, W., & Li, L. (2024). Grid-forming control for inverter-based resources in power systems: A review on its operation, system stability, and prospective. *IET Renewable Power Generation, 18*(6), 887–907. DOI: 10.1049/rpg2.12991

Kim, J.-S., & Sul, S.-K. (1996). A Novel Voltage Modulation Technique of the Space Vector PWM. *IEEJ Transactions on Industry Applications, 116*(8), 820–825. DOI: 10.1541/ieejias.116.820

Kirklin, S.. (2023). Data-driven materials discovery for next-generation batteries. *Nature Materials, 22*, 556–568.

Kizito, R., Li, X., Sun, K., & Li, S. (2020). Optimal distributed generator placement in utility-based microgrids during a large-scale grid disturbance. *IEEE Access : Practical Innovations, Open Solutions, 8*, 21333–21344. DOI: 10.1109/ACCESS.2020.2968871

Kostopoulos, G., Davrazos, G., & Kotsiantis, S. (2024). Explainable artificial intelligence–based decision support systems: A recent review. *Electronics (Basel), 13*(14), 2842. DOI: 10.3390/electronics13142842

Kumari, A., Gopal, Y., Dhaked, D. K., & Panda, K. P. Y.N. Vijaya Kumar, A single source five-level switched-capacitor based multilevel inverter with reduced device count, *e-Prime - Advances in Electrical Engineering, Electronics and Energy*,Volume 5,2023,100235,ISSN 2772-6711.

Kumari, A., Gopal, Y., Dhaked, D. K., Panda, K. P., & Kumar, Y. N. V. (2023). A single source five-level switched-capacitor based multilevel inverter with reduced device count. *E-Prime - Advances in Electrical Engineering. Electronics and Energy, 5*, 100235. DOI: 10.1016/j.prime.2023.100235

Kumar, R., Aggarwal, R. K., & Sharma, J. D. (2013). Energy analysis of a building using artificial neural network: A review. *Energy and Building, 65*, 352–358. DOI: 10.1016/j.enbuild.2013.06.007

Kumar, S. S., & Balakrishna, K. (2024). A novel design and analysis of hybrid fuzzy logic MPPT controller for solar PV system under partial shading conditions. *Scientific Reports, 14*(1), 10256. DOI: 10.1038/s41598-024-60870-5 PMID: 38704401

Kumar, V., Singh, N., & Pratap, B. (2017). Adaptive pitch control design for variable speed wind turbine using Chebyshev neural network. *In 2017 International Conference on Information, Communication, Instrumentation and Control (ICICIC)*, 1–6, IEEE. DOI: DOI: 10.1109/ICOMICON.2017.8279166

Kundu, S., Singh, M., & Giri, A. K. (2021). Design and Control of a Standalone Wind–Solar System with SEIG Feeding Linear/Nonlinear Loads. In: Mohapatro, S., Kimball, J. (eds) Proceedings of Symposium on Power Electronic and Renewable Energy Systems Control. Lecture Notes in Electrical Engineering, vol 616. Springer, Singapore. DOI: 10.1007/978-981-16-1978-6_28

Kundu, S., Singh, M., & Giri, A. K. Synchronization and control of WECS-SPV-BSS-based distributed generation system using ICCF-PLL control approach, Electric Power Systems Research, Volume 226, 2024, 109919, ISSN 0378-7796, DOI: 10.1016/j.epsr.2023.109919

Kundu, S., Giri, A. K., Kadiyan, S., & Singh, S. (2025). Power management and adaptive control method for a standalone distributed generation system under fault, DC-offset, and dynamic environments. *International Journal of Ambient Energy*, *46*(1), 2452245. Advance online publication. DOI: 10.1080/01430750.2025.2452245

Kundu, S., Giri, A. K., Singh, M., & Kadiyan, S. (2025). Adaptive control strategy for isolated renewable energy-based generation system with intermittency. *Electrical Engineering*, *107*(7), 8773–8785. DOI: 10.1007/s00202-024-02454-0

Kundu, S., Giri, A. K., Singh, M., & Singh, S. (2025). Effective Multiport Battery Charging Facility, Operation and Control in Hybrid Isolated Microgrid System. In Giri, A. K., & Singh, M. (Eds.), *Electric Vehicle Charging Infrastructures and its Challenges. Studies in Infrastructure and Control.* Springer., DOI: 10.1007/978-981-96-0361-9_11

Kundu, S., Singh, M., & Giri, A. K. (2022). Control algorithm for coordinated operation of wind-solar microgrid standalone generation system. *Energy Sources. Part A, Recovery, Utilization, and Environmental Effects*, *44*(4), 10024–10044. DOI: 10.1080/15567036.2022.2143943

Kundu, S., Singh, M., & Giri, A. K. (2023). Adaptive Control Approach-Based Isolated Microgrid System with Alleviating Power Quality Problems. *Electric Power Components and Systems*, *52*(7), 1219–1234. DOI: 10.1080/15325008.2023.2239222

Kundu, S., Singh, M., & Giri, A. K. (2023). Implementation of variable gain controller based improved phase locked loop approach to enhance power quality in autonomous microgrid. *International Journal of Numerical Modelling*, *36*(5), e3082. DOI: 10.1002/jnm.3082

Kundu, S., Singh, M., & Giri, A. K. (2024). SPV-wind-BES-based islanded electrical supply system for remote applications with power quality enhancement. *Electrical Engineering*, *106*(1), 279–294. DOI: 10.1007/s00202-023-01979-0

Kundu, S., Singh, M., Giri, A. K., Kadiyan, S., Chittora, P., & Shailly, D. (2025). Robust and fast control approach for islanded microgrid system and EV charging station applications. *Electrical Engineering*, *107*(7), 8385–8396. DOI: 10.1007/s00202-024-02291-1

Kundu, S., Sundriyal, N., Singh, N., Singh, S., & Ramirez, J. M. (2025). *Control strategy for a single-phase grid-connected PV system with bidirectional EV charging support. Modern computing technologies for EV efficiency and sustainable energy integration.* IGI Global Scientific Publishing., DOI: 10.4018/979-8-3373-2382-4.ch007

Lago, J., De Ridder, F., & De Schutter, B. (2021). Forecasting renewable generation with machine learning. *Renewable & Sustainable Energy Reviews*, *138*, 110688.

Lee, Y.-H., Kim, D.-H., & Hyun, D.-S. (n.d.). Carrier based SVPWM method for multi-level system with reduced HDF [harmonic distortion factor]. *Conference Record of the 2000 IEEE Industry Applications Conference. Thirty-Fifth IAS Annual Meeting and World Conference on Industrial Applications of Electrical Energy (Cat. No.00CH37129)*, 1996–2003. DOI: 10.1109/IAS.2000.882151

Leung, K. W., Fang, X. S., & Luk, K. M. (2000). Circularly polarized dielectric resonator antenna excited by dual conformal strips. *Electronics Letters*, *36*(8), 609–611.

Liao, Y., Yu, G., Chen, P., Zhou, B., & Li, H. (2023). Integration of decision-making and motion planning for autonomous driving based on double-layer reinforcement learning framework. *IEEE Transactions on Vehicular Technology*, *73*(3), 3142–3158. DOI: 10.1109/TVT.2023.3326548

Li, D., Zhao, D., Zhu, Y., & Xia, Z. (2015). Thermal comfort control based on MEC algorithm for HVAC systems. *2015 International Joint Conference on Neural Networks (IJCNN)*, 1–6. DOI: 10.1109/IJCNN.2015.7280436

Li, J., Wang, T., Zhang, D., Wu, Q., & Liu, J. (2025). Multi-agent deep reinforcement learning algorithms for distributed charging station management. *International Journal of Advanced Computer Science and Applications*, *16*(7), 535–543.

Li, J., Wang, Y., Lin, S., & Zhao, X. (2022). Nonlinear modelling and adaptive control of smart rotor wind turbines. *Renewable Energy*, *186*, 677–690. DOI: 10.1016/j.renene.2022.01.020

Lin, H., & Huang, J. (2022). Driver-behavior prediction for EV energy management. *Transportation Research Part C, Emerging Technologies*, *140*, 103671.

Lin, W. M., Hong, C. M., Ou, T. C., & Chiu, T. M. (2011). Hybrid intelligent control of PMSG wind generation system using pitch angle control with RBFN. *Energy Conversion and Management*, *52*(2), 1244–1251. DOI: 10.1016/j.enconman.2010.09.020

Li, P., Xue, Q., Zhang, Z., Chen, J., & Zhou, D. (2023). Multi-objective energy-efficient hybrid flow shop scheduling using Q-learning and GVNS driven NSGA-II. *Computers & Operations Research*, *159*, 106360. DOI: 10.1016/j.cor.2023.106360

Lipu, M. S. H., Hannan, M. A., Hussain, A., Saad, M. H. M., Ayob, A., & Blaabjerg, F. (2018). State of charge estimation for lithium-ion battery using recurrent NARX neural network model based lighting search algorithm. *IEEE Access : Practical Innovations, Open Solutions*, *6*, 28150–28161. DOI: 10.1109/ACCESS.2018.2837156

Li, T., Liu, X., Lin, Z., Yang, J., & Ioannou, A. (2024). A linear quadratic regulator with integral action of wind turbine based on aerodynamics forecasting for variable power production. *Renewable Energy*, *223*, 119605. Advance online publication. DOI: 10.1016/j.renene.2023.119605

Liu, C., Wang, J., Botterud, A., Zhou, Y., & Vyas, A. (2012). Assessment of impacts of PHEV charging patterns on wind-thermal scheduling by stochastic unit commitment. *IEEE Transactions on Smart Grid*, *3*(2), 675–683. DOI: 10.1109/TSG.2012.2187687

Liu, F., Wei, Z., Lin, Y., Huang, X., Li, Y., Huang, Y., & Lim, M. K. (2025). Vehicle-To-Grid technology acceptance for electric vehicle users, a systematic literature review and future research agenda. *International Journal of Consumer Studies*, *49*(3), 70065. DOI: 10.1111/ijcs.70065

Liu, H., Pang, A., Yin, J., Yi, H., & Mu, H. (2025). Collaborative optimization scheduling strategy for electric vehicle charging stations considering spatiotemporal distribution of different power charging demands. *World Electric Vehicle Journal*, *16*(3), 1–24. DOI: 10.3390/wevj16030176

Liu, J., Lin, G., Rehtanz, C., Huang, S., Zhou, Y., & Li, Y. (2022). Data-driven intelligent EV charging operating with limited chargers. *International Journal of Electrical Power & Energy Systems*, *141*, 108218. DOI: 10.1016/j.ijepes.2022.108218

Liu, Q., & Chen, R. (2023). Reinforcement learning for anomaly detection in cyber-physical systems. *IEEE Access : Practical Innovations, Open Solutions*, *11*, 32122–32134.

Liu, W., Placke, T., & Chau, K. T. (2022). Overview of batteries and battery management for electric vehicles. *Energy Reports*, *8*, 4058–4084. DOI: 10.1016/j.egyr.2022.03.016

Li, X., Wang, Q., Wen, H., & Xiao, W. (2019). Comprehensive studies on operational principles for maximum power point tracking in photovoltaic systems. *IEEE Access : Practical Innovations, Open Solutions, 7*, 121407–121420. DOI: 10.1109/ACCESS.2019.2937100

Li, X., Zhang, T., & Chen, H. (2021). Transformer fault diagnosis using deep CNN. *IEEE Transactions on Industrial Electronics, 68*(8), 7231–7242.

Li, Y., & Shi, R. (2015). An intelligent solar energy-harvesting system for wireless sensor networks. *EURASIP Journal on Wireless Communications and Networking, 2015*(1), 1–12. DOI: 10.1186/s13638-015-0414-2

Loh, P. C., Holmes, D. G., & Lipo, T. A. (2005). Implementation and control of distributed PWM cascaded multilevel inverters with minimal harmonic distortion and common-mode voltage. *IEEE Transactions on Power Electronics, 20*(1), 90–99. DOI: 10.1109/TPEL.2004.839830

Luk, K. M., & Leung, K. W. (2003). *Dielectric resonator antennas*. Research Studies Press.

Lu, R., Zhang, Y., Yang, T., Chen, Y., Wang, D., & Peng, X. (2024). A novel hybrid-action-based deep reinforcement learning for industrial energy management. *IEEE Transactions on Industrial Informatics, 20*(10), 1–15. DOI: 10.1109/TII.2024.3424529

Lu, Y. (2016). Industrial integration: A literature review. *Journal of Industrial Integration and Management, 1*(02), 1650007. DOI: 10.1142/S242486221650007X

Lyu, Z., Liu, Y., Sciazko, A., Komatsu, Y., Tao, J., Nakamura, A., Hara, T., Sun, K., Shikazono, N., & Han, M. (2025). Co-generation of electricity and chemicals from methane using direct internal reforming solid oxide fuel cells. *Advanced Energy Materials, 15*(10), 2403869. DOI: 10.1002/aenm.202403869

M'dioud, M., Bannari, R., & Elkafazi, I. (2022). Optimal DG allocations in the distribution networks using an improved PSO algorithm. [IJRER]. *International Journal of Renewable Energy Research, 12*(3).

M'dioud, M., Bannari, R., & Elkafazi, I. (2024). A novel PSO algorithm for DG insertion problem. *Energy Systems, 15*(1), 325–351. DOI: 10.1007/s12667-022-00546-z

Maasoumy, M., Pinto, A., & Sangiovanni-Vincentelli, A. (2011). Model-Based Hierarchical Optimal Control Design for HVAC Systems. *ASME 2011 Dynamic Systems and Control Conference and Bath/ASME Symposium on Fluid Power and Motion Control, Volume 1*, 271–278. DOI: 10.1115/DSCC2011-6078

Malik, S., Joshi, B. P., Bisht, V., Kharayat, P. S., & Goyal, S. (2024, May). Ambiguous Fuzzy Einstein Ordered Averaging Operator: Application to the classification of Power Generation methods. *In 2024 4th International Conference on Advance Computing and Innovative Technologies in Engineering (ICACITE)* (pp. 1853-1857). IEEE.

Mane, D. (2026). AI-driven optimization techniques for energy-efficient building systems. In *AI-Driven Approaches for Fully Automated Smart Engineering* (pp. 149–166). IGI Global.

Mannan, M., Mansor, M., Reza, M. S., Roslan, M. F., Ker, P. J., & Hannan, M. A. (2024). Recent development of grid-connected microgrid scheduling controllers for sustainable energy: A bibliometric analysis and future directions. *IEEE Access : Practical Innovations, Open Solutions*, *12*, 90606–90628. DOI: 10.1109/ACCESS.2024.3419443

Mao, M., Cui, L., Zhang, Q., Guo, K., Zhou, L., & Huang, H. (2020). Classification and summarization of solar photovoltaic MPPT techniques: A review based on traditional and intelligent control strategies. *Energy Reports*, *6*, 1312–1327. DOI: 10.1016/j.egyr.2020.05.013

Ma, R., Siaw, F. L., Thio, T. H. G., & Yang, W. (2024). New adaptive super-twisting extended-state observer-based sliding mode scheme with application to FOWT pitch control. *Journal of Marine Science and Engineering*, *12*(6), 902. Advance online publication. DOI: 10.3390/jmse12060902

Marquez, F. P. G., & Gonzalo, A. P. (2022). A comprehensive review of artificial intelligence and wind energy. *Archives of Computational Methods in Engineering*, *29*(5), 2935–2958. DOI: 10.1007/s11831-021-09678-4

Marugán, A., González, R., & Ruiz, E. (2018). Machine-learning techniques for solar energy forecasting. *Renewable & Sustainable Energy Reviews*, *81*, 2299–2313.

Marzouk, O. A. (2025). Wind speed Weibull model identification in Oman, and computed normalized annual energy production (NAEP) from wind turbines based on weather station data. *Engineering Reports*, *7*(3), e70089. DOI: 10.1002/eng2.70089

Masenge, I. H., & Mwasilu, F. A. (2024). Modeling and control dynamics of microgrid with renewable energy systems. In *Modeling and control dynamics in microgrid systems with renewable energy resources* (pp. 199–242). Academic Press., DOI: 10.1016/B978-0-323-90989-1.00015-4

Mathews, I., King, P. J., Stafford, F., & Frizzell, R. (2015). Performance of III–V solar cells as indoor light energy harvesters. *IEEE Journal of Photovoltaics*, *6*(1), 230–235. DOI: 10.1109/JPHOTOV.2015.2487825

Ma, Y., Matusko, J., & Borrelli, F. (2015). Stochastic Model Predictive Control for Building HVAC Systems: Complexity and Conservatism. *IEEE Transactions on Control Systems Technology*, *23*(1), 101–116. DOI: 10.1109/TCST.2014.2313736

Ma, Z., Callaway, D. S., & Hiskens, I. A. (2012). Decentralized charging control of large populations of plug-in electric vehicles. *IEEE Transactions on Smart Grid*, *3*(1), 351–359.

Ma, Z., Callaway, D. S., & Hiskens, I. A. (2013). Decentralized charging control of large populations of plug-in electric vehicles. *IEEE Transactions on Control Systems Technology*, *21*(1), 67–78. DOI: 10.1109/TCST.2011.2174059

Meena, G., Singh, V. P., Dixit, A., Mathur, A., & Bhatt, A. (2024, April). Steady-state models of ac microgrid for load flow solutions. *In 2024 1st International Conference on Innovative Sustainable Technologies for Energy, Mechatronics, and Smart Systems (ISTEMS)* (pp. 1-6). IEEE.

Meena, G., Singh, V. P., Dixit, A., Mathur, A., & Bhatt, A. (2024, April). Steady-state models of ac microgrid for load flow solutions. In 2024 1st International Conference on Innovative Sustainable Technologies for Energy, Mechatronics, and Smart Systems (ISTEMS) (pp. 1-6). IEEE. DOI: 10.1109/ISTEMS60181.2024.10560356

Meggers, F., Ritter, V., Goffin, P., Baetschmann, M., & Leibundgut, H. (2012). Low exergy building systems implementation. *Energy*, *41*(1), 48–55. DOI: 10.1016/j.energy.2011.07.031

Mishra, A., & Nandy, M. (2026). Human-AI collaboration in optimizing energy harvesting for sustainable wireless sensor networks. In *Multidisciplinary Advancements in Human-AI Augmentation* (pp. 225–242). IGI Global.

Mishra, P., & Singh, G. (2023). Energy management systems in sustainable smart cities based on the internet of energy: A technical review. *Energies*, *16*(19), 6903. DOI: 10.3390/en16196903

Mohamed, M. A. E. H., Kamel, S., & Meymand, H. Z. (2025). Intelligent integration of ANN and H-infinity control for optimal enhanced performance of a wind generation unit linked to a power system. *Optimal Control Applications & Methods*, *46*(1), 79–93. DOI: 10.1002/oca.3199

Moll, N., Barkoutsos, P., Bishop, L. S., Chow, J. M., Cross, A., Egger, D. J., & Gambetta, J. M. (2018). Quantum optimization using variational algorithms on near-term quantum devices. *Quantum Science and Technology*, *3*(3), 030503. DOI: 10.1088/2058-9565/aab822

Mongia, R. K., & Bhartia, P. (1994). Dielectric resonator antennas—A review and general design relations for resonant frequency and bandwidth. *International Journal of Microwave and Millimeter-Wave Computer-Aided Engineering*, *4*(3), 230–247. DOI: 10.1002/mmce.4570040304

Moshtaghi, P., Hajialigol, N., & Rafiei, B. (2025). A comprehensive review of artificial intelligence applications in wind energy power generation. *Sustainable Futures : An Applied Journal of Technology, Environment and Society*, *9*, 100638. Advance online publication. DOI: 10.1016/j.sftr.2025.100638

Mourtzis, D. (2024). Challenges and opportunities of the transition from Industry 4.0 to Industry 5.0. In *Manufacturing from Industry 4.0 to Industry 5.0* (pp. 97–131).

Mozer, M. C. (1998). The neural network house: An environment hat adapts to its inhabitants. *Cdn.Aaai.Org*. https://cdn.aaai.org/Symposia/Spring/1998/SS-98-02/SS98-02-017.pdf

Nabae, A., Takahashi, I., & Akagi, H. (1981). A New Neutral-Point-Clamped PWM Inverter. *IEEE Transactions on Industry Applications*, *IA-17*(5), 518–523. DOI: 10.1109/TIA.1981.4503992

NASA Langley Research Center. (n.d.). *NASA POWER Data Access Viewer*. Retrieved October 25, 2025, from https://power.larc.nasa.gov/data-access-viewer/

Natarajan, S., Mathur, S., Sidheekh, S., Stammer, W., & Kersting, K. (2024). Human-in-the-loop or AI-in-the-loop? Automate or collaborate? *arXiv*.

Natarajan, S., Mathur, S., Sidheekh, S., Stammer, W., & Kersting, K. (2025). Human-in-the-loop or AI-in-the-loop? Automate or collaborate? *Proceedings of the AAAI Conference on Artificial Intelligence*, *39*(27), 28594–28600. DOI: 10.1609/aaai.v39i27.35083

Ngo, Q. V., Yi, C., & Nguyen, T. T. (2020). The fuzzy-PID based-pitch angle controller for small-scale wind turbine. *International Journal of Power Electronics and Drive System*, *11*(1), 135–142. DOI: 10.11591/ijpeds.v11.i1.pp135-142

Nguyen, T. T. (2021). Enhanced sunflower optimization for placement distributed generation in distribution system. *International Journal of Electrical and Computer Engineering*, *11*(1), 107–113. DOI: 10.11591/ijece.v11i1.pp107-113

Nielsen, M. A., & Chuang, I. L. (2010). *Quantum computation and quantum information* (10th anniversary ed.). Cambridge University Press. DOI: 10.1017/CBO9780511976667

Nishad, D. K., Tiwari, A. N., Khalid, S., & Gupta, S. (2024). Power quality solutions for rail transport using AI-based unified power quality conditioners. *Discover Applied Sciences*, *6*(12), 651. DOI: 10.1007/s42452-024-06372-5

Olatunde, T. M., Okwandu, A. C., Akande, D. O., & Sikhakhane, Z. Q. (2024). Review of energy-efficient HVAC technologies for sustainable buildings. *International Journal of Science and Technology Research Archive*, *6*(02), 12–20. DOI: 10.53771/ijstra.2024.6.2.0039

Oldewurtel, F., Parisio, A., Jones, C. N., Gyalistras, D., Gwerder, M., Stauch, V., Lehmann, B., & Morari, M. (2012). Use of model predictive control and weather forecasts for energy efficient building climate control. *Energy and Building*, *45*, 15–27. DOI: 10.1016/j.enbuild.2011.09.022

Oldewurtel, F., Parisio, A., Jones, C. N., Morari, M., Gyalistras, D., Gwerder, M., Stauch, V., Lehmann, B., & Wirth, K. (2010). Energy efficient building climate control using Stochastic Model Predictive Control and weather predictions. *Proceedings of the 2010 American Control Conference*, 5100–5105. DOI: 10.1109/ACC.2010.5530680

Oldewurtel, F., Sturzenegger, D., & Morari, M. (2013). Importance of occupancy information for building climate control. *Applied Energy*, *101*, 521–532. DOI: 10.1016/j.apenergy.2012.06.014

Onyinyechukwu, C., Adetomilola, V. F., Valentine, I. I., Bright, N., Cosmas, D. D., & Kenneth, I. I. (2024). Integration of renewable energy in industrial operations: experiences from Canada, USA, and Africa.

Orús, R., Mugel, S., & Lizaso, E. (2019). Quantum computing for finance: Overview and prospects. *Reviews in Physics*, *4*, 100028. DOI: 10.1016/j.revip.2019.100028

Osman, A. M., & Alsokhiry, F. (2022). Sliding mode control for grid integration of wind power system based on direct drive PMSG. *IEEE Access : Practical Innovations, Open Solutions*, *10*, 26567–26579. DOI: 10.1109/ACCESS.2022.3157311

Ouali, S., & Cherkaoui, A. (2020). An improved backward/forward sweep power flow method based on a new network information organization for radial distribution systems. *Journal of Electrical and Computer Engineering*, *2020*, 5643410. Advance online publication. DOI: 10.1155/2020/5643410

Oussama, M., Abdelghani, C., & Lakhdar, C. (2022). Efficiency and robustness of type-2 fractional fuzzy PID design using salps swarm algorithm for a wind turbine control under uncertainty. *ISA Transactions*, *125*, 72–84. DOI: 10.1016/j.isatra.2021.06.016

Palizban, O., & Kauhaniemi, K. (2020). Microgrids and intelligent energy management. *Renewable & Sustainable Energy Reviews, 117*, 109499.

Panda, K. P., Anand, A., Bana, P. R., & Panda, G. (2018). Novel PWM control with modified PSO-MPPT algorithm for reduced switch MLI based standalone PV system. *International Journal of Emerging Electric Power Systems, 19*(5), 20180023. DOI: 10.1515/ijeeps-2018-0023

Pande, A. S., Soni, B. P., & Bhadane, K. V. (2023). Electrical models for EV's batteries: an overview and mathematical design of RC network. *SpringerAS Pande, BP Soni, KV BhadaneJournal of The Institution of Engineers (India): Series B, 2023•Springer, 104*(2), 533–547. DOI: 10.1007/s40031-022-00852-1

Pande, A. S., Soni, B. P., & Sharma, A. K. (2025). Modeling of one and two RC model and state estimation of Lithium-Ion Battery using Thevenin's equivalent Circuit Model. *Procedia Computer Science, 259*, 494–503. DOI: 10.1016/j.procs.2025.03.351

Pandey, A. K., Singh, V., & Jain, S. (2023). Maximum power point tracking algorithm based on fuzzy logic control using PV and IV characteristics for PV array. *IEEE Transactions on Industry Applications, 59*(4), 4572–4583. DOI: 10.1109/TIA.2023.3272536

Park, H., Kim, J. K., & Yi, S. C. (2023). Optimization of site utility systems for renewable energy integration. *Energy, 269*, 126799. DOI: 10.1016/j.energy.2023.126799

Parmar, D. B., & Giri, A. K. (2025). Integration of field-oriented and steady-state linear Kalman filter control in PMSG-based grid-connected system for improving voltage control and power balance operation. *Electrical Engineering, 107*(7), 8803–8817. DOI: 10.1007/s00202-024-02456-y

Patel, C., P. P., R., Dey, A., Ramchand, R., Gopakumar, K., & Kazmierkowski, M. P. (2012). Fast Direct Torque Control of an Open-End Induction Motor Drive Using 12-Sided Polygonal Voltage Space Vectors. *IEEE Transactions on Power Electronics, 27*(1), 400–410. DOI: 10.1109/TPEL.2011.2159516

Patel, V., Giri, V. K., & Kumar, A. (2024). Efficient power management strategies for AC/DC microgrids with multiple voltage buses for sustainable renewable energy integration. *Energy Informatics, 7*(1), 97. DOI: 10.1186/s42162-024-00377-5

Patterson, D., Gonzalez, J., Leong, M., & Dean, J. (2022). Carbon-emission efficiency in large-scale machine-learning training. *Communications of the ACM, 65*(8), 64–75.

Peruzzo, A., McClean, J., Shadbolt, P., Yung, M. H., Zhou, X. Q., Love, P. J., Aspuru-Guzik, A., & O'Brien, J. L. (2014). A variational eigenvalue solver on a photonic quantum processor. *Nature Communications, 5*(1), 4213. DOI: 10.1038/ncomms5213

Petersen, S., & Bundgaard, K. W. (2014). The effect of weather forecast uncertainty on a predictive control concept for building systems operation. *Applied Energy*, *116*, 311–321. DOI: 10.1016/j.apenergy.2013.11.060

Peterson, S. B., Whitacre, J. F., & Apt, J. (2010). The economics of using plug-in hybrid electric vehicle battery packs for grid storage. *Journal of Power Sources*, *195*(8), 2377–2384. DOI: 10.1016/j.jpowsour.2009.09.070

Petkovic, D., Cojbasic, Z., & Nikolic, V. (2013). Adaptive neuro-fuzzy approach for wind turbine power coefficient estimation. *Renewable & Sustainable Energy Reviews*, *28*, 191–195. DOI: 10.1016/j.rser.2013.07.049

Petosa, A., & Ittipiboon, A. (2010). Dielectric resonator antennas: A historical review and the current state of the art. *IEEE Antennas & Propagation Magazine*, *52*(5), 91–116. DOI: 10.1109/MAP.2010.5687510

Pham, T. D., Nguyen, T. T., & Dinh, B. H. (2021). Find optimal capacity and location of distributed generation units in radial distribution networks by using enhanced coyote optimization algorithm. *Neural Computing & Applications*, *33*(9), 4343–4371. DOI: 10.1007/s00521-020-05239-1

Plett, G. L. (2004). Extended Kalman filtering for battery management systems of LiPB-based HEV battery packs: Part 2. Modeling and identification. *Journal of Power Sources*, *134*(2), 262–276. DOI: 10.1016/j.jpowsour.2004.02.032

Pop, C., Cioara, T., Antal, M., & Salomie, I. (2023). Blockchain-based AI markets for peer-to-peer energy trading. *Energy Reports*, *9*, 132–146.

Pradabane, S., Narasimharaju, B. L., & Srikanth, N. V. (2014). Two-quadrant clamping inverter scheme for three-level open-end winding induction motor drive. *2014 IEEE International Conference on Power Electronics, Drives and Energy Systems (PEDES)*, 1–4. DOI: 10.1109/PEDES.2014.7041988

Pradabane, S., Srikanth, N. V., & Narasimharaju, B. L. (2014). A new alternate Fixedbias inverter SVPWM scheme for open-end winding induction motor drive. *International Review of Electrical Engineering*, *9*(1).

Pradheeba, P., Gopu, K., & Thanikasalam, A. (2025). *AI-enabled decision support systems for real-time industrial process automation and optimization.*

Prasad, P. S., & Parimi, A. M. (2025). Design of reinforcement learning based robust μ-synthesis controller for single phase grid-connected VSI. e-Prime-Advances in Electrical Engineering. *Electronics and Energy*, *12*, 100978.

Pratap, B., Singh, N., & Kumar, V. (2018). Robust control of variable speed wind turbine using quasi-sliding mode approach. *Procedia Computer Science, 125*, 398–404. DOI: 10.1016/j.procs.2017.12.052

Preskill, J. (2018). Quantum computing in the NISQ era and beyond. *Quantum : the Open Journal for Quantum Science, 2*, 79. DOI: 10.22331/q-2018-08-06-79

Priyadarshi, N., Padmanaban, S., Holm-Nielsen, J. B., Blaabjerg, F., & Bhaskar, M. S. (2019). An experimental estimation of hybrid ANFIS–PSO-based MPPT for PV grid integration under fluctuating sun irradiance. *IEEE Systems Journal, 14*(1), 1218–1229. DOI: 10.1109/JSYST.2019.2949083

Qi, L., Zhao, M., Wu, S., Zhang, X., Meng, P., Zhao, Y., & Deng, W. (2025). Adaptive torque feed-forward control for wind turbine MPPT considering predicted wind speed characteristics. *International Journal of Electrical Power & Energy Systems, 167*, 110598. Advance online publication. DOI: 10.1016/j.ijepes.2025.110598

Qiskit. (2023). *Qiskit documentation.* https://qiskit.org/

Quantum, I. B. M. (2023). IBM Quantum systems. https://quantum-computing.ibm.com/

Rahman, S., & Shrestha, G. B. (1993). An investigation into the impact of electric vehicle load on the electric utility distribution system. *IEEE Transactions on Power Delivery, 8*(2), 591–597. DOI: 10.1109/61.216865

Rajendran, S., Jena, D., Diaz, M., & Rodriguez, J. (2024). Design of modified complementary terminal sliding mode controller for wind turbine at region II using a two-mass model. *Results in Engineering, 24*, 103026. Advance online publication. DOI: 10.1016/j.rineng.2024.103026

Rao, M. S., Saraswathy, S., Neela, V., Arun, M. R., Gupta, S., & Taqui, S. N. (2023, November). Efficient Model Training in Decentralized Systems with Federated Learning. *In 2023 International Conference on Communication, Security and Artificial Intelligence (ICCSAI)* (pp. 900-904). IEEE. DOI: 10.1109/ICCSAI59793.2023.10421469

Rao, P., & Chowdhury, S. (2023). Reinforcement learning-based battery dispatch optimization. *Electric Power Systems Research, 213*, 108721.

Rappaport, T. S., Xing, Y., Kanhere, O., Ju, S., Madanayake, A., Mandal, S., Alkhateeb, A., & Trichopoulos, G. C. (2019). Wireless communications and applications above 100 GHz: Opportunities and challenges for 6G and beyond. *IEEE Access : Practical Innovations, Open Solutions, 7*, 78729–78757. DOI: 10.1109/ACCESS.2019.2921522

Ratnakaran, R., Rajagopalan, G. B., & Fathima, A. (2023). Artificial ecosystem optimized neural network controlled unified power quality conditioner for microgrid application. *Energy Informatics, 6*(1), 45. DOI: 10.1186/s42162-023-00301-3

Rawajfeh, M. A. A., & Gomaa, M. R. (2023). Comparison between horizontal and vertical axis wind turbine. *International Journal of Applied Power Engineering, 12*(1), 13–23. DOI: 10.11591/ijape.v12.i1.pp13-23

Refaat, M. M., Atia, Y., Sayed, M. M., & Fattah, H. A. (2020). Adaptive Fuzzy Logic Controller as MPPT Optimization Technique Applied to Grid-Connected PV Systems. In Eltamaly, A., & Abdelaziz, A. (Eds.), *Modern Maximum Power Point Tracking Techniques for Photovoltaic Energy Systems. Green Energy and Technology*. Springer., DOI: 10.1007/978-3-030-05578-3_9

Rekioua, D., Mokrani, Z., Kakouche, K., Oubelaid, A., Rekioua, T., Alhazmi, M., Ali, E., Bajaj, M., Dost Mohammadi, S. A., & Ghoneim, S. S. M. (2024). Coordinated power management strategy for reliable hybridization of multi-source systems using hybrid MPPT algorithms. *Scientific Reports, 14*(1), 10267. DOI: 10.1038/s41598-024-60116-4 PMID: 38704399

Rensource Energy. (2024). *Solar microgrid deployment report 2024*. Rensource Holdings.

Reyhan, B. G., & Coleri, S. (2025). Safe deep reinforcement learning for resource allocation with peak age of information violation guarantees. *IEEE Transactions on Communications*.

Richardson, D. B. (2013). Electric vehicles and the electric grid, a review of modeling approaches, impacts, and renewable energy integration. *Renewable & Sustainable Energy Reviews, 19*, 247–254. DOI: 10.1016/j.rser.2012.11.042

Romanova, O., & Kuzmin, E. (2024). From Industry 4.0 to Industry 5.0: Challenges and opportunities of evolutionary transition in Russia. In *The Future of Industry: Human-Centric Approaches in Digital Transformation* (pp. 55–72).

Roy, T. K., Ghosh, S. K., & Saha, S. (2023). Robust backstepping global integral terminal sliding mode controller to enhance dynamic stability of hybrid AC/DC microgrids. *Protection and Control of Modern Power Systems, 8*(1), 1–13. DOI: 10.1186/s41601-023-00281-2

Rufer, A., Veenstra, M., & Gopakumar, K. (1999). Asymmetric Multilevel Converter for High Resolution Voltage Phasor Generation. *Proceedings of the 1999 European Conference on Power Electronics and Applications (EPE 1999)*.

Saeed, H., Zubair, M., & Alomainy, A. (2022). Terahertz antennas for 6G communication systems: Challenges, design strategies, and applications. *IEEE Access : Practical Innovations, Open Solutions, 10*, 375–390.

Sah, D. K., & Amgoth, T. (2020). Renewable energy harvesting schemes in wireless sensor networks: A survey. *Information Fusion, 63*, 223–247. DOI: 10.1016/j.inffus.2020.07.005

Sahoo, S. (2024). Enhancing pitch angle control in variable speed doubly fed induction generator based wind energy conversion system using advanced adaptive neuro-fuzzy approach with particle swarm optimization. *Wind Engineering, 49*(2), 444–461. DOI: 10.1177/0309524X241275145

Sahoo, S., Subudhi, B., & Panda, G. (2019). Torque and pitch angle control of a wind turbine using multiple adaptive neuro-fuzzy control. *Wind Engineering, 44*(2), 125–141. DOI: 10.1177/0309524X19849825

Sayed, E. (2025). Decision intelligence in the age of industrial transformation: A survey from Industry 4.0 to Industry 7.0. *Sustainable Machine Intelligence Journal, 12*, 20–40. DOI: 10.61356/SMIJ.2025.12565

Schuld, M., & Petruccione, F. (2018). *Supervised learning with quantum computers.* Springer., DOI: 10.1007/978-3-319-96424-9

Sekhar, K. R., & Srinivas, S. (2013). Discontinuous Decoupled PWMs for Reduced Current Ripple in a Dual Two-Level Inverter Fed Open-End Winding Induction Motor Drive. *IEEE Transactions on Power Electronics, 28*(5), 2493–2502. DOI: 10.1109/TPEL.2012.2215344

Sele, D., & Chugunova, M. (2023). Putting a human in the loop: Increasing uptake, but decreasing accuracy of automated decision-making. *PLoS One*, •••, 19. PMID: 38335162

Selim, A., Kamel, S., & Jurado, F. (2020). Efficient optimization technique for multiple DG allocation in distribution networks. *Applied Soft Computing, 86*, 105938. DOI: 10.1016/j.asoc.2019.105938

Semwal, S., Kant, S., Bandyopadhyay, K., Asim, M., Garg, R., & Goyal, S. (2024, October). Implicit Control of DC-DC Converter based DC Motor for the Charging of Electric Vehicle using Fuzzy Logic Controller. *In 2024 5th IEEE Global Conference for Advancement in Technology (GCAT)* (pp. 1-6). IEEE.

Semwal, S., Kant, S., Bandyopadhyay, K., Asim, M., Garg, R., & Goyal, S. (2024, October). Implicit Control of DC-DC Converter based DC Motor for the Charging of Electric Vehicle using Fuzzy Logic Controller. In 2024 5th IEEE Global Conference for Advancement in Technology (GCAT) (pp. 1-6). IEEE. DOI: 10.1109/GCAT62922.2024.10923979

Sen, P., Sahoo, A. K., Panda, K. P., & Jha, V. (2023). Single-phase switched-capacitor boost multilevel inverter interfacing solar photovoltaic system. *E-Prime - Advances in Electrical Engineering. Electronics and Energy*, *6*, 100350. DOI: 10.1016/j.prime.2023.100350

SGCC (State Grid Corporation of China). (2025). *Energy Brain annual white paper 2025.* State Grid Corporation of China.

Shankar, R., Chaudhary, B. P., Mehraj, H., Gupta, S., Arslan, F., Srivastava, A. K., & Goswami, C. (2023). Impact of node mobility on the DL based uplink and downlink MIMO-NOMA network. *International Journal of Information Technology : an Official Journal of Bharati Vidyapeeth's Institute of Computer Applications and Management*, *15*(6), 3391–3404. DOI: 10.1007/s41870-023-01362-z

Shao, Y., Liu, J., Huang, J., Hu, L., Guo, L., & Fang, Y. (2022). The implementation of fuzzy PSO-PID adaptive controller in pitch regulation for wind turbines suppressing multi-factor disturbances. *Frontiers in Energy Research*, *9*, 828281. Advance online publication. DOI: 10.3389/fenrg.2021.828281

Sharma, H., Haque, A., &Jaffery, Z. A. (2018). Modeling and optimisation of a solar energy harvesting system for wireless sensor network nodes. *Journal of sensor and Actuator Networks,* *7*(3), 40. https://doi.org/DOI: 10.3390/jsan7030040

Sharma, K., Gupta, C., & Gupta, S. (2019, July). Infant weeping calls decoder using statistical feature extraction and gaussian mixture models. *In 2019 10th International Conference on Computing, Communication and Networking Technologies (ICCCNT)* (pp. 1-6). IEEE. DOI: 10.1109/ICCCNT45670.2019.8944527

Sharma, A., Adhikari, A., Adhikari, B., Pandey, P., & Tamrakar, I. (2024). Active cell balancing approach for efficient battery management system. *KEC Journal of Science and Engineering*, *8*(1), 23–26. DOI: 10.3126/kjse.v8i1.69260

Sharma, R. K., Saran, V. B., Kanaujia, S., & Gupta, S. (2024, May). Evaluation of mppt controller performance using fuzzy logic design with wireless sensor node. *In 2024 International Conference on Smart Systems for applications in Electrical Sciences (ICSSES)* (pp. 1-6). IEEE. DOI: 10.1109/ICSSES62373.2024.10561365

Sharma, R. K., Sharma, A., & Kanaujia, S. (2024). Smart Maintenance of Railway Infrastructure Using Wireless Sensor Networks. *International Journal of Experimental Research and Review, 46*, 113–126. DOI: 10.52756/ijerr.2024.v46.009

Sheng, H. W., & Jun, D. Y. (2013). Research on new pitch control algorithm of wind power generators. *Power Electronics, 47*(2), 53–54.

Shiau, J. K., & Ma, C. W. (2013). Li-ion battery charging with a buck-boost power converter for a solar powered battery management system. *Energies, 6*(3), 1669–1699. DOI: 10.3390/en6031669

Shukla, R. D., Tripathi, R. K., & Gupta, S. (2010, November). Power electronics applications in wind energy conversion system: A review. *In 2010 International Conference on Power, Control and Embedded Systems* (pp. 1-6). IEEE. DOI: 10.1109/ICPCES.2010.5698663

Siemens, A. G. (2024). *MindSphere industrial AI energy platform report 2024.* Siemens.

Singh, D., Singh, A., Chowdhury, I. R., Gupta, R., Ram, P., & Gupta, N. (2025). Government of India, Ministry of Statistics and Programme Implementation, National Statistics Office. *Energy statistics India 2025.* https://www.mospi.gov.in

Singh, M., & Giri, A. K. (2022, June). Implementation of in-loop filter based DSC-FLL control approach for power quality improvement in isolated microgrid. In 2022 IEEE 31st international symposium on industrial electronics (ISIE) (pp. 516-521). IEEE.

Singh, N., & Pratap, B. (2025). Robust observer design for variable speed wind turbine using a two-mass model. *In Proceedings of the International Conference on Systems, Control and Automation (ICSCA 2023) Lecture Notes in Electrical Engineering, 1260*, 171–180, Springer. DOI: 10.1007/978-981-97-7384-8_15

Singh, S. P., Kumar, M., Kumar, A., & Saxena, R. (2022). UWB antenna for ITS application. In A. Dhawan, V. S. Tripathi, K. V. Arya, & K. Naik (Eds.), *Recent trends in electronics and communication (VCAS 2020)* (Lecture Notes in Electrical Engineering, Vol. 777). Springer. DOI: 10.1007/978-981-16-2761-3_102

Singh, N., Pratap, B., & Swarup, A. (2016). Design of robust control for wind turbine using quantitative feedback theory. *IFAC-PapersOnLine, 49*(1), 718–723. DOI: 10.1016/j.ifacol.2016.03.141

Singh, N., Pratap, B., & Swarup, A. (2017). Discrete-time robust control of variable speed wind turbine using quantitative feedback theory. *In 2017 IEEE India Council International Conference (INDICON)*, 1–6, IEEE. DOI: DOI: 10.1109/INDICON.2017.8487583

Singh, N., Pratap, B., & Swarup, A. (2018). Robust control design of variable speed wind turbine using quantitative feedback theory. *Proceedings of the Institution of Mechanical Engineers. Part A, Journal of Power and Energy, 232*(6), 691–705. DOI: 10.1177/0957650917751463

Singh, N., Pratap, B., & Swarup, A. (2019). Design of dynamic surface controller for robust performance of variable speed wind turbine. *IET Renewable Power Generation, 13*(9), 1511–1521. DOI: 10.1049/iet-rpg.2018.5922

Singh, N., Pratap, B., & Swarup, A. (2019). Nonlinear robust observer based adaptive control design for variable speed wind turbine. *Journal of Engineering Research, 7*(3), 258–285. DOI: 10.1016/S2307-1877(25)00618-2

Sinha, R., Patel, M., Gupta, S., & Sinha, K. K. (2024, June). Performance Analysis of Breast Cancer Predictor using Machine Learning Techniques. In 2024 15th International Conference on Computing Communication and Networking Technologies (ICCCNT) (pp. 1-5). IEEE. DOI: 10.1109/ICCCNT61001.2024.10724436

Sinha, R., Sinha, K. K., Patel, M., Gupta, S., & Priya, S. (2024, July). Detection of leukemia disease using convolutional neural network. *In 2024 5th International Conference on Image Processing and Capsule Networks (ICIPCN)* (pp. 451-456). IEEE. DOI: 10.1109/ICIPCN63822.2024.00080

Sinha, R., Kaur, N., Gupta, S., & Thakur, P. (2023, November). Diagnosis of Parkinson's Disease using Hybrid Ensemble Technique. In *2023 International Conference on Ambient Intelligence, Knowledge Informatics and Industrial Electronics (AIKIIE)* (pp. 1-5). IEEE. DOI: 10.1109/AIKIIE60097.2023.10390458

Sioshansi, R., & Denholm, P. (2010). Emissions impacts and benefits of plug-in hybrid electric vehicles and vehicle-to-grid services. *Environmental Science & Technology, 43*(4), 1199–1206. DOI: 10.1021/es802324j PMID: 19320180

Široký, J., Oldewurtel, F., Cigler, J., & Prívara, S. (2011). Experimental analysis of model predictive control for an energy efficient building heating system. *Applied Energy, 88*(9), 3079–3087. DOI: 10.1016/j.apenergy.2011.03.009

Sivakumar, S., Sathik, M. J., Manoj, P. S., & Sundararajan, G. (2016). An assessment on performance of DC–DC converters for renewable energy applications. *Renewable & Sustainable Energy Reviews, 58*, 1475–1485. DOI: 10.1016/j.rser.2015.12.057

So, C. (2020). Human-in-the-loop design cycles: A process framework that integrates design sprints, agile processes, and machine learning with humans. *arXiv: Human–Computer Interaction*. DOI: 10.1007/978-3-030-50334-5_9

Somasekhar, V. T., Srinivas, S., & Gopakkumar, K. A space vector based PWM switching scheme for the reduction of common-mode voltages for a dual inverter fed open-end winding induction motor drive, *Power Electronics. Specialists Conference*, 2005. PESC '05. IEEE 36th, 2005, pp. 816-821. DOI: 10.1109/PESC.2005.1581721

Somasekhar, V. T., Srinivas, S., & Kumar, K. K. (2008). Effect of Zero-Vector Placement in a Dual-Inverter Fed Open-End Winding Induction Motor Drive With Alternate Sub-Hexagonal Center PWM Switching Scheme. *IEEE Transactions on Power Electronics, 23*(3), 1584–1591. DOI: 10.1109/TPEL.2008.921170

Song, J., Zhang, H., Zhang, J., Hu, P., & Feng, J. (2024). Observer-based adaptive fuzzy backstepping control of pitch angle of wind turbine. *2024 IEEE 2nd International Conference on Electrical, Automation and Computer Engineering (ICEACE)*, 887–892. DOI: 10.1109/ICEACE63551.2024.10898733

Sonwane, P., Yadav, M., Bunker, N., Sonwane, S., & Shirsath, V. (2024). Economical Design Perspective for Aircraft by Optimizing Airfoil S1223. *E-Prime - Advances in Electrical Engineering. Electronics and Energy*, *8*, 100531. DOI: 10.1016/j.prime.2024.100531

Sorte, P. K., Panda, K. P., & Panda, G. (2021). Current reference control based MPPT and investigation of power management algorithm for grid-tied solar PV-battery system. *IEEE Systems Journal, 16*(1), 386–396. DOI: 10.1109/JSYST.2021.3052959

Sovacool, B. K., & Dworkin, M. H. (2022). *Energy justice: Re-balancing equity and technology*. Oxford University Press.

Srikakolapu, J., Arya, S. R., Giri, A. K., Kundu, S., & Singh, M. (2024). DSTATCOM components modeling and control. custom power devices for efficient distributed energy systems, 21-42.

Srinivas, S., & Ramachandra Sekhar, K. (2013). Theoretical and Experimental Analysis for Current in a Dual-Inverter-Fed Open-End Winding Induction Motor Drive With Reduced Switching PWM. *IEEE Transactions on Industrial Electronics, 60*(10), 4318–4328. DOI: 10.1109/TIE.2012.2209615

Srinivas, S., & Somasekhar, V. T. (2008). Space-vector-based PWM switching strategies for a three-level dual-inverter-fed open-end winding induction motor drive and their comparative evaluation. *IET Electric Power Applications*, *2*(1), 19–31. DOI: 10.1049/iet-epa:20070134

Stemmler, H. (1994). High-power industrial drives. *Proceedings of the IEEE*, *82*(8), 1266–1286. DOI: 10.1109/5.301688

Strubell, E., Ganesh, A., & McCallum, A. (2019). Energy and policy considerations for deep learning. *Proceedings of ACL*, 3645–3650. DOI: 10.18653/v1/P19-1355

Sudha, G., Natha, A., Rajendiran, M., Vanitha, U., Kiran, V., Patil, H., & Rajaram, A. (2024). Optical oasis harnessing energy efficiency in wireline, optical, and wireless communications. *Journal of Environmental Protection and Ecology*, *25*(4), 1179–1187.

Sun, B., Huang, Z., Tan, X., & Tsang, D. H. K. (2018). Optimal scheduling for electric vehicle charging with discrete charging levels in distribution grid. *IEEE Transactions on Smart Grid*, *9*(2), 624–634. DOI: 10.1109/TSG.2016.2558585

Sundriyal, N., Kundu, S., Gupta, S., Bhatt, A., & Singh, N. (2025). *Hybrid EV Charging Systems: A Study Report on Integration of Fuel Cell with EV for Clean Solution. Hybrid Renewable Power Infrastructure for Sustainable Electric Vehicle Development. Studies in Infrastructure and Control*. Springer., DOI: 10.1007/978-981-95-1323-9_9

Sundriyal, N., Ramirez, J., & Corrochano, B. (2023). A comparison of geometric algebra and harmonic domain for linear circuit analysis. *Mathematics for Applications*, *12*(1), 115–127. DOI: 10.13164/ma.2023.08

Sun, J., Wu, X., Palade, V., Fang, W., & Shi, Y. (2015). Random drift particle swarm optimization algorithm: Convergence analysis and parameter selection. *Machine Learning*, *101*(1–3), 345–376. DOI: 10.1007/s10994-015-5522-z

Sun, L., Yang, P., & Guo, S. (2023). Natural-language processing for predictive maintenance in smart grids. *IEEE Transactions on Industrial Informatics*, *19*(6), 8020–8031.

Sun, Y., & Han, W. (2025). A review of enhancing wind power with AI: Applications, economic implications, and green innovations. *Digital Economy and Sustainable Development*, *3*(11), 11. Advance online publication. DOI: 10.1007/s44265-025-00059-4

Sutikno, T., Subrata, A. C., & Elkhateb, A. (2021). Evaluation of fuzzy membership function effects for maximum power point tracking technique of photovoltaic system. *IEEE Access : Practical Innovations, Open Solutions*, *9*, 109157–109165. DOI: 10.1109/ACCESS.2021.3102050

Sutton, R. S., & Barto, A. G. (2018). *Reinforcement learning: An introduction* (2nd ed.). MIT Press.

Taibi, E., Gielen, D., & Bazilian, M. (2012). The potential for renewable energy in industrial applications. *Renewable & Sustainable Energy Reviews*, *16*(1), 735–744. DOI: 10.1016/j.rser.2011.08.039

Talaei Khoei, T., Ould Slimane, H., & Kaabouch, N. (2023). Deep learning: Systematic review, models, challenges, and research directions. *Neural Computing & Applications*, *35*(31), 23103–23124. DOI: 10.1007/s00521-023-08957-4

Tang, Z., Yang, Y., & Blaabjerg, F. (2021). Power electronics: The enabling technology for renewable energy integration. *CSEE Journal of Power and Energy Systems*, *8*(1), 39–52.

Tan, K., Wang, X., & Li, B. (2022). Reinforcement learning for vehicle-to-grid energy management. *Applied Energy*, *308*, 118344.

Tarascon, J.-M., & Armand, M. (2021). (Referenced for battery chemistry in the storage section.)

Taylor, J., Maitra, A., Alexander, M., Brooks, D., & Duvall, M. (2009). Evaluation of the impact of plug-in electric vehicle loading on distribution system operations. *2009 IEEE Power & Energy Society General Meeting*, 1–6.

Tesla Inc. (2022). *Opticaster AI energy optimization system*. Tesla.

Thandlam, A. K., Vignesh, T., Saravanan, A., Subramani, P., Marimuthu, R., & Gupta, S. (2024, April). Next-gen vehicle safety: The futuristic approach to auto-stop in modern vehicles. In 2024 10th International Conference on Communication and Signal Processing (ICCSP) (pp. 265-270). IEEE. DOI: 10.1109/ICCSP60870.2024.10543666

Thirunavukkarasu, G. S., Seyedmahmoudian, M., Jamei, E., Horan, B., Mekhilef, S., & Stojcevski, A. (2022). Role of optimization techniques in microgrid energy management systems—A review. *Energy Strategy Reviews*, *43*, 100899. DOI: 10.1016/j.esr.2022.100899

Truong, K. H., Nallagownden, P., Elamvazuthi, I., & Vo, D. N. (2020). A quasi-oppositional-chaotic symbiotic organisms search algorithm for optimal allocation of DG in radial distribution networks. *Applied Soft Computing*, *88*, 106067. DOI: 10.1016/j.asoc.2020.106067

Tu, N., Fan, Z., Pang, X., Yan, X., Wang, Y., Liu, Y., & Yang, D. (2024). A multi-objective scheduling method for hybrid integrated energy systems via Q-learning-based multi-population dung beetle optimizers. *Computers & Electrical Engineering*, *117*, 109223. DOI: 10.1016/j.compeleceng.2024.109223

Tyagi, M., Gupta, S., & Khan, M. I. (2024, May). Reduction of voltage instability in complex power networks through the utilization of D-STATCOM. *In 2024 4th International Conference on Advance Computing and Innovative Technologies in Engineering (ICACITE)* (pp. 1662-1666). IEEE. DOI: 10.1109/ICACITE60783.2024.10616856

Tyagi, M., Meena, D. C., & Kundu, S. "CBPF-FLL-Based Operational Coordination of SPV, BESS, and Wind-Driven SEIG Hybrid Systems for Distributed Generation," 2025 IEEE International Conference on Computer, Electronics, Electrical Engineering & their Applications (IC2E3), Srinagar Garhwal, India, 2025, pp. 1-6, DOI: 10.1109/IC2E365635.2025.11166884

Ukoba, K., Olatunji, K. O., Adeoye, E., Jen, T. C., & Madyira, D. M. (2024). Optimizing renewable energy systems through artificial intelligence: Review and future prospects. *Energy & Environment, 35*(7), 3833–3879. DOI: 10.1177/0958305X241256293

UNDP. (2025). *Global AI for Energy Partnership report.* United Nations Development Programme.

UNESCO. (2022). *Recommendation on the ethics of artificial intelligence*. UNESCO.

United Nations. (2024). *Sustainable Development Goals progress report 2024.* United Nations.

van der Meer, S., et al. (2024). (Referenced in blockchain-enabled V2G case.)

Verma, M., Bhambri, S., Gupta, S., & Buduru, A. B. (2019). Making smart homes smarter: Optimizing energy consumption with human in the loop. *arXiv:1912.03298.*

Vu, D. H., Muttaqi, K. M., & Sutanto, D. (2020). An integrated energy management approach for the economic operation of industrial microgrids under uncertainty of renewable energy. *IEEE Transactions on Industry Applications, 56*(2), 1062–1073. DOI: 10.1109/TIA.2020.2964635

Wang, J., Li, X., & Zhao, Q. (2023). Multi-agent reinforcement learning for microgrid coordination. *Applied Energy, 339*, 120933.

Wang, J., Sheng, Y., Ge, H., Bai, X., Su, J., Guo, Q., & Sun, H. (2024). Charging pricing for autonomous mobility-on-demand fleets based on game theory. *Journal of Modern Power Systems and Clean Energy, 12*(6), 2006–2016. DOI: 10.35833/MPCE.2024.000139

Wang, Z., & Li, H. (2023). Adaptive scheduling for EV charging using reinforcement learning. *Transportation Research Part C, Emerging Technologies, 151*, 104054.

WEC (World Energy Council). (2024). *World Energy Trilemma Index 2024.* World Energy Council.

Wei, T., Wang, Y., & Zhu, Q. (2017). Deep reinforcement learning for building HVAC control. *Dl.Acm.OrgT Wei, Y Wang, Q ZhuProceedings of the 54th Annual Design Automation Conference 2017, 2017•dl.Acm.Org, Part 128280*. DOI: 10.1145/3061639.3062224

Wei, T., Zhu, Q., & Maasoumy, M. (2014). Co-scheduling of HVAC control, EV charging and battery usage for building energy efficiency. *2014 IEEE/ACM International Conference on Computer-Aided Design (ICCAD)*, 191–196. DOI: 10.1109/ICCAD.2014.7001351

Welchko, B. A. (2005). A double-ended inverter system for the combined propulsion and energy management functions in hybrid vehicles with energy storage. *31st Annual Conference of IEEE Industrial Electronics Society, 2005. IECON 2005.*, 6 pp. DOI: 10.1109/IECON.2005.1569110

Wen, H., Liu, Y., Lee, S. M., & Park, J. H. (2023). Optimal power regulating for wind turbine with LQR and disturbance accommodation pitch control. *9th International Conference on Control, Automation and Robotics (ICCAR)*, 262–267. DOI: DOI: 10.1109/ICCAR57134.2023.10151773

WHO. (2025). *Tracking SDG 7: Energy Progress Report 2025*. World Health Organization & World Bank.

World Economic Forum. (2024). *Global Cyber Resilience in Energy 2024*. World Economic Forum.

Xie, J., Dong, H., & Zhao, X. (2023). Data-driven torque and pitch control of wind turbines via reinforcement learning. *Renewable Energy*, *215*, 118893. Advance online publication. DOI: 10.1016/j.renene.2023.06.014

Yang, L., Nagy, Z., Goffin, P., & Schlueter, A. (2015). Reinforcement learning for optimal control of low exergy buildings. *Applied Energy*, *156*, 577–586. DOI: 10.1016/j.apenergy.2015.07.050

Yin, X. X., Lin, Y. G., Li, W., Liu, H. W., & Gu, Y. J. (2015). Adaptive sliding mode back-stepping pitch angle control of a variable-displacement pump controlled pitch system for wind turbines. *ISA Transactions*, *58*, 629–634. DOI: 10.1016/j.isatra.2015.07.006

Yu, C., Guo, Z., Pan, Z., Zhou, Y., Zhang, H., Chen, B., Tan, P., Guan, W., & Zhong, Z. (2025). Performance and stress analysis of flat-tubular solid oxide fuel cells fueled with methane and hydrogen. *Acta Mechanica Solida Sinica*, *38*(3), 402–414. DOI: 10.1007/s10338-024-00514-3

Yu, L., Xie, D., Huang, C., Jiang, T., & Zou, Y. (2019). Energy Optimization of HVAC Systems in Commercial Buildings Considering Indoor Air Quality Management. *IEEE Transactions on Smart Grid, 10*(5), 5103–5113. DOI: 10.1109/TSG.2018.2875727

Zamchiy, A. O., & Baranov, E. A. (2022). Polycrystalline silicon thin films for solar cells via metal-induced layer exchange crystallization. *Coatings, 12*(12), 1926. DOI: 10.3390/coatings12121926

Zhang, C., & Mi, C. (2022). AI-based optimization for electric-vehicle integration. *IEEE Transactions on Industrial Electronics, 69*(10), 10188–10199.

Zhang, C., & Plestan, F. (2021). Adaptive sliding mode control of floating offshore wind turbine equipped by permanent magnet synchronous generator. *Wind Energy (Chichester, England), 24*(7), 754–769. DOI: 10.1002/we.2601

Zhang, L., Brown, T., & Samuelsen, G. S. (2011). Fuel reduction and electricity consumption impact of different charging scenarios for plug-in hybrid electric vehicles. *Journal of Power Sources, 196*(15), 6559–6566. DOI: 10.1016/j.jpowsour.2011.03.003

Zhang, L., Ji, Y., Li, X., Huang, Z., Cui, D., Chen, H., Gong, J., Breer, F., Junker, M., & Sauer, D. U. (2025). Multi-objective charging scheduling for electric vehicles at charging stations with renewable energy generation. *Green Energy and Intelligent Transportation, 4*(4), 100283. DOI: 10.1016/j.geits.2025.100283

Zhang, R., Han, T., & Feng, Y. (2023). Graph-neural-network-based fault location in power-distribution systems. *IEEE Transactions on Smart Grid, 14*(1), 777–789.

Zhang, S., Chen, J., & Tang, X. (2022). Multi-objective control and energy management strategy based on deep Q-network for parallel hybrid electric vehicles. *International Journal of Vehicle Performance, 8*(4), 371–386. DOI: 10.1504/IJVP.2022.125943

Zhang, Z., Xiao, Y., Ma, Z., Xiao, M., Ding, Z., Lei, X., Karagiannidis, G. K., & Fan, P. (2019). 6G wireless networks: Vision, requirements, architecture, and key technologies. *IEEE Vehicular Technology Magazine, 14*(3), 28–41. DOI: 10.1109/MVT.2019.2921208

Zhao, F., Guo, Y., & Chen, B. (2024). A review of lithium-ion battery state of charge estimation methods based on machine learning. *World Electric Vehicle Journal, 15*(4), 131. DOI: 10.3390/wevj15040131

Zheng, W., Wang, M., & Li, S. (2022). Digitalization and carbon emissions, how to achieve the dual goals of China's energy conservation and emission reduction. *Energy Reports*, *8*(10), 10923–10935.

Zhenhua, L., Yaming, G., & Qiang, L. (2025). Asymptotic tracking stability control method for grid-connected wind turbine systems based on quantitative trajectory constraints. *IEEE Access : Practical Innovations, Open Solutions*, *13*, 32248–32259. DOI: 10.1109/ACCESS.2025.3542774

Zhou, Y., & Brown, R. (2022). Load forecasting with interpretable machine learning. *Energy and AI*, *9*, 100171.

Zhou, Z., Wu, J., & Yan, Y. (2024). Artificial intelligence in grid optimization and control: A review. *Renewable & Sustainable Energy Reviews*, *178*, 113244.

Zobaa, A. F., & Bansal, R. C. (2011). *Handbook of renewable energy technology*. World Scientific. DOI: 10.1142/7489

About the Contributors

Gunjan Gupta (nominee for prestigious HERS SA Young achievers Award), received B.Tech (Electronics & Telecommunication) and M. Tech (VLSI) degrees from India in 2006 and 2011 respectively, DEng in Renewable Energy from Cape Peninsula University of Technology, Cape Town, in 2018. She started her career in 2006, an academic experience of more than 17 years and supervised more than 200 Undergraduate students. She worked as a Post Doctoral Researcher (Satellite Systems Design) at French South African Institute of Technology for three years (2020-2023). Currently, She is a Senior Lecturer in the Department of Electrical, Electronics and Computer Engineering, member of African Space Innovation Centre (ASIC), F'SATI at the Cape Peninsula University of Technology (CPUT), Bellville, South Africa. She has published numerous technical papers in accredited Journals, peer-reviewed conference proceedings and book chapters. Her research focusses on LoRa Networks (Reliability and accuracy), LPWAN Technologies, LoRa based Air quality IoT devices, LoRa WAN in satellite communication (CUBESATS), Satellite systems mission and design. Her current research activities include Integration of LoRaWAN and 5G for IoT, LoRa Direct-to-Satellites, IoT ground station. She lectured various subjects including Solid State Devices & Circuits, Digital Systems-(1 & 2), Communication Systems, Electronic Communication, Physics, Electromagnetic Field Theory, etc. She is a recipient of numerous awards which includes Best adopting faculty, merit award for batch topper, Student activity council member, best paper awards in national and international conferences.

Sandeep Gupta received his B.Tech. Degree in Electrical and Electronics Engineering from UPTU Lucknow in the year 2006. He has completed his Ph.D. from NIT Allahabad in the area of control and power systems with AI techniques. He is presently working as an Associate Professor at Graphic Era (Deemed to be University), Dehradun (India). He has a total of 15 years of experience. His areas of interest in research are the application of artificial intelligence to power system control design, FACTS devices, renewable energy, power electronics, and stability of power systems with machine learning. He is a member of numerous professional organizations, including IE, IRED, IDES, IAENG, IAER, and the Society of Professional Engineers. He has been the author or co-author of 150 papers published in international journals and presented at different national and international conferences. He is currently guiding the research scholars under Machine Learning, Control & Power Systems, Wireless Technology, and Internet of Things.

Tarun Varshney received his BE in Electronics and Instrumentation Engineering from Rajasthan University, Jaipur, India, in 2001, M.Tech in Control and Instrumentation from MNNIT, Allahabad, India, in 2007, and Ph.D from MNNIT Allahabad in 2014. Currently he is working as HOD & Professor in Department of Electrical, Electronics and Communication Engineering, SET, Sharda University, Greater Noida, UP, India. His research interests include intelligent control systems, multi-variable systems modelling and neural networks. He is reviewer of SCI, E-SCI and Scopus indexed journals. He is Senior Member of IEEE, Life member of ISTE, Member of ISA and Fellow of IE(I).

Tianyi Han received his B.S. degree in mechanical engineering from Xi'an Jiaotong University, China, in 2016, and his Ph.D. degree in mechanical engineering from Tsinghua University, China, in 2021. He was a joint Ph.D. student at University of Montreal, Canada, from November 2019 to August 2020. From March 2022 to March 2024, he was a post-doctor in the State Key Laboratory of Tribology at Tsinghua University. He is currently an assistant professor at Tsinghua University, China. His research interests include liquid superlubricity, hydration lubrication, intermolecular and surface forces.

***

Omkar Biramane is a final-year B.Tech student at Vishwakarma Institute of Information Technology (VIIT), Pune. I have a deep and active interest in computer science, focusing on web development with technologies like Next.js and React, as well as fundamental concepts in programming and operating systems.

Grishma Yadav Bobhate holds Bachelor degree and Masters degree in Computer Engineering from Sant Gadge Baba Amravati University and Savitribai Phule Pune University in 2011 and 2014. She is currently pursuing a Ph.D from Sandip University, Nashik. Her research focused on Machine Learning, Deep Learning, Natural Language Processing and Internet of Behaviors. Her skills interest in Front End Technologies with ReactJS and python programming. Currently, she is working as an Assistant Professor in the Department of Computer Science and Engineering specializing in Artificial Intelligence and Machine learning at Vishwakarma Institute of Information Technology, Pune.

Surinder Chauhan is an academician and researcher specializing in Artificial Intelligence, Smart Grids, and Optimization Algorithms. He is currently a faculty member at the University of Petroleum and Energy Studies (UPES), School of Computer Science. Prior to UPES, He also served as a faculty member at Chandigarh University and NIT Kurukshetra. He earned his Ph.D. from the National Institute of Technology, Kurukshetra, focusing on optimal placement of micro-Phasor Measurement Units (μPMUs) and AI-based fault detection. He also holds an M.E. in Power Systems & Electric Drives from Thapar University and a B.Tech. in Electrical Engineering (Software specialization) from Kurukshetra University. Dr. Chauhan has over six years of teaching experience and more than two years of industry experience with Tata Consultancy Services (TCS). He has published 11+ SCI/SCOPUS indexed research papers and presented at leading IEEE international conferences including PIICON, SEFET, and ASIANCON. A recipient of the MHRD Fellowship and a member of IEEE, he also serves as a reviewer for reputed international journals. His technical expertise includes C, C++, Data Structures, AR & VR, Python, MATLAB, Java, and OPAL-RT, with a strong focus on advancing AI-driven solutions for smart energy systems.

Ram Raghotham Rao Deshmukh is a Professor in the Department of Electrical and Electronics Engineering at SR University, Warangal. He holds a Ph.D. in Electrical Engineering from Cardiff University, UK, and brings over 20 years of academic and industry experience, including key engineering roles in the UK. His research interests include electrical machines, electric vehicles, and artificial intelligence applications in power systems. He currently serves as the Senior Director Admissions and Marketing at SR University. ORCID 0000-0002-8836-3943

Shadab Hussain is a global technology leader shaping the future of intelligence systems. As Lead Engineer at MathCo and Co-founder of Quantum Computing India, he explores the frontier of AI, Quantum, and Ethics. A Google Developer Expert and Davos Council contributor, Shadab champions human-centered innovation and the global movement toward trustworthy, future-ready AI.

Gaurav Jain has completed PhD in renewable Engineering from university college of Engineering, Kota in 2020 located in the state of Rajasthan, India. He has done M. Tech from Guru Nanak Dev Engineering College Ludhiana in 2012 from the state of Panjab and B. Tech from University of Rajasthan Jaipur in 2009. He also did MBA in Power Management from University of Petroleum and Energy Studies

Dehradun in 2024. His research interest includes in Power System, Deregulation, power management, Power Plant Engineering and Transmission pricing etc. He has published more than 45 publications in conferences and journals in various areas. He delivered several expert talks and two patent filed. Currently he is associated with Poornima college of Engineering, Jaipur

Sandeep Kanaujia is pursuing his PhD from United University Rawatpur Jhalwa, Prayagraj, India. He is the assistant Professor in United Group of Institution and teaching in various department of engineering since 2011. He has published over many papers in reputed journals and conference and chapters also .He has many numbers of National and international Patent design, copyrights and grants. Attended many Conferences, Workshops and Seminar. Work as Reviewer and organiser for many conferences. Received best paper award as well as key note speaker.

Vinod Kumar D.M. is a Professor in the Department of Electrical and Electronics Engineering at SR University, Warangal. He holds a Ph.D. in Electrical Engineering from IIT Kanpur, M.Tech in Power System Engineering & B.E in Electrical Engineering from Osmania University, Hyderabad and brings over 41 years of academic experience, including key engineering roles in the India. His research interests include Power System Operation and Control, AI applications in Power Systems, Power System Security, Power System State Estimation, Evolutionary Algorithms. ORCID 0008-0002-8078-8058

Sombir Kundu received his B.Tech in Electrical Engineering from the Maharishi Dayanand University, Rohtak, Haryana, India, in 2011, M.Tech in Electrical Engineering with a specialization in Power System from the DCRUST, Murthal, Sonepat, Haryana, India in 2014 and PhD from the Department of Electrical Engineering, Delhi Technological University, New Delhi, India in 2023. In August 2014, he joined the Department of Electrical Engineering, School of Engineering and Technology (a unit of Ganga Technical Campus), Bahadurgarh, Haryana as an Assistant Professor. In May 2022, he joined the Department of Electrical Engineering, Ganga Institute of Technology and Management, Jhajjar, Haryana, India as an Assistant Professor. In August 2024, he joined the Department of Electrical Engineering, MMEC, Maharishi Markandeshwar (deemed to be University), Mullana, Ambala, Haryana, India as an Assistant Professor. He has authored and co-authored of 20 research papers in reputed journals such as Elsevier, Wiley, Tailor Francis, Springer, etc. He has also authored and co-authored 5 book chapters which are published in Elsevier Publication and CRC Press. He has served as a session chair for the International Conference on Intelligent Computing, Simulation and Optimization, December 08-10, Goa. He was awarded with the

Research & Innovation Excellence Award at Delhi Technological University, New Delhi in 2024. He is also serving as an editor of Scientific Reports, Springer Nature Journal. His areas of research interests include power electronics, renewable energy, microgrids, power quality, and the application of adaptive and robust control techniques in microgrids.

Duli Chand Meena received his B.Tech in Electrical Engineering from the National Institute of Technology, Kurukshetra, Haryana, India. M.Tech in Electrical Engineering with a specialization in Power System from the National Institute of Technology, Kurukshetra, Haryana and PhD from the Department of Electrical Engineering, Delhi Technological University, New Delhi, India. Currently, he working in the Department of Electrical Engineering, Delhi Technological University, New Delhi, India as an Associate Professor. He has authored and co-authored of several research papers in reputed journals. His areas of research interests include Power Systems, Control of Electrical Machines, Self-Excited Induction Generator(SEIG), Power Quality and Renewable Energy.

Nikita Rawat received her B.Tech degree in Electrical and Electronics Engineering from UTU in 2013 and her M.Tech degree in Control Systems from GEU in 2015. She completed her Ph.D. from GEU in 2022. Her research interests include power quality disturbances, renewable energy systems, and solar PV parameter estimation.

Vishal Sharma received his B.Tech in Electrical Engineering from I.E.T, Dr. Rammanohar Lohia Avadh University, Ayodhya, and is pursuing his M.Tech in Power & Energy System at NIT Delhi. His work focuses on smart and sustainable renewable energy systems, including AI-based MPPT control, hybrid optimization, digital twin-enabled EMS, game-theoretic coordination, multi-agent control, EV integration, and advanced battery management.

Ajay Sharma with more than 23 years of teaching and research experience is a core researcher in the field of Bio Medical Instrumentation, Advanced Embedded Systems & Vital signal processing. Has a number of national and International publications related to this field. By profile he is an Associate Professor in the department of Electronics and Communication, apart from this he is heading over the "Foreign relations Cell" of United Group, Prayagraj, India. In addition to this he is the "Cultural Chief" of United group of Institutions. A very dynamic, hardworking and goal centric academician with his unflinching attitude he took his organization to the newer heights. He is also the member of reviewer panel of various "Scopus" & "SCI" indexed international journals. For various Internships, trainings and conferences he visited more than half a dozen countries including

United States of America, Spain, Thailand, Singapore, Malaysia etc. He is a brilliant speaker, motivator and an expert in personality & holistic development of students.

Prabhat Chandra Shrivastava is working as an assistant professor at J.K. Institute of Applied Physics & Technology, University of Allahabad, India.

Anand Shukla completed his Ph.D. in mathematics with AI techniques from NIT Allahabad. He has 14 years of experience and is currently working at Wollega University, Ethiopia. His areas of interest in research are the application of artificial intelligence with machine learning. He is a member of numerous professional organizations. He has been the author or co-author of 40 papers published in international journals and presented at different national and international conferences.

Surya Karan Singh received his B.Tech in Electrical Engineering from MAKAUT, India (2018), and M.Tech in Power & Energy Systems from NIT Silchar (2021). He is currently a researcher at NIT Delhi, focusing on intelligent and sustainable energy technologies. His work spans AI-driven energy management, federated learning-based power system forecasting, peer-to-peer energy trading, dynamic tariff optimization, and multi-criteria decision-making frameworks. His broader interests include distributed generation, electric mobility, hybrid energy storage, predictive maintenance, and resilient smart energy networks.

Amit Kumar Singh received his B.Tech in Electrical Engineering from UPTU, India (2007), and M.Tech in Power & Energy Systems from NIT Silchar (2009). He earned his Ph.D. in Electrical Engineering from IIT Patna in 2014. He served as a faculty member at IITRAM Ahmedabad (2015–2017) and is currently with the Department of Electrical Engineering, NIT Delhi (since 2017). His research interests include power system restructuring, electricity markets, distributed generation, renewable energy integration, electric vehicles, fault detection, and smart energy network operation and control.

Navdeep Singh received his B.Tech in Electrical Engineering from the Kurukshetra University, Kurukshetra Haryana, India, in 2012, M.Tech with a specialization in Renewable Energy Systems from National Institute of Technology (NIT) Kurukshetra, Haryana, India in 2014 and PhD from the School of Renewable Energy and Efficiency from NIT Kurukshetra, Haryana, India in 2019. In September 2019, he joined the Department of Electrical Engineering, MMEC, MMDU, Mullana, Ambala, Haryana, as an Assistant Professor. He has authored and co-authored 11 research papers in reputed journals and International conferences such as IET Renewable Power Generation, Proceedings of the Institution of Mechanical

Engineers, Part A: Journal of Power and Energy, Journal of Engineering Research, International Federation of Automatic Control (IFAC), INDICON, Elsevier, Procedia Computer Science, Springer Nature Link, etc. He reviewed several International Journals and IEEE Conferences. He also served as a session chair for two International Conferences in 2023 and 2024. His research interests include renewable energy, applications of control techniques to Wind Turbines, and Energy efficiency.

Prashant Singh is a Senior Scientist at VTT Technical Research Centre in Finland, where he specializes in batteries and charging systems. His work focuses on energy management and control for Hybrid Energy Storage Systems, which include batteries and supercapacitors. He conducts research on Hardware-in-the-Loop (HIL) testing for battery management systems (BMS), with a strong emphasis on power and energy storage applications. Prashant has advanced proficiency in MATLAB/Simulink and extensive hands-on experience implementing real-time HIL simulations using Speedgoat and OPAL-RT platforms.

Bhanu Pratap Soni is Assistant Professor and Head of Electrical Engineering Department at CETVET, Fiji National University. He holds a Ph.D. in Intelligent Power Systems from MNIT Jaipur and M.Tech/B.Tech (Hons) in Electrical Engineering. He received the Young Scientist Award (SERB-DST, India) in 2018 and the Eminent Young Researcher Award in 2020. As a Senior Member of IEEE, Chartered Engineer (IEI), and member of Engineers New Zealand and Engineers Australia, Dr. Soni has over 60 publications. His research interests include AI/ML in power systems, smart grids, and optimization. He also serves as an editor and reviewer for reputed journals.

Pravin Sonwane had completed Post graduation in Integrated Power system from VNIT, Nagpur in 2005 and PhD from Savitribai Phule Pune University in 2016. His research area is optimal power system. Intelligent systems for Electrical Power Grids, Monitoring and control of earth pit parameters, artificial prosthetic leg and swarm intelligence for drone system in navigation for defense purpose. He is also interested to extend his specialization in both generation and transmission, which need development considering the fundamental understanding of the phenomena in electrical sector in addition to financial aspect such as minimizing energy cost, power system planning and growth by implementing intelligent systems. In addition to this he is extending work for development of advanced sensors and actuators to achieve the targeted task within the systems in future areas of research specifically in power electronic semiconductor devices. He is intended to solve complex engineering issues through collaborative efforts and apply his theoretical and mathematical skills in the development of useful and novel technologies. Total publications are 55, fetched multiple funding's from various sectors for research projects, organized more than 20 national events, delivered several expert talk (one international talk in Brazil) and three patent filed. He is dedicated faculty and instructor for the student community and follow all student centric process.

Vaishali Sonwane is the research scholar at Sandip University Nashik and currently working in the faculty of Computer Science, Poornima University, Jaipur. She completed her Masters in Digital Systems from Pune University and B. Tech in Electronics Engineering in 2005 and PhD from Sandip University, Nashik. Her interest of research is in wind energy, sustainable energy, computer systems, processing units, aircraft design etc. She has published around twenty publications in various areas. Her work on artificial prosthetic leg is appreciated in the E-PGcon, Pune in 2017 and DC Motor control project was appreciated by Savitribai Phule Pune University. She has received fellowship of five lack rupees from MJPRTI, Nagpur. Currently she is working on full stack developer and entrepreneurship essentials.

Surasi Sharan Kumar is a Full-Time Research Scholar in the Department of Electrical and Electronics Engineering at SR University, Warangal. He holds M.Tech. in Power System Control and Automation & B.Tech in Electrical and Electronics Engineering from JNTU Hyderabad. His research interests include Electric Vehicles, Machine Learning, Quantum Computing and Artificial Intelligence applications in Battery Management System. ORCID 0009-0008-4188-9850

Mohit Tyagi received his B.Tech in Electrical & Electronics Engineering from the Uttar Pradesh Technical University University, Lucknow, Uttar Pradesh, India, in 2010, M.Tech in Electrical Engineering with a specialization in Power Power & Energy Systems from the Mahamaya Technical University, Noida, Uttar Pradesh, India in 2012 and Pursuing PhD from the Department of Electrical Engineering, Delhi Technological University, New Delhi, India.. In August 2010, he joined the Department of Electrical & Electronics Engineering, H.R. Institute of Technology, Ghaziabad, Uttar Pradesh as an Assistant Professor. In February 2014, he joined the Department of Electrical & Electronics Engineering, Vidya College of Engineering, Meerut, Uttar Pradesh, India as an Assistant Professor. In January 2019, he joined the Department of Electrical Engineering, R.D. Engineering College, Ghaziabad, Uttar Pradesh, India as an Assistant Professor. His areas of research interests include power electronics, renewable energy, microgrids, power quality, and the application of adaptive and robust control techniques in microgrids.

Ram Kumar Yadav has a total of 18 years of experience. His research interests include intelligent control systems, Renewable Energy with SDGs Control. He is reviewer of SCI, E-SCI and Scopus indexed journals.

Index

P

Q

R

S

T

U

V

W

www.ingramcontent.com/pod-product-compliance
Lightning Source LLC
LaVergne TN
LVHW062329300926
849161LV00007B/1633